Metaplasticity in Virtual Worlds:
Aesthetics and Semantic Concepts

Gianluca Mura
Politecnico di Milano University, Italy

INFORMATION SCIENCE REFERENCE
Hershey • New York

Senior Editorial Director:	Kristin Klinger
Director of Book Publications:	Julia Mosemann
Editorial Director:	Lindsay Johnston
Acquisitions Editor:	Erika Carter
Development Editor:	Michael Killian
Publishing Assistant:	Natalie Pronio, Jamie Snavely & Deanna Zombro
Typesetter:	Casey Conapitski, Natalie Pronio & Deanna Zombro
Production Editor:	Jamie Snavely
Cover Design:	Nick Newcomer

Published in the United States of America by
Information Science Reference (an imprint of IGI Global)
701 E. Chocolate Avenue
Hershey PA 17033
Tel: 717-533-8845
Fax: 717-533-8661
E-mail: cust@igi-global.com
Web site: http://www.igi-global.com

Copyright © 2011 by IGI Global. All rights reserved. No part of this publication may be reproduced, stored or distributed in any form or by any means, electronic or mechanical, including photocopying, without written permission from the publisher. Product or company names used in this set are for identification purposes only. Inclusion of the names of the products or companies does not indicate a claim of ownership by IGI Global of the trademark or registered trademark.

Library of Congress Cataloging-in-Publication Data

Metaplasticity in virtual worlds : aesthetics and semantic concepts / Gianluca
Mura, editor.
    p. cm.
  Includes bibliographical references and index.
  ISBN 978-1-60960-077-8 (hardcover) -- ISBN 978-1-60960-079-2 (ebook) 1.
Human-computer interaction. 2. Virtual reality. 3. New media art. 4.
Communication and technology. I. Mura, Gianluca, 1965-
  QA76.9.H85M48 2010
  006.8--dc22
                                2010036722

British Cataloguing in Publication Data
A Cataloguing in Publication record for this book is available from the British Library.

All work contributed to this book is new, previously-unpublished material. The views expressed in this book are those of the authors, but not necessarily of the publisher.

Editorial Advisory Board

Paul Catanese, *Columbia College Chicago, USA*
Félix Francisco Ramos Corchado, *CINVESTAV GDL, Mexico*
Everardo Reyes Garcia, *Tecnologico de Monterrey Campus Toluca, Mexico*
Olga Sourina, *Nanyang Technological University, Singapore*
Daniel Thalmann, *EPFL VRLab École Polytechnique Fédérale de Lausanne, Switzerland*
Hidenori Watanave, *Tokyo Metropolitan University, Japan*

Table of Contents

Preface xi

Acknowledgment xiv

Chapter 1
Virtual Metaplasticity (Ars Metaplastica) 1
    Gianluca Mura, Politecnico di Milano University, Italy

Chapter 2
The Metaplastic Virtual World Theory 27
    Gianluca Mura, Politecnico di Milano University, Italy

Chapter 3
The Tangiality of Digital Media 58
    Paul Catanese, Columbia College Chicago, USA
    Joan Truckenbrod, School of the Art Institute of Chicago, USA

Chapter 4
Pervasive Virtual Worlds 78
    Everardo Reyes-García, Monterrey Tech at Toluca, Mexico

Chapter 5
Semantic Entities in Virtual Worlds: Reasoning Through Virtual Content 90
    Vadim Slavin, Lockheed Martin Space Systems, USA
    Diane Love, Lockheed Martin Information Systems & Global Services, USA

Chapter 6
Humanness Elevated Through its Disappearance 102
    Jeffrey M. Morris, Texas A&M University, USA

Chapter 7
The Mediation Chamber: Towards Self-Modulation 121
    Chris Shaw, Simon Fraser University, Canada
    Diane Gromala, Simon Fraser University, Canada
    Meehae Song, Simon Fraser University, Canada

Chapter 8
A Behavioral Model Based on Personality and Emotional Intelligence for
Virtual Humans 134
    Héctor Orozco, Centro de Investigación y de Estudios Avanzados del I.P.N., México
    Félix Ramos, Centro de Investigación y de Estudios Avanzados del I.P.N., México
    Daniel Thalmann, École Polytechnique Fédérale de Lausanne, Switzerland
    Victor Fernández, Centro de Investigación y de Estudios Avanzados del I.P.N., México
    Octavio Gutiérrez, Centro de Investigación y de Estudios Avanzados del I.P.N., México
        & Grenoble Institute of Technology, France

Chapter 9
The Virtual World of Cerberus: Virtual Singer using Spike-Timing-Dependent
Plasticity Concept 158
    Jocelyne Kiss, LISAA, University Paris East, France
    Sidi Soueina, Strayer University, USA
    Martin Laliberté, LISAA, University Paris East, France
    Adel Elmaghraby, University of Louisville, USA

Chapter 10
Learning from Baroque 167
    Carola Moujan, Université Paris 1 LETA/CREDE, France

Chapter 11
Synthetic Worlds, Synthetic Strategies: Attaining Creativity in the Metaverse 182
    Elif Ayiter, Sabanci University, Turkey

Chapter 12
Kritical Artworks in Second Life 198
    Dew Harrison, University of Wolverhampton, UK
    Denise Doyle, University of Wolverhampton, UK

Chapter 13
Digital Media and the Quest for the Spiritual in Art 217
    Ina Conradi Chavez, Nanyang Technological University, Singapore

Chapter 14
Plastika [Totipotenta] 228
    Catherine Nyeki, Plurtimedia Artist "plasticienne", France

Chapter 15
Spatial Design and Physical Interface in Virtual Worlds 240
    Hidenori Watanave, Tokyo Metropolitan University, Japan

Chapter 16
The City of Abadyl 251
    Michael Johansson, PRAMnet, Sweden & China

Chapter 17
Social and Citizenship Competencies in a Multiuser Virtual Game 266
    Germán Mauricio Mejía, Universidad de Caldas, Colombia
    Felipe César Londoño, Universidad de Caldas, Colombia
    Paula Andrea Escandón, Universidad de Caldas, Colombia

Compilation of References 281

About the Contributors 300

Index 306

Detailed Table of Contents

Preface xi

Acknowledgment xiv

Chapter 1
Virtual Metaplasticity (Ars Metaplastica) 1
    Gianluca Mura, Politecnico di Milano University, Italy

This chapter introduces the definition of new metaplastic discipline. It also defines the metaplastic virtual medium used in different research areas with application in virtual art and virtual heritage.

Chapter 2
The Metaplastic Virtual World Theory 27
    Gianluca Mura, Politecnico di Milano University, Italy

Conceptualization of the metaplastic virtual world is explained with its artistic archetypes and it leads to the composition of a new theoretical model with examples of application.

Chapter 3
The Tangiality of Digital Media 58
    Paul Catanese, Columbia College Chicago, USA
    Joan Truckenbrod, School of the Art Institute of Chicago, USA

This chapter introduces the concept of Tangiality by illustrating how it is represented in the contemporary interdisciplinary, hybrid media and multi-modal artists' definitions examples.

Chapter 4
Pervasive Virtual Worlds 78
    Everardo Reyes-García, Monterrey Tech at Toluca, Mexico

This chapter describes the notion of contemporary society of media which virtualizes human senses and capabilities integrated within a collective, pervasive hyperspace of information. It is demonstrated with three different examples of digital media applications.

Chapter 5
Semantic Entities in Virtual Worlds: Reasoning Through Virtual Content 90
    Vadim Slavin, Lockheed Martin Space Systems, USA
    Diane Love, Lockheed Martin Information Systems & Global Services, USA

This chapter explores how a parallel semantic knowledge base describing the virtual world can improve the utility of virtual world environment by enabling virtual agents to interact and behave like their peer human participants.

Chapter 6
Humanness Elevated Through its Disappearance 102
    Jeffrey M. Morris, Texas A&M University, USA

This chapter gives consideration to the different types of presence manifested in various communication formats, stage presence in technology-mediated performance, and several artworks that bring new light to the artist's approach to virtual worlds as a kind of counterpoint with reality.

Chapter 7
The Mediation Chamber: Towards Self-Modulation 121
    Chris Shaw, Simon Fraser University, Canada
    Diane Gromala, Simon Fraser University, Canada
    Meehae Song, Simon Fraser University, Canada

This chapter describes scientific methods of evaluations and findings and it discusses first-person phenomenological aspects of biofeedback technologies and their possible future directions.

Chapter 8
A Behavioral Model Based on Personality and Emotional Intelligence for
Virtual Humans 134
    Héctor Orozco, Centro de Investigación y de Estudios Avanzados del I.P.N., México
    Félix Ramos, Centro de Investigación y de Estudios Avanzados del I.P.N., México
    Daniel Thalmann, École Polytechnique Fédérale de Lausanne, Switzerland
    Victor Fernández, Centro de Investigación y de Estudios Avanzados del I.P.N., México
    Octavio Gutiérrez, Centro de Investigación y de Estudios Avanzados del I.P.N., México
       & Grenoble Institute of Technology, France

This chapter presents a behavioral model based on personality and emotional intelligence for Virtual Humans. This model allows virtual humans performing actions based on their current affective state, beliefs, desires and intentions.

Chapter 9
The Virtual World of Cerberus: Virtual Singer using Spike-Timing-Dependent
Plasticity Concept 158
    Jocelyne Kiss, LISAA, University Paris East, France
    Sidi Soueina, Strayer University, USA
    Martin Laliberté, LISAA, University Paris East, France
    Adel Elmaghraby, University of Louisville, USA

The Cerberus is an avatar singer model that performs his musical style and emotion using competitive learning rule within artificial neural networks methodologies.

Chapter 10
Learning from Baroque.. 167
    Carola Moujan, Université Paris 1 LETA/CREDE, France

This chapter discusses from a cultural perspective, philosophical and aesthetic issues related to mixed reality spaces. It proposes a perceptual analysis of early Baroque architecture for bringing aesthetic depth and relevance into mixed reality installations.

Chapter 11
Synthetic Worlds, Synthetic Strategies: Attaining Creativity in the Metaverse................................. 182
    Elif Ayiter, Sabanci University, Turkey

This chapter underlines theoretical premises of an immersive learning approach to the visual arts implementation into online three dimensional synthetic worlds.

Chapter 12
Kritical Artworks in Second Life.. 198
    Dew Harrison, University of Wolverhampton, UK
    Denise Doyle, University of Wolverhampton, UK

This chapter outlines the curatorial process of creative practices for showing artworks exhibitions in the Kriti SL island. It indicates new reading guidelines of Art exhibition practice in Second Life

Chapter 13
Digital Media and the Quest for the Spiritual in Art.. 217
    Ina Conradi Chavez, Nanyang Technological University, Singapore

This chapter presents a virtual art concept of immersive 3D installations for social interaction through sensation and physical engagement.

Chapter 14
Plastika [Totipotenta] 228
    Catherine Nyeki, Plurtimedia Artist "plasticienne", France

This chapter describes biological creatures "living" in their virtual world. It composes a virtual "vivarium" with diverse digital interactive pieces inspired with latest biotechnology and cellular biology.

Chapter 15
Spatial Design and Physical Interface in Virtual Worlds 240
    Hidenori Watanave, Tokyo Metropolitan University, Japan

This chapter proposes an architecture spatial model design methodology to create physical human experience in 3D virtual worlds.

Chapter 16
The City of Abadyl 251
    Michael Johansson, PRAMnet, Sweden & China

This chapter introduces a complex digital space that can be utilized as an open world in order to generate scenarios and producing new artifacts for the co-creators of Abadyl.

Chapter 17
Social and Citizenship Competencies in a Multiuser Virtual Game 266
    Germán Mauricio Mejía, Universidad de Caldas, Colombia
    Felipe César Londoño, Universidad de Caldas, Colombia
    Paula Andrea Escandón, Universidad de Caldas, Colombia

This chapter discusses findings of theory and models about effects of video gaming in education and proposes an e-learning virtual game.

Compilation of References 281

About the Contributors 300

Index 306

Preface

Concepts of virtual worlds are strongly related to the current innovations of the Internet and new media communications. Nowadays, the research areas of virtual worlds include different scientific fields coming from Science, Technology, Arts, Design and Digital Cultures.
This book explores the interdisciplinary development of Virtual world's aesthetics and semantics through many approaches of metaplastic conceptualizations. The term metaplasticity, generally in neuroscience, defines mnemonic and adaptive synaptic properties. Metaplasticity conceptualization, differently in plastic arts, is a process of creation and transformation.

From these definition qualities, the goal of this book becomes finding new definitions of the metaplasticity term for Digital media and Virtual Worlds fields. It would be proposed diverse interdisciplinary meanings of digital media researches for the purposes of this book.

This volume starts with *Virtual Metaplasticity* (*Ars Metaplastica*) chapter where the analyses of Modern Art history and the definition of new Metaplastic discipline are based on the intersection of plastic formalism with digital technology. The Metaplasticity concept defined within an interdisciplinary research field introduces the characteristics of a metaplastic virtual media model and their possible applications to different research fields.

The *Metaplastic Virtual World Theory* chapter describes how the metaphor of artistic machine finds its new realization within the metaplastic virtual world, where the machine metaphor itself becomes aesthetic expression of the virtuality. The methodology explains the becoming of new virtual worlds typologies made through abstract art languages and artificial intelligence.

Joan Truckenbrod and Paul Catanese in *Tangiality of Digital Media* illustrates the immateriality as a conceptual materiality of virtual world. It makes an interdisciplinary analysis of many contemporary experiences that use this concept in their examples of hybrid and multimodal artworks.

Everardo Reyes Garcia continues with *Pervasive Virtual Worlds* experience where virtual media are explained through different social media definitions that virtualize human senses and capabilities. These notions are expressed with three different digital installations.

Vadim Slavin and Diane Love in their chapter *Semantic Entities in Virtual Worlds* describe how semantic knowledge base could define virtual world architecture richness by giving their virtual entities behavior and their relationships with human participants.

Jeffrey Morris in his *Humanness Elevated through its Disappearance* argues on human's different types of presence and the blurring boundaries between real and virtual in art experiences. His artistic approaching to virtual world is seen as a counterpoint with reality.

Diane Gromala, Chris Shaw and Meehae Song describe in their *Meditation Chamber: Towards Self-Modulation* an immersive virtual environment, which analyzes human feelings including stress, anxiety and pain and their biofeedback technologies. They discuss results and new findings on possible applications in neurosciences.

Daniel Thalmann, Félix Ramos, Héctor Orozco, Victor Fernández and Octavio Gutiérrez in *A Behavioral Model based on Personality and Emotional Intelligence for Virtual Humans* developed a behavioral model for Virtual Humans and implemented it in a system that uses various calculus formalism driven by emotions and try to define virtual human behaviors for each situation they experience in the environment.

Jocelyne Kiss, Sidi Soueina, Martin Laliberté and Adel Elmaghraby in *Virtual World of Cerberus* present an avatar singer application that performs different sounds and express their correspondent facial emotions. This study applies the metaplasticity notion differently, as used in neuroscience, in an avatar model to enhance memory functions and simulated feelings related to a virtual context.

Carola Moujan in *Learning from Barouque*, gives a historical and cultural perspectives through phylosophical studies implications with proposing mixed realities artistic installations. This essay indicates a radical change into artist's view for creating meaningful experience and participation of spectators in artworks. It explains the Baroque architecture as an relevant early example of virtuality in Art.

Elif Ayiter in her chapter *Synthetic Worlds, Synthetic Strategies: Attaining Creativity in the Metaverse* attempts to define theoretical premises and definition of an immersive learning approach pertaining to visual arts to be implemented in an online synthetic worlds. The author proposes recent educational approaches as well as an examination of creative practices into the formulation of an virtual world learning strategy.

Dew Harrison and Denise Doyle in *Kritical Artworks in Second Life*, introduce their SL island where they experiment creative practice and give curatorial view in virtual world. They introduce various case studies of artist's exhibition in Second Life.

Ina Conradi in *Digital Media and the Quest for the Spiritual in Art*, presents her public virtual art exhibition and defines a virtual world's aesthetics based on spectator's oneiric virtual world realization. The virtual media accomplishes its aesthetics within different mental states with a connected user.

Catherine Nieky in her *Plastika Totipotenta* presents her biotope virtual "vivarium", where biological creatures "live" into a simulacrum and the concept of metaplasticity is applied from natural and nanosciences into biotechnological virtual world.

Michael Johansson with his *City of Abadyl* generates an interactive virtual scenario of an open work project of the city of Abadyl. It is a database that contains information which is continuously updated and this interactive updating it is called "fieldasy".

Hidenori Watanave in his *Spatial Design and Physical Interface in Virtual Worlds* proposes a new spatial model, a contents oriented architectural space design methodology to create spatial experience in a 3D virtual world.

Germán Mauricio Mejía, Felipe César Londoño, Paula Andrea Escandón in their *Social and Citinzenship Competencies in a Multiuser Virtual Game* describe educational gaming that shows a metaphor of collective challenge and indicates an overview of everyday life complexity and learning transferability. The authors discuss findings according to proposed theories and models about effects of video games in education and behavior.

The chapters collected an in-depth coverage of the state-of-the-art of virtual worlds experiences. It has traced some of research tendencies and paths, regarding virtual worlds aesthetics and semantics, and grouped them into the following areas:

- knowledge communication in virtual worlds as social media;
- development of methodologies dedicated more precisely to human perceptiveness, with greater involvement of human participants into different and inter-crossed levels of realities;
- intersection of relations and roles between dynamics of virtual world and representation, and virtual participant and co-creator;
- cultural strategies analysis of creative practice, education, exhibition and communication within virtual worlds;
- virtual worlds concepts between architectures, new territories, new biotype creatures into real-virtual societies. In conclusion, the experiences of contemporary virtual worlds indicate a necessity of changing the actual methodologies and practices toward new languages dedicated to humans and solving their cultural and social issues.

Gianluca Mura
Editor

Acknowledgment

I would like to gratefully thank all the Editorial Board members and particularly Paul Catanese, Félix Francisco Ramos Corchado, Everardo Reyes-García, Olga Sourina, Daniel Thalmann, and Hidenori Watanave.

I would also like to thank all the Authors that have generously and patiently contributed to this book with their important and precious work, that has made this book possible, particularly Elif Ayiter, Ina Conradi, Denise Doyle, Adel S. Elmaghraby, Paula Andrea Escandón, Victor Fernández Cervantes, J.Octavio Gutierrez-Garcia, Diane Gromala, Dew Harrison, Michael Johansson, Jocelyne Kiss, Martin Laliberté, Felipe César Londoño, Diane J.Love, Germán Mauricio Mejía, Jeffrey M.Morris, Carola Moujan, Catherine Nyeki, Héctor Rafael Orozco Aguirre, Chris Shaw, Vadim Slavin, Meehae Song, Sidi Soueina and Joan Truckenbrod.

I wish to thank all IGI Global for their kind and precious help during all the time for developing and supporting this book project, particularly thanks to Jan Travers, Elizabeth Ardner, Dave DeRicco, Mike Killian and Erika Carter. Their professional assistance and guidance have made this collaboration a great enriching experience.

I want to gratefully acknowledge the support and collaboration of Nataša Duboković, for all the time and dedication to this book.

Thank you.

Gianluca Mura
Editor

Chapter 1
Virtual Metaplasticity (Ars Metaplastica)

Gianluca Mura
Politecnico di Milano University, Italy

ABSTRACT

This chapter defines a new metaplastic discipline through analysis of Digital art history and their relations among artistic and scientific achievements. Virtual Realities found in Art and Science a new modality of definition based on technological codifying with language philosophies that interpret new digital aesthetics. It would be introduced new metaplastic virtual media used in different research areas with examples of applications.

INTRODUCTION

Art and virtuality discover their common origins within the poetic vision of artists, for example imaginary worlds of Kandinsky, Klee, Miro'. Their worlds are, in fact, inhabited with organisms similar to cells, bacteria and other microorganisms in fantastic and indefinite spaces. Artists (re)create new realities with systems of symbols that express the representation of their concepts. At the same mode, computer's virtual realities are patterns of information. The description of artists' imaginary worlds occur with representations of basic elements, combinations of elements in sequence, structures, surfaces and kaleidoscopes of images. These expressions narrate of new visual and musical languages modalities with their peculiar characteristics. Their research processes of new shapes and disciplines established archetypes of virtual reality languages. According to Jaron Lanier, a pioneer of the virtual reality, artists should create new worlds:

"Instead of communicating symbols like letters, numbers, pictures, or musicals' notes, you are creating miniature universes that have their own internal states and mysteries to be discovered."

In the last decades of the 20th Century, technological innovations in different research areas have developed new interdisciplinary interests between art, design, science and technology. From the 1950s, Computer Art is a term used to define artistic electronic data processes for creating new aesthetics. In the same period, the mathematician Norbert Wiener in his essay "Cybernetics and Society" (1950) presented Cybernetics which refer to the general analysis of control systems and communication in living organisms and machines. The exhibition "Cybernetic Serendipity" (1968) at the Institute of Contemporary Arts in London was important because there were presented various artworks of computer graphics, of sound and luminous spaces, "intelligent" robots which could be considered precursors of contemporary artistic installations.

Some artworks were dedicated to machine's aesthetics and its transformations, other artworks created audio visual motives and automation's poetry, like "open systems" installation where spectator acts as a part of a programmed system. From the 1960s, the researches within new technologies created new research areas in arts and sciences. In that period, important transdisciplinary studies were developed by artists as Andy Warhol, Robert Rauschenberg, Jean Tinguely, John Cage and others at the Osaka Universal Exposition, Pepsi Cola's pavilion in 1970. (Christiane Paul, 2004 p.16). The artistic creative process is modified with new technologies, generating new methodologies and different disciplinary areas in Digital Art.

In this phase, new creation methodologies became with following the simulations of the preceding technique. This simulation property gives to digital systems the unprecedented capacity of blurring the traditional distinct boundaries, by subjecting all techniques of information to the creation of hybrid art forms and to becoming a hybrid art system itself. This is the first hybridization typology of creation methodologies.

The new aesthetics of Digital Arts are founded with this continuous antithesis of blurring the boundaries and resisting to this new artistic becoming. This is the second hybridization typology of creation methodologies. According to Christiane Paul, one of these new methodologies of creation is the combinatorial processes that obey to the rules of Dadaist Poetry, which were resumed by the members of OULIPO (Ouvroir de Littérature Potentielle) and their creation poetics became part of combinatorial loop of conceptual gaming. The Fluxus artists have specific rules when they stage their performances. Vera Molnar uses mathematical functions to create new drawing methodologies for computer graphics. In these research areas, the Permutation Art, defined by Abraham Moles in 1971 found its aesthetics. It is created with combinations of limited set composed with simple elements, which gave representations of infinite perceptual games.

A different method forms of "self-production" has been suggested by mathematician H. Conway with his famous "Game of Life"(1970) which has inspired the future development of Generative Arts theories. Chistiane Paul in his book Digital Art indicates various artistic experience that have been developed in the 1960s and 1970s period, and in the following part will be given their description. Michael Noll in his artwork "Computer Composition with Lines"(1964) uses a graphic table to draw combinations of lines and other pictorial representations. Manfred Mohr composes his artworks with computer as exercises on parametric variations of basic graphic repertoire. This methodology originated the "generative drawing". Mohr is the first plastic artist that had contributed to the theorizing of criteria and methodologies of creation new aesthetics with computers. The first generation of Algorists in the 1960's and 1970's included Herbert Franke, Manfred Mohr, Frieder Nake, George Nees, Harold Cohen and Roman Verotsko. The term "Algorithm" descended from the name of an Arabian mathematician who was active around 820 AD in Baghdad. It is believed that his surname, *al-Khowarazmi* is the source for the term *algorism*. According to Roman Verotsko (2009):

"Artists engaging new computing and visualizing technologies had to either collaborate with engineers for programming their ideas or else create their own procedures (algorithms). The Algorists are artists who create art using algorithmic procedures that include their own algorithms".

AESTHETICS OF INTERACTION

The notion of interactivity means the possibility of real time interaction with digital artwork. Art of Virtual Environments integrates informatics devices and modifies its interaction with the Audience. It cannot be reduced only to technicist manipulations. This is the third hybridization typology of creation methodologies.

John Cage's artworks have strongly anticipated interactive arts experience with introducing participation of spectators through objects, combinations and their casual instructions.

In the same way, Duchamp stated that artwork becomes physical and objective through its Interactivity. Nowadays, researches on "spectator's participation" were used in different areas of the Media Art as Performances, Kinetic Art, Conceptual Art, Body Art and other forms of art.

Aesthetics of Participation found its new redefinition within actual digital technologies.

Technological progresses have conducted to the creation of the first interactive media for Internet with reintroducing and extending on the Web experiences based on relations between digital artwork, audiences and authors. Digital creative process becomes the result of complex collaborations among different co-authors from various levels and matter of research in interdisciplinary areas. Interactive artworks contain dialogs with their spectators that are more than simply observers, they have an acting function.

The interactive artwork is created with two actors. The first actor that originates or defines programming rules for spectator's conditions and the second actor-spectator that introduces the progress of artwork with the goal of acting its potentiality, differently from the traditional spectator that has no possibility of interaction. The artwork is therefore, constituted of two different semiotic objects: the actor that is the computer Program and the other object, the spectator with the role of co-authoring or co-acting. The computer within its interactivity logic is capable of virtual objects reproductions that don't act as "things" anymore with forms and immutable properties, but as artificial "beings" more or less sensibly, more or less lively, more or less autonomously, more or less intelligently. These achievements came from research area of Artificial Life and Cognitive Sciences. Researches in the fields of Artificial Intelligence were going toward non programmed behavior by utilizing genetic algorithms' properties. There is a basic principle within this studies: interactivity in its highest level of complexity between basic elements of Artificial Life (genes or neurons) and their configurations correspond to the production of these emergent phenomena.

This disciplinary field is derived from the first developments of Cybernetics. New aesthetics of hybrid forms between artwork and spectator led Contemporary Art to its conceptual and physical dematerialization. The phenomenon of dematerialization entered the arts of showing with digital choreographies where dancers interacted with their virtual doubles projected on scenes. Digital Arts restore their lost technique with introducing a new way of automatic use of information. The computer is a hybrid machine and it is the first machine that uses an interpretive language for its functioning.

VIRTUAL REALITIES

The virtual has its plain reality because it's virtual.
Gilles Deleuze, Difference and Repetition. (1994)

The definition of Virtual Reality is confusingly expressed between one concept and its contrary. In fact, reality and virtuality aren't in opposition: Virtual, from Latin *virtus* that means strength, of what exists powerfully in reality, it's essential and existential condition. The fundamental dualism is between ours' perception and what really exists, and between what has been induced in ourselves and what exists outside of ourselves. These phenomena that solicit our senses are always real: they are physical phenomena. On the contrary, the cognitive representations induced by physical phenomenon in our brain, could be an object that really exists or created by our imagination, so it doesn't exist: in the first case the object is real; in the second case the object is defined as virtual. The Virtual Reality is often defined as an expression of meaning, deduced from an immersive and interactive simulacra created by computer, which provides the user with the sensation of being completely immersed within artificial environment.

For understanding the growth of virtual realities concepts, it's important to introduce briefly the fundamental phases of their history. The article entitled "As We May Think" written by Vannevar Bush and published by the Atlantic Monthly in 1945, had a revolutionary role for the history of communications. He described a machine called Memex that had a form of a desk with a translucent plane, where the user could read his documentations. The Memex's content was full of books, journals and images translated to microfilms in a readable format. This important project is considered to be an archetype of all hypertextual systems. Memex, was essentially an analogical machine. The ENIAC (Electronic Numerical Integrator and Computer) was the first machine with alphanumerical code that was commercialized. During the 1940s, Wiener's cybernetic theories served as base for researches on "symbiosis" between human and machines, concept which would be largely explored later on by many multimedia artists. In the 1960s, there were important developments of information science; the effects of that period was largely based on art and technology improvements. The concepts of Vannevar Bush was taken and developed by Theodor Nelson into basic notions of hypertext and hypermedia which indicate a space of reading and writing of text, image and sound electronically doubled with the possibility of connection to a "Docu-verse" (universe of documents).

Nelson's hypertext isn't linear but layered and it allowed to its readers/writers personal information paths. His notions anticipate future files' transfers and Internet messages which will start with the Web, in 1990. Political environment of the 1960s with the cold war conditions, initiate the ARPA project in the USA for creating dominant positions in technology. Further goals of this group was the development of a communication system. This communication system was called ARPANET and accomplished in 1969. The successive phases of technological development happened at the end of that decade contributed with two fundamental notions of information science: the information space and the interface. Douglas Engelbart, from Augmentation Research Center of the Stanford University, introduced a revolutionary idea of windows' direct bit-mapping in 1968. Every pixel was assigned a bit memory unit with a value of 0 or 1. Engelbart with his invention gave a possibility of direct space manipulation with the introduction of interface devices like the mouse and optical pen, user's hand prosthesis. During the 1970s, Alan Kay and his research group of the Xerox PARC in Palo Alto, widened these concepts by introducing the GUI (Graphic User Interface) through the introduction of the Desktop metaphor, which became very popular with the Apple Computers. From the ARPANET built in the 1960s, it has been developed the World Wide Web by Tim Berners-Lee and the CERN (European Council of Nuclear Research) which would become a multimedia information system for public audience in 1990. The Web has a protocol of hypertext transfer (http, HyperText Transfer Protocol) that allows access to documents written with HTML(HyperText Markup Language). The first multiple user environments MUD(Multiple User Dungeons) are essentially textual interfaces and founded on the model of the first electronic role video games. The MOO (Muds Object-Oriented) are software more similar to video conference systems than to role games. Visual virtual worlds created by users, have started to spread on the Web with the 3D programming language named VRML (Virtual Reality Modeling Language). Tony Parisi and Mark Pesce at the first conference on the World Wide Web in 1994, presented Labyrinth, the first prototype of interface 3D that demonstrated the need to create a language for describing 3D scenes on the Net. At that point, users of virtual environments needed a movable image that represented themselves in this fictitious reality: the avatar.

During the 1980s, new research area was founded: the area of virtual reality. The Virtual Reality is usually defined as a result of particular technological systems. Those systems include a computer with suitable computational capacities, controlled by a set of linked devices. The following paragraph introduces some examples of Virtual Reality definitions:

- "The terms virtual worlds, virtual cockpit, and virtual workstations were used to describe specific projects... In 1989, Jaron Lanier, coined the term virtual reality to bring all of the virtual projects under a single rubric. The term therefore typically refers to the three-dimensional realities implemented with stereo viewing goggles and reality gloves" (Krueger, 1991);
- "Virtual Reality is electronic simulations of environments experienced via head mounted eye goggles and wired clothing enabling the end user to interact in realistic three-dimensional situations" (Krueger, 1991);
- "Virtual Reality is an alternate world filled with computer-generate images that respond to human movements. These simulated environments are usually visited with the aid of an expensive data suit which features stereophonic video goggles and fiberoptic data gloves"(Biocca,1992).

These definitions of Virtual Reality include the notion of electronically simulated environments and systems that need "goggles and gloves" for accessing into them. This way of seeing them is limiting because it doesn't give the right explanation of these system's opportunities. The natural consequences of these definitions cause behaviors for this kind of realities that assign crucial part only to important "hardware" or "software". But this conduct should be changed with adding another level of definitions. What is the Virtual? Pierre Levy gives a brief response to this question by proposing a wider vision of the phenomenology with the following definition:

"...it is the transformation from a being modality to another being modality." Therefore, "...simply, one of the possible being modalities, opposed not to "Reality" but to "Actuality".....". It is a part of human experiences and in these terms we should consider the best possible way of understanding it. The goal of Virtual Reality is resumed within creating experience which should be of remarkable quality as real world's experience, or an unusual situation "proposed differently" or impossible to experience in Reality (Levy,1997)..

The philosopher Karl Popper defines three kind of world's experience of society:

"First, there is the physical world – the universe of physical objects...this is I will call World 1. Second, there is the world of mental states; this is I will call World 2. But there is also a third such world, the world of the contents of thoughts, and, indeed of the products of the human mind; this I will call World 3" (Popper, 1977).

The aim of the virtual reality is also to create a conceptual "world-in mind" (World3) experience through high technology systems. The human perceptions and cognition are the links between the real world (World1) and this conceptual worlds (World3) (Kunii, 2005). The artists, as we have seen it in the previous paragraph, with their own artworks create conceptual "world-in mind" realities like many "World3". The conceptual world modeling requires "to integrate art, science and technology to create integrated culture".

Therefore, in what terms is it possible to define more precisely Virtual Reality without referring to particular computer hardware conditions? The fundamental and perceptible experience within these artificial environments is the sensation of "Being There" in this unknown World.

In 1992, Jonathan Steuer argued his preferences in indicating virtual reality through notions of "Presence" and "Tele-presence" where "Tele-presence" was defined as "a mediated perception of *an environment*" (Steuer, 1992). Gibson stated his concept:" ...the essence of being in an *environment*" (Gibson, 1992). Sensorial immersion in simulated worlds exclude the user from real world with inducing psychological phenomena of disembodiment. This phenomenology announces the possibility of freeing the body and becoming another living being: a Cyborg. The concept of cyborg, of extended body and post humanity are frequently used in Digital Art. Stelarc is a performance artist that made numerous artworks which propose different systems of prostheses-interfaces which unify humans to machines as Exoskeleton (1998).

It is a compressed air system with six telescopic legs that the artist uses for his movements within its installation. Stelarc did other experiments for the development of a muscular stimulation system that permitted to the audience a remote control of his body. Moreover, Virtual Reality is defined within our individual consciousness; in fact it's related to various factors needed to the creation of a sense of presence, which differs in every individual. Perceptual states are influenced with different contents of mediated environment, its entities' types and interactions levels. The term "Presence" identifies personal impressions induced by complex expressions of sensorial paths, unified with previous memories of the same experiences (Gibson, 1996). This terminology is used to describe a mediated experience within a physical environment. Shapiro and McDonald (1995) distinguished between "reconstructed reality" created with lots of personal information through presentations of experienced events and "constructed reality" where everything is founded on the way humans accept the presentations of events as reality. Heeter (1992) describes three distinct presence typology or "Being There": personal subjective experience, social presence and personal objective experience. The emotional situation described as "Presence" refers to natural perceptions induced by virtual environment, Steuer, instead, intended "telepresence" a mediated condition. The concept of "telepresence" refers to every medium that induces in the self a sense of presence. This concept has been introduced by Marvin Minsky (1980) with his referring to teleoperating systems, on remote manipulation of objects. The previous concept of telepresence has been widened with redefining "Virtual Reality" without any need of referring to hardware systems:"Virtual Reality is defined as physical or simulated environment where is perceived a telepresence experience"(Marvin Minsky, 1980). Nowadays, this kind of situations are "normal" for many users. There is a possibility to talk to somebody who is not physically present in the same room. The movie "Being There" (Ashby & Kosinsky) reproduced it in 1979. We need other concepts for giving a definition to every characteristic of this phenomenology .Multimedia artist Michael Naimark defines these characteristics as "realness" and "interactivity". Brenda Laurel (1990) and Rheingold (1991) made similar distinctions. These are some of the principal definitions:

- **Vivacity:** This term defines a medium's capacity to produce sensorially rich environments;
- **Interactivity:** This term defines a range of interaction possibilities at user's disposal, capable to create relations and to influence form and content of the mediated environment.

Sutherland (1965) indicated that virtual reality is the ultimate simulation. He enthusiastically described his physical immersion in the electronic space which he called "mathematical wonderland". This kind of Reality couldn't be reproduced in its physical form in actual three-dimensional space, "There's no there there" but only in the Place, difficult-to-define within the computer, which William Gibson [1984] a science fiction writer defined Cyberspace.

There are many studies that try to define individual factors which contribute to a forming of perceptions, experienced in the virtual environment. Most of the researches are related to a sense of presence or self-consciousness of being in virtual environment. It becomes a place where mediated communication happens. Discussions and trials searched for elements from natural environments which should be reproduced or modeled to facilitate our consciousness of "Being Present" (Lombard & Ditton, 1997). The advances in scientific areas dedicated to mediated experiences are closely related to the technological improvements and to the Human Computer Interface developments. Steuer with the definition of Vividness and Interactivity indicated them as the basic elements which contribute to the immersion level in a CMC communication channel(Computer Mediated Communication)(Steuer,1995).

It's useful to introduce and understand some meaningful examples of immersive environments typologies. Char Davies's installation "Osmose" is important because the spectator-"immerse-ant" is equipped with head-mounted video device and a jacket with sensors that registered breathing and body movements. The breathing and the participant's movements create their own paths within a forrest representation. Davies's artwork is very impressive because of its realization with transparencies and luminous fluxus and it gives an oneiric dimension to the staged virtual space. The spectator-"immerse-ant" achieves his new breathing function that makes him navigate in this artificial environment and returns to real world.

Jeffrey Shaw, Agnes Hegedus and Bernard Lintermann installation "conFIGURING the CAVE"(1996) uses an environment called CAVE(Cavern, acronym of Cave Automatic Virtual Environment) developed by researchers of the Electronic Visualization Laboratory (EVL) at the University of Illinois, Chicago. CAVE environment simultaneously represents the virtual scene on four surrounding and transparent walls, staged with an audio system. The name of this artificial environment refers to one of the best-known allegory of the Cave in Plato's Republic (Book VII). This allegory has been developed on the concepts of reality, representation and their perception through the image of the Prisoner detained within a cavern, where his perception of the world's reality is re-assumed with Shadows on the Walls of Dancing Flames. ConFIGURING the CAVE creates an immersive environment with projecting the synthetic images on the walls. The user with moving his head and body, explores and modifies the elements of audio-visual space.

Installation Memory Theater VR (1997) of Agnes Hegedus, uses the virtual reality from a different, more conceptual and historical point of view. The visitor controls the projection onto a large, cylindrical screen of four interactive film scenes, that represent the history of imaginary space visualizations. The installation's idea explicitly relate to the ancient concept of the memory theater, re-adapted to actuality with a new consideration of architectural space as a digital information space. The concept of information space returns to the originary idea of Memory Palaces or Method of Loci and other mnemonic arts technique used in ancient history. Cicero in 2 century BC imagined to associate diverse discourse elements to different parts of an imaginary city and go from one place to another of that city while he was making a speech. This mnemonic technique is founded on the idea that our memory functions with a spatial analogy.

In the XVI century, a system of signs and physical structures stopped to give storing indications about a transcendent wisdom of the world. Giulio Camillo (1480-1544) started with the idea of a Palace of Memory and he built a Memory Theater structure exhibited in Paris and Venice. The building was composed of columns with images, characters and ornaments on them that should have represented all the Universe's Wisdom, with allowing the visitors to have all their questions answered with the help of Cicero's wisdom. In the VR Memory Theater, Hegedus used this example from the ancient history to recreate it in the contemporary imaginary space and understand the history of Virtual Reality.

John Klima's 3D virtual world Glasbead (1999) uses the model of multi-user gaming. It is a multi-user virtual environment, made in spherical form and composed with musical instruments that could be played from distance and composes many environments of sounds. Virtual environment "The File Room" (1994) of Antoni Muntadas uses its three-dimensional space to create an evolutive encyclopedia realized through continuous collaboration, by following the model of "work in progress" with the goal of free information's diffusion necessity. TeleGarden (1995) installation made by Ken Goldberg and exposed at Linz's Center Ars Electronica, explores the theme of telepresence with connecting the spectator's real space with installation's remote space. The connected "internauts" interact with plants in a garden and they could see the consequences of their action in real time. The phenomenon of dematerialization enters the arts of staging with digital choreographies where dancers interact with their virtual double. Bill Seaman in World Generator (1996-97) with Gideon May offered to its participants the possibility of construction and exploration of virtual worlds. The World Generator created a complex space of semantics combinations with 3D objects, images, poetic phrases, film, sounds selected by users. The concept of this artwork installation is directly linked to the OULIPO's art movement. Maurice Benayoun in World Skin (1997) places the spectators in the center of a battle, with a camera that registered their experiences and reflected on representations acts. The interaction of physical and virtual architecture happens in Hani Rashid and Lise-Anne Couture Asymptote (1987) with trying to transfer virtual properties into reality and enriching the perception in physical space. Architect Marcos Novak (1998) with his virtual worlds, interprets cyberspace as a liquid architecture with all programmable structures, transcending real space and making it capable of intelligent interactions with the spectator.

THE VIRTUAL EXPERIENCE

Interactive environment design analyzes cognitive involvement of user's experiences during the last decades. The human user or participant, as virtual performer, emerges in his corporal, emotional and social entity. The complexities of these entities are used in actual socio-technical networks and through their intersections, they establish cultural networks of meanings. Nowadays, subjective experiences of user's interaction in virtual environments have been investigated also in psychological research fields. These research studies use the concept of "Presence" perception in virtual reality. This psychological human state is generally defined as a user's subjective sensation of "being there", a part of the scene developed by the medium (Barfield W., 1995). The concept of presence is relevant for designing and evaluating the virtual media. It is the key for understanding virtual reality's experience, but it isn't the only parameter. The user's emotions are essential for understanding how humans interpret virtual worlds and how to obtain important implications for conceptual understanding of virtual experience (Hang, 1999). The process of adaptation to artificial environments provides us with knowledge of biological features, creation of interaction, building artifacts, behavioral rules and development of new culture. The parameters of mediated experience compose the structure of Virtual Medium. Virtual Environment in its general composition function could be defined with its peculiar characteristics, which differs from other media, like Communication Systems. Communication systems are composed with a group of functions that are identified with a main interface, communication channels and organization infrastructures. McLuhan (1964) clearly defines that this is a prototype of Communication System and the physical world is his content. "All Media are human sensorial extensions" (McLuhan,1964).Researchers of Virtual Reality have proceeded into developing new sensorial extensions and their interfaces. Confronting Virtual Reality developments with other communication media like Television, Computers and Telephone, VR emerges as a relevant meta medium (Kay & Goldberg, 1997). "It isn't only a technology but it is a destination" (Biocca, 1995).

The final goal is, in fact, making a "full immersion of the human sensorimotor channels into a vivid *experience*" (Biocca,1995). In this ideal system, human body is immersed in the communication process with Information. Other media also include us in their environment: radio and television stimulate our imagination and observation, but they don't immerse our senses inside their environments.

THE VIRTUAL MEDIUM

Communication Media are classified regarding to their level of interaction. Interactivity parameter is an extension that is used to evaluate user's participation level under which he could modify the forms and contents of media environments in real time. Laurel (1990) explicitly discussed Interface Design, emphasizing the importance of natural experience of ours' interaction with technological media (Laurel, 1990). She describes a medium in terms of Mimesis, imitation or representation of the sensible world aspects, especially human actions, in literature and art, as relations between user and technology from acting to gaming. The Engagement, emotional state described by Laurel, serves as a critical factor in personal relations. Virtual environments have been diffusely represented in science fiction literature rather than elsewhere. Cyberspace, an electronic reality was described by William Gibson (1984) in his Neuromancer, where he offers a slightly different vision of an interactive and multisensory environment. He unifies the "real reality" with "synthetic reality" within an information matrix directly linked through a neural system to the media environment's hardware. This way a mediated experience links body's sensorial organs, where given stimuli receive responses directly from the brain, causing extreme sensations of fullness and profoundness. Gibson distinguishes one kind of cyberspace's experience from another non-interactive, and he defined it "simstim", also experienced through neural interfaces, but passive as traditional media. The phenomenology of Virtual Realities could be critically analyzed through a definition of gaps, expressed in three different modalities. The first examination offers a point of view based on the technology which suggests that the best way of understanding a "VR system" is to find out the presence or absence of a minimum hardware requirements. This modality of definition doesn't identify a conceptual analysis unit necessary for virtual reality. If the virtual reality consists with a set of machines, how could we identify a singular virtual reality? The third problem is: when and how VR is losing its theoretical dimensions? When and what are the conditions under VR cannot exist?

When a system that has the minimum required hardware is a VR system and when it isn't? In lacking of a clear definition, probably the most efficient solution would be the changing of actual definition to a broaden theoretical vision. The actual terminology has all possible uses in academic and popular fields where every "non-real" aspect or situation becomes "virtual reality". This attitude induces even more confusion and impoverishment in understanding virtual reality's phenomenology. The actual tendency is to define VR in terms of "experience" instead of "set of hardware". The definability of Virtual Reality in these terms allows a better possibility of analysis where a set of dimensions indicates a range within virtual reality could change its meaning related to other types of mediated experiences. This chapter indicates the research areas for developing new definitions of virtuality in relation to interdisciplinary areas of Metaplastic Design. One of the goals for VR is to create a sense of realness within these research areas by combining high resolution and complex computer graphics. It is displayed with particular devices capable of objects' manipulations and navigating within created environment. Previous experiences from all over the world largely remain with unsolved issues. How to efficiently use virtual reality and moreover what is the language typology that should be used within these artificial realities?

ABSTRACT LANGUAGES

The Art is a form for public expression of emotions and feelings from deep inside of our subjectivity. Art arises emotions into autonomous space-time entities, which induce us emotions and it becomes a high quality and subjective experience (Lévy, 1997).

Lévy continues with these words:

"With researching safety and control, we are looking for The Virtual because it leads us towards ontological regions which are not threatened by common risks."

Art follows this tendency and therefore "virtualizes the virtualization" because Art traces its path "Here and Now" with emphasizing sensuality. The Abstract Art is considered to be a form of language (Levi-Strauss, 2007).

One of the principal models of Abstract Art has been music and language in a very close relation. Art languages through their own plastic signs compose a visual semiology through abstraction from figurative art representation. Even within Reason's philosophers, the explanation of the abstraction phenomenology could be introduced from their *Encyclopedie* as follows:

"Abstraction is a Spiritual Operation which we use to describe Impressions on external objects, or of our Sensorial experiences within our Minds, where we develop our particular images of a specific Figure and which we separately store and maintain in our Minds[...] and because we cannot share our thoughts with others but verbally describing them, this need and ours use to give specific names to real objects, we use it even to define metaphysical concepts with our own words" (L'Encyclopedie, Diderot).

The poetic model of abstract Art within its relations with natural, human language substitutes a denotative function of words with expressive and emotional forces of plastic images.

As the Poetry represents language in its purity, so does abstract art and its pioneers with the "Art Purity" within abstract art conception. Form and color are used only to express their pure properties, not to describe figures or objects (Kandisky, 1968). The Russian's Cubism-Futurism Art Movement is very interesting because of its particular mixture of poets and painters. They unified Poetry and Painting in a unique language, a specific model for abstract art pioneers. They elaborated a methodology for construction of an abstract art grammar composed with an alphabet, a syntax and a semantics. There are opinions, even nowadays, that abstract art cannot contain signs, because it doesn't denote a meaning and so it doesn't have the value of signs (Bru, 1955, p.43). According to these opinions, abstraction is a process of non-figurative definition. Umberto Eco (1976) argued by adding a process of a relation between three elements: Signifier, Signified (meaning) and its Referent. He indicates the eventuality of signs' existence without any referents like Pegasus a mythological animal figure. We could observe that the object or referent it's not absolutely necessary to have its sign. De Mauro (1967, pg.83-84) underlines that: "...the linguistic sign doesn't unify an object and its name, but a concept and its acoustical image....", or respectively a Signified(meaning) to its Signifier. Jacobson (1960), on the matter, indicated the importance of the links between Plastic arts and language which suggested the art experimentation of Picasso with his important semiotic models. Levi-Strauss (1979) disagreed by arguing that without an actual referential object the reasoning becomes an Academic speculation on the signifier. Disagreements stand within the idea of considering the artwork's meaning crucial in identifying actually represented object. Eco (1976) suggested his solution by solving the ambiguity of languages with defining them Codes of expression. The verbal language is considered to be a particular code with two articulations and abstract art language is considered a code with a simple articulation. Nelson Goodman[2003] suggests that the relation between image and object it's just one of the possible forms of Symbolizations called Denotation. The Rhetoric Groupe μ (1992), from the University of Liegi, within the studies of Visual Semiotics' advancements refused the idea of the visual sign, which uses plastic elements identified as significant and links the icon and to its Signified(meaning). They substitute the visual sign with two real signs, the iconic sign and the plastic sign, both constituted with a signifier and a signified (meaning) (the Expression level and the Content level). They are two independent signs, but often in interaction. The merit of the Groupe μ (1992) is the disambiguation of the iconic sign from the plastic sign, by making plastic signs a new typology of signs. This means that it exists the relation between expression and content, and signifier and signified(meaning). Because we are talking of signs, the plastic significant is related to its signified(meaning) of plastic significant (which can vary). It is impossible to anticipate their meaning, but we have to analyze other elements all together: form, color, structures to understand their meaning, and this process should be relational and topological through oppositions (Calabrese, 1985). The Semiotics demonstrates that a code is valid even if it is unstable. In conclusion, from this researches it arises that a plastic sign is a new type of signs. The plastic signs are unstable, not predetermined but directly linked to oppositions which every artwork determines (Mounin, 1968). By analyzing them as indexes and symbols, their expressive value is independent from their iconic content. Feelings, emotions, notions could be expressive in the sense of the concept of their plastic meaning:

"The primary attention to the Significant Dimension in the proper plane of the plastic image, or organization of forms, colors, lines, which determine the composition of Artwork should be carefully considered with operation of Mise en Discourse by visual description and interactive simulacra with the Spectator. Klee chose another way: asked the Painting to demonstrate its semiotic independence." (Roque, 2004).

ABSTRACT VIRTUAL SPACE

"The virtual space is the calculated conquer of Measure"

Virtual Space described with the poetics of Paul Klee, found new possibilities of content expression through the attempt of world's reconstruction with his qualities values. These results could be obtained with the operation of information's extraction and their transformation, qualification, a "qualification" of quantity. The search for the World of Quality it's an intense and dynamic activity to obtain. The Form of these spaces are obtained, as Klee in (Edschmid, 1920) narrates:

"...The World of Qualities, which opens more deeply when you enter it, but it isn't a World of preexisting Forms, they change in fieri, in the course of execution, in Gestaltung; it is the World of Infinite and Organic relations, which rise from Real encounters and their measure, from the Effective Force which every image develops in its particular Space-Time position..."

Every infinitesimal element of Space is naturally, fundamentally defined in the Point. A non dimensional Point, neither Warm nor Cold, neither White nor Black, but a generic element of the artificial space of Pure abstraction: the Grey Point . It is the principal law of movement and figuration elaborated through Pure Energy in determined directions. These processes constitute The Idea of All Principles (Cherchi, 1978). Humans in Virtual Space are Dynamic Centers surrounded with Up-Down-Front-Back-Left-Right (to be noted similarities with the philosophies of Husserl and Heidegger) spaces. Objects, within their nature of continuous transforming and dynamics in these Spaces become:

"... Object isn't certain anymore, it could have existed or not, it could still exist, to be or not to be: Being, definitively, only the encounter of Coordinates, a luminous point within obscure extension of Possible Space-Time, could always mutate into another Subject, which trajectory could pass through that Point." Paul Klee in (Edschmid, 1920)

To ourselves, the metaphor of our Existence will be passing in that Point, and we could Become that "Being" in other space-time conditions, part of That Object (Kafka's The Metamorphosis). The images of virtual space cannot be separated from reality and every part of it becomes a moment of being there and existence. In the World of Qualities, the images are not only representations but they acquire Vitality by Conservation the Memory of Human Beings.

FROM CREATION TO THE SYSTEM

Paul Klee's "Creative Confession" (Edschmid, 1920) is the first document that contains an open declaration about his poetics where Human-Artist, Designer use themselves as a cause through Art. Paul Klee, with the artist's Confession, had intended to be world's Consciousness that becomes in the artist's creations and in his artwork's becoming. Art that "doesn't repeat visible things, but renders visible" (Edschmid, 1920) imposes to the Artist a necessity to construct Everything from Nothing.

"For Klee, being able to recognize himself as artist, had meant to reuse all the elementary forms of vision to justify with their genetic process of becoming. Here, in his painting which is becoming, in the universe that is becoming in its physiognomy, converge on hidden reasons of "being" and with its Intelligence cannot be given with irrelevance or with gratuity."....." There is no Universe of Forms to decompose, with syllogistic geometry deduction a predefined drawing of combinations, but exists a necessity of becoming to primordial elements from which Constructively, with spontaneous movement, creates Physiognomy of the World which enfolds us." (Cherchi, 1978).

Klee's reality is an experimented reality through a qualitative translation both in a conscious and unconscious way. His analysis of the plastic syntax would never be Pure and Morphologically distant description of its types and technical conditions, property of Painters, but it would always be those types and conditions as vital interaction of the artists and their experiences of the world.

"A sort of Logical Formalism seems to be imposed as Equilibrium in Chaos of Sensations, at least in a sense that they are significant choices and predisposed to a verified effect or to be verified." (Cherchi, 1978).

The Figuration or *Gestaltung* is a "tool" used by Klee to configure, because apparent emptiness of Chaos is becoming a world of Form's phenomenology. Following the theories of Paul Klee, the action of vital forces that animate Universe, act in a universal state of tension within polarity of Chaos-Cosmos. Cosmo-logically, that polarity is explained as a duality of "being or not being" of All, which follows physical and biological laws. Authentic chaos figuratively corresponds to the concept of the Point (not mathematically) and the concept of the grey color: the grey point. The a-dimensionality of the point and the neutrality of grey, live their critical equilibrium, balanced between Being and Non-being.

"Existing has a sense that etymologically resembles to Klee's cosmogony hypothesis: Ex-sistere, to exit the chaos towards an essential order. Because of that assumption the general tension of ambivalence that initiates and concludes in authentic chaos, it contra-poses the chaos as antithesis which prefigures the perfectible possibility of Cosmos as order. This antithesis indicates the need of a process which couldn't be developed without being involved into dramatic choice, the active or passive selection: certain forms are assumed not others, this or that form is chosen, but the important thing is the becoming of being explained as a conscious duty of order. This order has no need for explanation of organic mediation as a need that the artist realizes between natural order and artificial order." (Boulez, 1984).

The dispute between two Orders is recursive with Klee, and the order between "visible" and "invisible" (Edschmid, 1920) contains it very clearly: that dichotomy is constantly analyzed. The artificial order seems reasonable compared with rational geometry that seems untruthful: but it's a contradiction, because Klee uses geometry and underlines its practical value (Cherchi, 1978). The natural order doesn't represent a positive category versus a negative category of artificial order. This discussion converges more on speculations on the nature of speculations than making deductions about order.

"If our senses are linked to the condition of finiteness, the natural order is the world of Invisible, the world of Genesis, of relativity, of movement: every sensorial definition or definition with practical proofs, couldn't be schematized and lose its richness by achieving clarity of the artificial order." (Edschmid, 1920)

These Klee's specific evaluations are necessary to be introduced because they refer to the Metaplastic Dynamic Forming which will be introduced in the chapter "Metaplastic Virtual World" that follows within this book. According to Klee, when natural order determines its internal flows, its dualities as movement and anti movement reciprocally counterbalanced, the artificial order fixes its dynamical flows and computation in a gradual quantification. On one side, there is Self-consciousness, but oriented into formalistic adventure which is progressively distanced from the organic content to be applied to. On the other side, it consists with needs for ordered Self-actualization, immediately and livingly, to follow the "Creative Confession" of *Form In Fieri*. "Creative Confession" means in fact, organic consciousness of natural order, and by understanding it and controlling it through analytical experience, its polyvalent states and Cosmic forces surpass the limits of *Erlebnis (lived non-conceptualized)* (Edschmid, 1920). These polarities within their alternating meanings, often determine contradictorily emotional states in human mind. Klee was even more sensible on issues linked with Visual Language codification, by dictating his rules in theoretical essays, and as a theorist of *Gestaltung* he felt the need for a construction of a system that would be useful for the Images' perception. He translated images into Representations through rationality of tools for Communication and rendered them intelligible with its peculiar aesthetic experience.

METAPLASTICA

"Here is learned how to organize movement in logical relations" Klee

The earlier plastic art movements characterized the conceptual and structural basis of the virtual media with the spatial and kinetic methodologies activated by the user during its interaction processes. The conceptualization of Klee's logical-plastic formalism indicates the possibility for a definition of virtual space theoretical Metaplastic model. It is defined through composition rules of abstract art languages. The plasticity, concept in opposition with elasticity, in digital media terms is a characteristic of the user's activities that within its own interaction process can create, modify and perform every form and content of the newer virtual media. The increased plasticity of the post Web 2.0 digital media include the social dimension as another level of potentiality to extend human communicative and creative possibility for the new virtual communities. The term *metaplasticity* is defined within the neuroscience or in an algorithmic sense of plasticity which are different point of view from a definition that would be given in the metaplastic discipline. The *Metaplastic discipline* is composed with different transdisciplinary fields coming from Art, Design, Architecture, Cybernetics, Psychology, Semiotics, Artificial Intelligence and Computer Sciences.

Metaplastic discipline defines proper goals characteristics with derivation from its originary disciplines as:

- *Interdisciplinarity* of existing relations between Art, Design, Science and Technology;
- *Dematerialization* of artworks and its processes from their disciplines;
- *Hybridization* between Aesthetics and Technology(hybridization type 1) becomes re-definition of sensible forms production practices; it becomes in socialization through artworks and in different levels of interaction modalities: between artwork, author and spectator(hybridization type 3) and between society, science and technology (hybridization type 2);
- *Interactivity* as a fundamental paradigm of dynamic relations occurred among author, spectator and artwork;
- *Synaesthetic immersivity* of spectator through his sensorial and psychological involvement coordinated in the interactive representation;
- *Communication of Wisdom* as a cultural goal to be obtained through the creation of new metaplastic media.

Digital Metaplasticity describes plastic qualities of digital media configurations and its expressions through the applications of abstract art languages and methodologies to computational symbolic systems. The metaplastic media, one of discipline's objects, within its own aesthetic and semantic codes define a new culture of the representation. Interaction processes defined with metaplastic codes, trace behaviors and plastic multisensorial qualities.

Metaplastic languages or codes are methodologies based on abstract art languages rules applied to digital symbolic systems needed for the construction of Metaplastic media Entities. The following definitions introduce some other necessary levels of theoretical definitions:

- *Metaplastic Entities* are complex objects of any present and future typology, which directly act and interact within the applied metaplastic metalanguage rule configurations;
- *Metaplastic Virtual Worlds* are digital spaces typology of Metaplastic Entities;
- *Metaplastic Metaspace* is a cyberspace composed with a network of Metaplastic Virtual Worlds.

METAPLASTIC VIRTUAL MEDIUM

The development of a theoretical meta-model for virtual media definition is based on the metaplastic ontology (Figure 1). This model is a conceptual map described within three fundamental study fields: the information field for the "Construction of knowledge"; the sensorial field for "Emotional involvement" and the area of "Social Participation". Every field mentioned before, was defined with more subfields. The parameters were placed on the conceptual map axis, each for every field through a qualitative evaluation to define a digital communication media. The two external zones were defined, respectively, as the "Private space" and the "Social space", within the use of media and its cultural content. The internal circle's width, visually indicated the conditions of being part of the media included or not included in the social spaces. Defining that media property, we defined different reality conditions: virtual reality, extended reality or mixed reality.

The level that described the "emotional involvement" area, indicated which conditions had improved to create an immersive reality. The application of the methodology to the laboratory projects, defined their quality and made it visible for different categories and various relations levels. The resultant models of the virtual media, offered useful indications about the shared informations; the definition of the user's activity; the interaction modalities with the information and with other connected users. The visual language model had confirmed the importance of the aesthetic value as a meaning element and to emphasize the emotional intensity of the medium.

Figure 1. Ontologies: (Left) Metaplastic discipline; (Right) Metaplastic Media. ©2010 Gianluca Mura

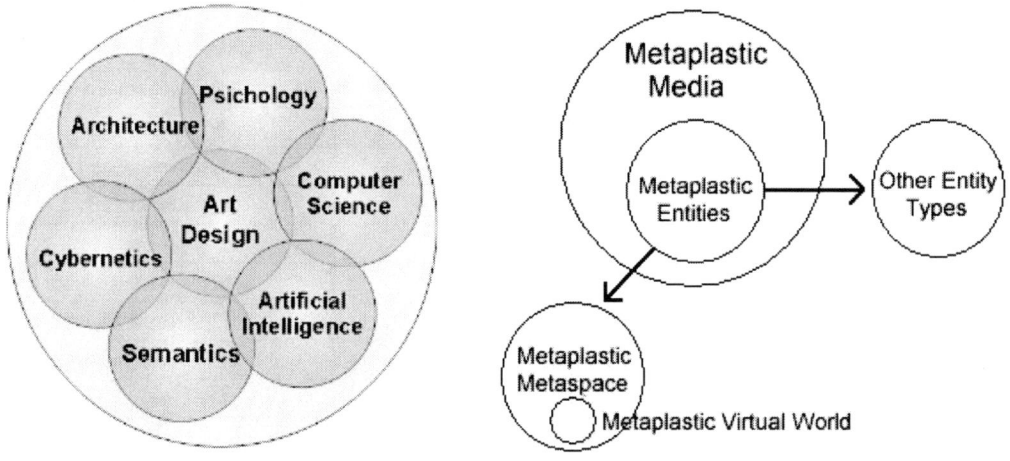

THE METAPLASTIC MEDIA MAP PLANNING TOOL

This paragraph describes a visual methodology for the definition of metaplastic media. The method is shown with valid parameters of a virtual environment project. Each value in the Media characteristics map, determines its polar coordinates (value, angle) within the Figure 2. This way, it is defined media's different characteristics and their values of planned media hypothesis. It is useful to obtain every media's quality value with the application of following function:

Metaplastic Media f = (S, St, P, Iv, Cg, If, Sm, A, E, K, SK, SE);

where are used the following properties: S = user's social participation; Cg = cognitive property; St = space-time ubiquitous connectivity (mixed and augmented reality); P = playfulness; Iv = emotional involvement; If = information contents

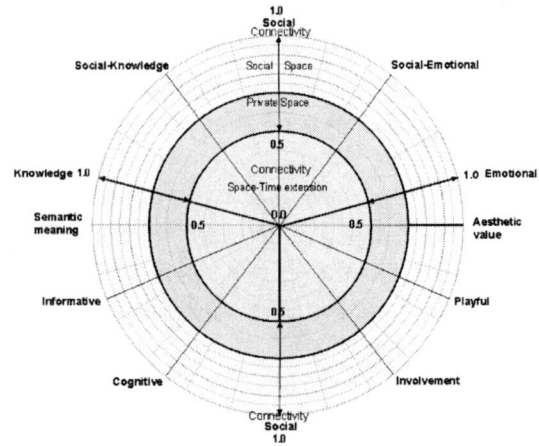

Figure 2. The Metaplastic Media Model. Use this map for media planning. ©2010 Gianluca Mura.

fruition; Sm = semantic meaning; A = aesthetic value; E = emotional content; K = structured information contents; SK = Social-Knowledge content value; SE = Social-Emotional content value; X,Y = map values in polar coordinates. (Table 1)

Table 1. Metaplastic media form. Use for calculating metaplastic virtual media hypothesis.

Metaplastic Media Planning Form Metaplastic Media f = (S, St, P, Iv,Cg,If,Sm,A,E,K,SK,SE);			
Value	X in [0..4] grade evaluation (Low/High) grades: 0 = Low .... 4 = High	Value	Y in [0.. 360°]
St Space-time connectivity	Grade [0..4] .... * 0,25 = ...	Y =	90° (social axis)
P Playful	Grade [0..4] .... * 0,25 = ...	Y =	337,5° (playful axis)
Iv Involvement	Grade [0..4] .... * 0,25 = ...	Y =	307.5° (involvement axis)
Cg Cognitive	Grade [0..4] .... * 0,25 = ...	Y =	232,5° (cognitive axis)
If Informative	Grade [0..4] .... * 0,25 = ...	Y =	202,5° (informative axis)
Sm Semantic meaning	Grade [0..4] .... * 0,25 = ...	Y =	180° (semantic axis)
A Aesthetic values	Grade [0..4] .... * 0,25 = ...	Y =	0° (aesthetic axis)
S Social Connectivity	(St + ...) / 2 = ...	Y =	90° (social axis)
E Emotional	(P + Inv + ...) / 3 = ...	Y =	15° (emotional axis)
K Knowledge value	(Cog + Inf + ...) / 3 = ...	Y =	165° (knowledge axis)
SK Social-Knowledge	(S + K) / 2 = ...	Y =	112,5° (social knowledge axis)
SE Social-Emotional	(S + E) / 2 = ...	Y =	22,5° (social emotional axis)
Space	Social space **S** <= 0.5 (Yes=4 No = 0)		(social axis)
	Private space **S** > 0.5 (Yes=4 No = 0)		(social axis)

Every coordinate pair (x,y) of calculated values are obtained in the Media planning form, Table 1. The obtained values are drawn on respective radial scale and it is unified to give proper map fields that illustrate desired media's properties (see example in Figure 3).

The "Virtuality in Arts and Design": Virtual Exhibition Projects' Experience

The workshop discussed themes linked with creation and application of the virtual reality in the Arts and Craft for producing and communicating cultural content. The event gave an occasion to express and discuss different combinations of art languages and industrial design planning. The activity was developed in two different but interactive levels: the student laboratory and the discussion forum about inherent themes. The meetings were held between the professors and students at the Ecole du Louvre with Professor Xavier Perrot and Gianluca Mura of the Politecnico di Milano University, Industrial Design Faculty (Perrot, Mura, 2005).

Simultaneously through a web blog there was a continuous debate and elaboration on the proposed projects. As results, the workshop offered different exposition models with significant examples developed through the media map methodology within the student laboratory. The complexity of digital media for expositions, requested a particular attention to describe the relations between visitor and interactive devices. The analysis of those relations solved conceptual problems about the communication of cultural content. For that purpose, it was proposed that the designer-planner should have suitable competences and high sensibility, regarding elements for discussing cultural complexity. The international workshop offered an important occasion for interdisciplinary collaboration and analysis of the phenomena at various levels with consequential elaborations of methods and tools on related themes. The "Virtuality in Arts and Design Workshop", during its sessions took into consideration different media typologies and their inherent problems. The workshop had experienced the encounter of museology and museography disciplines with industrial design, using theoretical instruments to develop unifying project practice. The final results were considered

Figure 3. Metaplastic Media Map example of a virtual environment project. ©2010 Gianluca Mura.

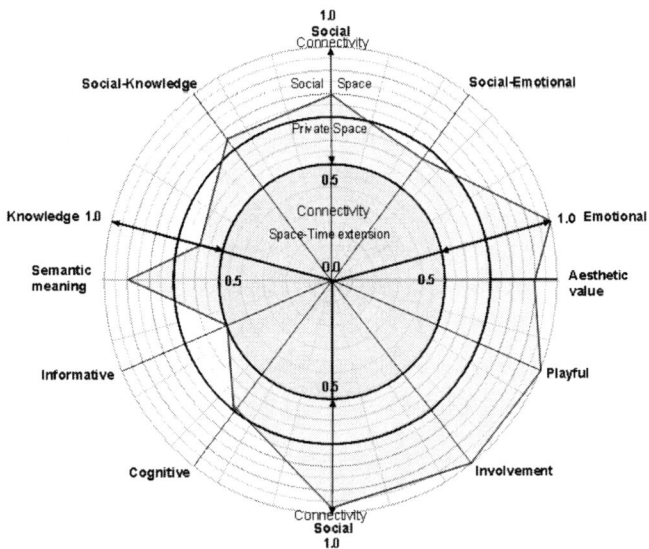

helpful as a comparison base for successive didactic improvement (workshop website at http://workshop.artsmachine.com). The workshop had obtained the High Patronage of the Italian Ministry of Foreign Affairs for cultural exchange.

DESIGNING POETICS IN IMAGINARY SPACE

The conceptual activity of designing Metaplastic Virtual Environments (VE) poses a number of problems to solve in different ways, through a direct relations between the visitors' psychological being and both the embodying interface and the mediated content. The design's process complexity involves several fields of knowledge like technical and scientific information, social, cultural and psychological factors. The abstraction levels and the construction of meaning for multidisciplinary analysis requires the development of a multidimensional knowledge system (Mura, 2007).

"The medium of 'immersive virtual space', or virtual reality as it is generally known has intriguing potential as an arena for constructing metaphors about our existential being-in the world and for exploring consciousness as it is experienced subjectively, as it is felt" (Davies, 2001, pp 293-300).

The research about the human hybridization with the virtual space has been developed in various artistic currents like the Body Art, Fluxus and the most recently NetArt. From these analyses emerge the need for suitable communication space. Canadian artist Davies realized it with her work a new kind of spaces.

"Through which our minds may float among three-dimensionally extended yet virtual forms in a paradoxical combination of the ephemerally immaterial with what is perceived and bodily felt to be real" (Davies, 2001)

The philosopher Gaston Bachelard in his book "The poetics of space" examines the potential of psychological transformation of "real" places like deserts, valleys and deep sea, open spaces different from urban spaces to which we have used to:

"By changing space, by leaving the space of one's usual sensibilities, one enters into communication with a space that is psychically innovating. For we do not change place, we change our nature" (Bachelard, 1966).

Other significant examples could be found in the work of artists as Jeffrey Shaw and Petra Gemeinboek with "Uzume" and many others. Open virtual environments offer individual and collective access with different types of experience construction which stimulate emotions, introspection, reflection and consideration on surrounding space. Virtual reality immersive systems need new definitions of concepts and space movements, with logical test and aesthetic values defined within user's interactions. Planning of dynamic processes within virtual environment require a particular sensibility for dialog construction between user's extended senses and his surrounding immaterial space. This metaphysical space is a metaphor of the post-modern society which is completely immersed into globalized information fluxus. Metaplasticity tries to give an opportunity of redefinition of "user-human scale" condition, becoming an independent actor of its personal political and social knowledge dimension.

Fluid immersive multidimensional space is, in fact, a landscape of knowledge where the user could make his abstract surrounding experience. The interactor, user or Virtual Performer redefines his role within its abstract space immersion. This enables him to interact with the environment and make new forms of data creations, also with other connected users. The user in his relation with the synthetic environment realizes different level of participation/inclusion within the system, which changes his role from spectator to interactor and at the end "immersant", defined with Char Davies's words in her installation "Osmose" (Davies, 2001).

The volume forms develop into a system maps and has a route orientation function. The orientation in the system is possible through signs which indicates to users the possible options. The remote presence of the user's body and his extended senses, like analogue/opposite field polarity in the immaterial-materiality of sound, light, form and colour produce various space aesthetic effects. The user in these new plastic spaces could explore and interact with information and contribute to change, delete or recreate it in different conceptual landscapes. User's results give important information about the conceptual expression of his artwork and resulting knowledge maps, and their dynamics within the virtual environments. Interactions in performance, produce aesthetic effects of the virtual environment with transforming the entire system into a new artwork (Figure 4).

IMAGINARY MUSEUMS

Technologies of Virtual Reality have irreversibly changed a sense of historical, philosophical, and technological role within museum's institutions, not only by changing their technologies, but with changing institutional structures and cultural organization. They have permitted museum's to open doors for mass media society of the Global Village(Mcluhan, 1962), with realizing a visionary model of "museums without walls"(Malraux, 1957). For better understanding a phenomenology relative to these developments, we have to resume museum's definitions. The official museum definition is the following:

"A museum is a non-profit making, permanent institution in the service of society and of its development, and open to the public, which preserve, acquires, conserves, researches, communicates and exhibits, for purposes of study, education and enjoyment, material evidence of people and their environment. (...)" (ICOM, 2009).

Another definition:

"Museums are gathering places, places of discovery, places to find quiet, to contemplate and to be inspired. They are our collective memory, our chronicle of human creativity, our window on the natural and physical world"(Ambrose, 1993).

The role and knowledge transformations happened within museums not only because of technology. The transformations have their origins inside the same causes of museum's becoming, in fact:

"Museums have their roles in our relations with artworks, which we think that don't exist, have never existed, where is ignored or was ignored the modernity of European civilization; which ex-

Figure 4. Metaplastic abstract spaces. Xense project. ©2010 Gianluca Mura.

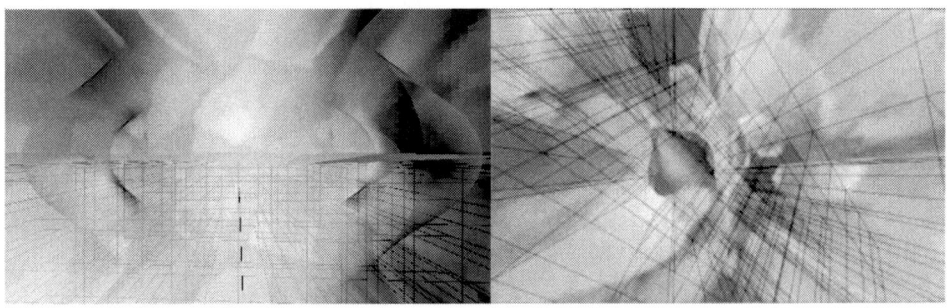

ists for less than two centuries. The 19th century lived of museums; we still live with them, but we forget that they have imposed to the spectators a new relation with artworks" (Malraux, 1957).

Malraux stated that the phenomenology of museums' changing functions and meaning are strictly related with artwork's exposition and to museum's spectators-visitors roles. The act of sharing artifacts have transformed their aesthetic value. New possibilities of communication allow museum's spectators-visitors the construction of personal paths of museum's visit adapted to their own knowledge levels.

"the technological museum is able to become an augmentation system of human capabilities, helping men to facing with the complexity of the actual society"(Engelbart, 1963) "and with complex problems in order to understand them"(Perrot, 1993).

Therefore, the definition of Cultural Heritage would be:

"Any concept or thing, natural or artificial, which is considered to have aesthetic, historical, scientific or spiritual significance"(ICOM, 2004).

Recently, there has been further developments of Museology's Disciplines. In fact, the ICOM has included specific conditions:"Museums has to have a Collection, which has to be properly stored, conserved and archived". Exceptions of those conditions are:

"Museum with a Unique Artwork, Naval Museums or House-Museums with culturally linked objects; Art Museums which apparently haven't acquired a Permanent Collection, Science Museums or Children's Museums without Collections and Virtual Museums with Virtual Artifacts"(ICOM, 2004).

This definition excludes most of Institutions that don't acquire, conserve or research materials despite they might have a service to Society's functions and its development. These Institutions and their evolution could be linked with research/ studies motives, educational and ludic functions, physically or virtually accessible for the Public. Virtual Museums with these new redefinitions in museum's cultural areas has been accepted as a new museum's typology. D. Tsichritzis and Gibbs (1991) were among the first to use the term "Virtual Museum"(McKenzie,1994) even if the first conceptual description was made by Malreaux (1957) with the introduction of the concept "Museum without Walls" or "Imaginary museum". Within actual research context has been accepted this definition:

"Virtual Museum is a collection of digital artifacts and information resources and virtually staged with a possibility of virtual visits and fruition by the visitors-users". (Malraux, 1957)

The concept of Virtual Museum has two slightly different definitions:

1. Indicates a digital version of existing collections, which could be defined as *digital museum*;
2. Indicates an *imaginary museum* which doesn't have real museum correspondent.

The difficulty of exact functional and topological classification for Virtual Museums is founded in their characteristics of immateriality and adaptability. Initially, a diffused opinion was that Internet would be used only to attract new interest for real museum's collections and that it would be used only for marketing purposes. The characteristics of the surrogate visit doesn't replace a real visit because the real museum visit is on the contrary amplified, extended, enriched.

If it is true that Virtual Museums cannot replace real artifacts experience, they should fulfill the following aspects described by Valentino and Mossetto (AAVV, 2001):

- Stimulate the public to go and see the original artwork in museums;
- Include other sensory experiences other than vision: sound, touch, or even olfactory, which should be inappropriate in a real museum (http://elteatrocampesino.com, 2010);
- Give a sense of orientation in order to easily find exposed contents;
- Contextualize the artifacts;
- Visualize techniques used by artist for their artworks;
- Provide restoring and virtual reconstruction of artworks and sights of cultural interest;
- Create new representations in imaginary space of artifacts that could be only virtually unified when a museum cannot expose them together (Malraux, 1957);
- Offer museum's and its exhibitions history in virtual space;
- Exhibit collections to wider, global audience;
- Show sights not exposed to public.

Virtual museum visit is more than a series of instructions' execution and extends a limit of a simple display of photography. By looking at computer monitor, we can walk through artifacts in a gallery, look closer to its details and we could save it and print it for personal use or obtain further information on artist's bibliography. In most cases, it's possible to participate to a guided tour, accompanied with voices and sounds. Photographs or digital reproductions aren't original artifacts, but copies on different digital devices. Malraux never thought of Imaginary Museum as a substitute for Real Museum, but as its particular extension, with proper specific functions and important historical researches. Museums on the Web offer new opportunities for eliminating accessibility barriers for disabled visitors. It is an accessible museum for all, and a precious occasion to learn and educate, as in for example SFMoma with their "Learning how to look at a picture", where artworks are explained with high educational methods. Education is the most important goal and function within a museum and it assumes diverse forms related to the place and circumstances. Virtual Museum gives a possibility to researchers with their researches divulgation. This new cultural phenomenon has an important area within museum activities.

For example, ICOM (International Council of Museums) and UNESCO uses many resources to investigate and report these new activities. This process is irreversible, but we shouldn't forget that many museums don't have their websites. Moreover, most museums don't have important international supporters, for example. Most of them have their website similar to a brochure and others are investing more resources in creation of their website contents. The Hermitage in St. Petersburg, which has one of the best museum's website of the world is created by IBM. An excellent example within their website is a Visual Search Engine named QBIC(Query by Image Content) (http://www.hermitagemuseum.org,2010). With choosing a criteria of research, color for example, a selection of artwork appears where that color is used. This search engine permits a particular kind of search by shape/form visual quality. The actual tendency within museums community is to experiment new types of interaction and exposition, trying to link and involve the largest audience possible. Another example to consider is the virtual space project for the Guggenheim museum: the Guggenheim Virtual Museum. This project shows an augmented environment for the exploration and the development of cultural experiences within the cyberspace.

"The museum deeply changes with interactive processes and becomes a "multiple museum" where

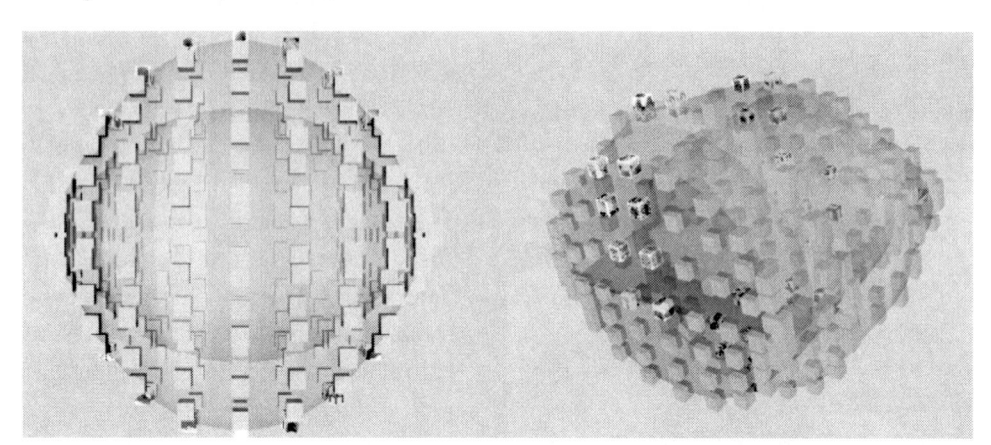

Figure 5. Metaplastic Virtual Museum. Front and side views. ©2010 Gianluca Mura

multimedia allow a superior degree of mediation with the artworks" (Perrot, 1999).

According with Perrot (1999), the role of multimedia within museums should be different from the real museum exhibition. Digital multimedia should offer a different criteria for museum's visits, by avoiding that the virtual museum becomes a simple reproduction of the real museum. The virtualization of the museum shouldn't start from a conceptual opposition between real and virtual museum but it should contain both spaces simultaneously within the same institution. Therefore virtual museum becomes a media with augmented artworks' documentation. Malraux, at last, imagined the possibility of creation of a unique global "Museum without Walls" that nowadays could be seen as the Net of nets (Internet). Social networks platforms extend interactive cultural mediation with artworks through the creation of self media networks.

These new digital networks follow the idea of Paul Valéry (1928) in "The conquest of ubiquity" ("La conquête de l'Ubiquité") where he anticipated long before nobody could even thought about the new possibilities of cultural improvements that will be offered by Internet and its social networks half a century after his essay. The main structures of virtual museum project has new goals with this particular typology:

- The *System of Interpretations* has the function of exhibition setting with the criteria of staged digitalized artifacts collection;
- The *System of Relations* provides a feedback function to visitors and among visitors;
- The *System of Coordination* provides a function of ordering for staging the collection.

This virtual museum typology adds new structures: archiving, exhibition setting, digital collection and social function. The following properties and roles define new system's parts:

- **Digital Collection:** coded information of collected artifacts;
- **Ordering:** museological criteria for classification of exhibition collection (Figure 7);
- **Exhibition Design:** museographic criteria for exposition composing;
- **Museum Experience Design:** identifies visitor's cultural and social interactive experience level (Figure 6).

Figure 6. Visitor experience model ©2005 Xavier Perrot. Used with permission

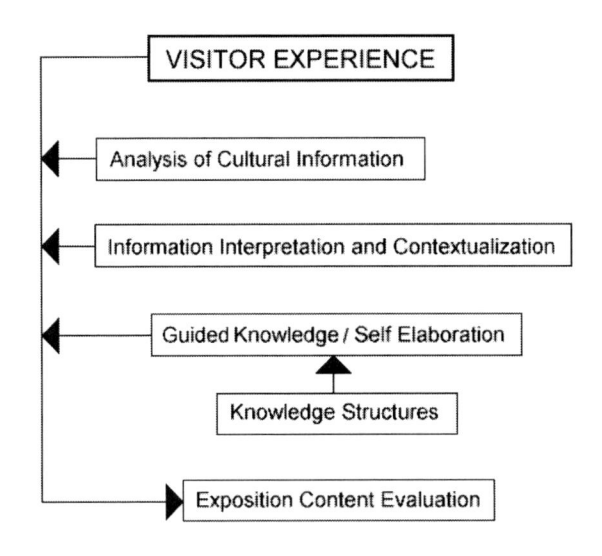

This scheme offers several combinations possibilities which would be used to create combinations of Virtual Museums. These configurations will be used within the Metaplastic language. The *Metaplastic Virtual Museum* project (Figure 5) is a particular metaplastic virtual world model built as a web museum (Gianluca Mura, 2008) which offers a complete immersive experience into cultural information contained within it. The project is a virtual space that contains cultural objects as "information-avatar", which represents 3D-artwork signs linked with its own remote data. The visitor can navigate and explore the metaplastic virtual museum and get the cultural information simply by touching the represented objects. The virtual environment guides the visit with audio-visual signs to create a high comfortable museum's exploration. The project has been exposed before at the "Museums and the Web" ICHIM International Conference at the Ecole du Louvre in Paris.

CONCLUSION

The Metaplastic discipline re-discover its fundamentals within artistic and scientific experiences. Definitions of virtual worlds are based on their artistic archetypes in art theorisations. Within the interdisciplinary encounter between artistic research and digital technologies, it is determined the possibility for metaplastic languages to construct new visions and methodologies applied to digital media developments. It has been shown the first hypotheses of metaplastic media develop-

Figure 7. Virtual Museum. Museological Path of Design objects could be explored within different theme paths, even in mixed navigation of various themes©2010 Gianluca Mura.

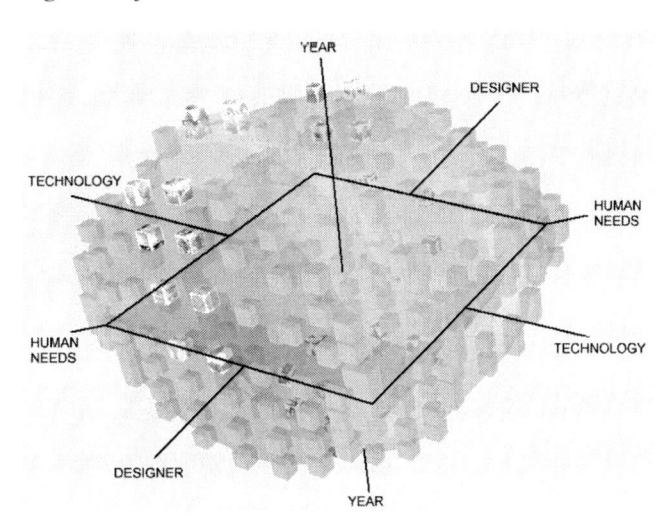

ment with starting from virtual heritage and their multimedia applications. These typologies of virtual media models conducted to the definition of the metaplastic virtual worlds and they would be widely described within the next chapter.

ACKNOWLEDGMENT

This chapter is based on the doctoral dissertation of Gianluca Mura,"Un modello di spazio virtuale del Design" developed at the Politecnico di Milano University, Italy.

I would like to thank Dr. Nataša Duboković for her collaboration in writing this chapter.

REFERENCES

Artsmachine Media Lab. (2010). Retrieved from http://www.artsmachine.com

Bachelard, G. (1966). *The poetics of space* (p. 203). Boston, MA: Beacon Press.

Barfield, W., Zelter, D., Sheridan, T. B., & Slater, M. (1995). Presence and Performance within Virtual Environments. In W. Barfield & T.A. Furness (Eds,), *Virtual Environments and advanced interface design*. Oxford: Oxford University Press

Benedict, M. (Ed.). (1991). *Cyberspace First Step*. Cambridge, MA: MIT Press.

Biocca, F., & Levy, M. R. (1995). *Communication in the age of virtual reality*. Hillsdale, NJ: Lawrence Erlbaum.

Boccioni, U. (1971). *Manifesto tecnico della scultura futurista*. Milano, Italy: Feltrinelli.

Boulez, P. (1984). *Forma in Punti di riferimento*. Torino, Italy: Einaudi.

Bru, C. (1955). *L'estetique de l'abstraction*. Paris: Presses Universitaires de France.

Bush, V. (1945). As We May Think. *Atlantic Monthly*, *176*(1), 101–108.

Calabrese, O. (1985). *Il luogo dell'Arte*. Milano, Italy: Bompiani.

Cherchi, P. (1978). *Paul Klee teorico*. Bari, Italy: De Donato.

Coates, G. (1992). *A virtual show. A multimedia performance work*. San Francisco.

Couchot, E., & Hillaire, N. (1991). *L'Art Numerique*. Paris: Ed. Flammarion.

Davies, C. (2001). *Multimedia: from Wagner to Virtual Reality* (pp. 293–300). New York: W.W. Norton & Company.

Davies, C. (2001). Retrieved May 5, 2010 from http://www.immersence.com.

De Mauro, T. (Ed.). (1967). *Corso di linguistica generale*. Roma-Bari, Italy: Laterza.

Deleuze, G. (1994). *Difference and Repetition*. London: Routledge.

Eco, U. (1962). *Opera aperta*. Milano, Italy: Bompiani Editore.

Eco, U. (1976). *A theory of Semiotics*. Bloomington, IN: Indiana University Press.

Edschmid, K. (1920), Paul Klee Schopferische Konfession. In *Tribune der Kunst und Zeit* (Vol. XIII). Berlin, Germany.

Gibson, W. (1984). *Neuromancer*. New York: Ace Books.

Goodman, N. (2003). *I Linguaggi dell'Arte. L'esperirenza estetica: rappresentazione e simboli, trad. I* (Saggiatore, I., Ed.). Milano, Italy.

Grau, O. (2003). *Virtual Art*. Cambridge, MA: MIT Press.

Groupe μ, (1992). *Traité du signe visuel. Pour une rhétorique de l'image*. Paris: Le Seuil.

Hang, H. P., & Alessi, N. E. (1999). Presence as an emotional experience. In *Medicine meets virtual reality: The Convergence of Physical and Information Technologies options for a new era in Healthcare* (pp. 148–153). Amsterdam: IOS Press.

Holtzman, S. R. (1995). *Digital Mantras*. Cambridge, MA: MIT Press.

Jakobson, R. (1960). Linguistics and poetics. In Sebeok, T. A. (Ed.), *Style in Language* (pp. 350–377). Cambridge, MA: MIT Press.

Kandinsky, W. (1968). *Punto Linea Superficie*. Milano, Italy: Adelphi Edizioni.

Krueger, M. (1991). *Artificial Reality II*. London: Addison Wesley.

Kunii, T. L. (2005). *Cyberworld modeling – integrating Cyberworld, the Real World and Conceptual World. CyberWorlds 05, CW2005* (pp. 3–11). IEEE.

Lanier, J. (1989). Interview with Jaron Lanier. *Whole Earth Review*, *64*, 108–119.

Laurel, B. (1990). *The Art of Human-Computer Interface Design*. Reading, MA: Addison-Wesley.

Levi-Strauss, C. (1979). *Myth and Meaning*. Berlin, Germany: Schocken Books.

Levi-Strauss, C. (2007). *Anthropology and Aesthetics(Ideas in Context)*. Cambridge, UK: Cambridge University Press.

Lévy, P. (1992). *Le tecnologie dell'intelligenza, trad. it*. Milano, Italy: Ed.A/Traverso.

Lévy, P. (1997). *Qu'est-ce que le virtuel? Il Virtuale, it.translation*. Milano, Italy: Raffaello. Cortina Editore.

Malraux, A. (1957). *Il Museo dei musei*. Milano, Italy: Mondadori.

McKenzie, J. (1994). Retrieved March 2010 from http://www.bham.wednet.edu/muse.htm

McLuhan, M. (1964). *Understanding Media, trad. it*. Milano, Italy: Garzanti.

Moles, A. A. (1990). *Art et Ordinateur*. Paris: Ed. Blusson.

Mounin, C. (1968). *Introduction a la semiologie*. Paris: Minuit.

Mura, G. (2007). The Metaplastic Virtual Space. In Wyeld, T. G., Kenderdine, S., & Docherty, M. (Eds.), *Virtual Systems and Multimedia* (pp. 166–178). Berlin, Germany: Springer.

Mura, G. (2008). The meta-plastic cyberspace: A network of semantic virtual worlds. In *ICIWI 2008 WWW/Internet International Conference*, Freiburg, IADIS.

1993Museums in Education. Towards the End of the Century. In Ambrose, T. (Ed.), *Education in Museums, Museums in Education*. Edinburgh, Scotland: HMSO.

Nelson, G. (2003). *I Linguaggi dell'Arte. L'esperienza estetica: rappresentazione e simboli, trad. it* (Saggiatore, I., Ed.). Milano, Italy.

Nelson, T. H. (1965). *A File Structure for the Complex, The Changing and the Indeterminate*. Cambridge, MA: MIT Press.

Paul, C. (2004). *L'Art Numerique*. Paris: Thames & Hudson.

Perrot, X. (1993). Applications in Museums. In D. Lees (Ed.), *Museums and Interactive Multimedia. Proceedings of the Sixth International Conference of the MDA and the Second International Conference on Hypermedia and Interactivity in Museums (ICHIM '93)*. Cambridge, UK, September 20-24, Archives & Museum Informatics, Pittsburgh, PA, USA.

Perrot, X. (1999). L'avenir du musée à l'heure des médias interactifs. In B. Darras, D. Chateaux (p.149) *Arts et multimédia*. Paris: Publication de la Sorbonne.

Perrot, X., & Mura, G. (2005). VR Workshop 2005 "Virtuality in Arts and Design" Virtual Exhibition Projects. Archives&Museum Informatics. Retrieved from http://www.archimuse.com/publishing/ichim_05.html

Popper, F. (1970). *L'arte cinetica*. Torino, Italy: Einaudi.

Popper, K. R., & Eccles, J. C. (1977). *The self and its brain*. New York: Springer International. QBIC (Query by Image Content). Retrieved March 2010 from http://www.hermitagemuseum.org

Ralston, A., & Reilly, E. D. (1983). *Encyclopedia of Computer Science and Engineering*. New York: Van Nostrand Reinhold.

Rheingold, H. (1991). *Virtual Reality*. London: Secker & Warburg Limited.

Roque, G. (2004). *Che cos'è l'Arte astratta?* Roma, Italy: Donizelli.

Steuer, J. (1992). Defining virtual reality: Dimensions determining Telepresence. *The Journal of Communication*, 42.

Sutherland, I. (1965). *The ultimate Display*. Paper presented at the IFIP Congress, *65, 506-508*, Information Processing.

Tsichritzis, D., & Gibbs, S. (1991, October). Virtual Museums and Virtual Realities. In *Proc. Intl. Conf. on Interactivity and Hypermedia in Museums*, Pittsburgh, USA.

Valéry, P. (1928). La Conquete de l'ubiquite. In *De La Musique avant toute chose*. Paris: Editions du Tambourinaire.

Wiener, N. (1948). *Cybernetics or Control and Communication in the Animal and the Machine*. Cambridge, MA: MIT Press.

Woolley, B. (1993). *Virtual Worlds*. Oxford, UK: Blackwell Publishers.

KEY TERMS AND DEFINITIONS

(Digital) Metaplasticity: Describes plastic qualities of digital media high configurations and its expressions through the applications of abstract art languages and methodologies to computational symbolic systems.

Digital Museum: A digital multimedia version of existing collection or museum.

Imaginary Museum: A Virtual Museum. Metaplastic Virtual Museum.

Metaplastic Interactivity: Interaction process defined with metaplastic codes, traces behaviors and plastic multisensorial qualities of the virtual Metaplastic space.

Metaplastic Language: Conceptual model based on abstract art languages rules applied to digital symbolic systems.

Metaplastic Metaspace: A conceptual cyberspace composed by a network of web metaobjects (Metaplastic Virtual Worlds). The dynamic relations between each own network nodes and the user establish the metaspace meaning representation through its shapes and behavior.

Metaplastic Virtual Museum: A Metaplastic Metaspace/Virtual World as Virtual Museum.

Metaplastic Virtual World: A complex digital space which directly interact and act behaviors within the applied metaplastic metalanguage rule configurations.

Museum Without Walls: Virtual Museum from Malraux's definition.

Virtual Museum: A collection of digital artefacts and information resources and virtually staged with a possibility of virtual visits and frui-

Chapter 2
The Metaplastic Virtual World Theory

Gianluca Mura
Politecnico di Milano University, Italy

ABSTRACT

This chapter introduces a methodology of metaplastic discipline for the realization of new virtual world. It explains the theoretical and artistic background of metaplastic virtual worlds evolutions, from their archetypes to their definition. The union between the plastic elements and the fuzzy logic systems, found its expression with new metaplastic aesthetic values. The virtual media acquire form and meaning through its process of conceptual interpretation. The metaphor of the artistic machine finds its new realization where the machine itself becomes aesthetic expression of the virtuality. The following paragraphs use metaplastic definitions within Art and Sciences application fields. In the conclusion of this chapter, two practical examples will be introduced.

ARCHETYPES OF VIRTUAL MEDIUM

The poetics of artistic machine indicated different hypotheses of evolution within the Avantgarde movements' ideas from the beginning of the 20th Century. Therefore, it is necessary to introduce several historical moments of artistic and scientific thoughts that have conducted to actual media developments. This paragraph on Art History would introduce and follow a conceptual path of different notions exploration about artistic machines visions. It is also a search for aspects of artistic machines which had led to the definition of original characteristics of the metaplastic virtual worlds. The Manifest of Futurist Reconstruction of the Universe wrote by Balla and Depero in 1910, emphasized the particular interests of the Futurists' movement which suggested a future fusion of Art and Science. Boccioni, one of the major Futurist's exponents, aimed to represent the Space of Soul and he has tried to decompose the object-painting to explore the Pure Becoming of Perception. He places the spectator at the center of the painting as in his artwork "The Matter" (1912), where the idea of plans interpenetration, reaches up remarkable levels. With that painting, he obtained the objective that would made the spectator and the entire scene eternal, nearly a possible virtual space-time (Boccioni, 1971, reprinted). It is the light that gives Life to the Shape, for artists and in this concept he seems to relate to secret relations that exist today between digital images and coded artifacts, which regulate movement in the computer space. The Constructivism becoming and diffusion, coincided with the Futurist's-Machinist's aesthetics development, which assumed their own inspiring models from the technological and the Industrial worlds' achievements. In the Bauhaus artistic movement, we find correspondence with these topics, in the theories of Laszlo Moholy-Nagy expressed in "Theater der Totalität" and those of Walter Gropius in his article "Totaltheater" of 1927. Moholy-Nagy wrote:

"It is time to develop activities which will not allow the masses to remain spectators, which will not only move them inwardly, but seize them, make them participate, and in the highest transports of ecstasy, allow them to enter the action on stage" (Moholy-Nagy, 1922).

Moholy-Nagy reinterpreted the ideas of Richard Wagner, famous composer, with reducing the importance of the spoken word in a synthesized vision of space, movement, sound, light, composition and increased different artistic expressions with technical equipment. He demanded complete mobilization for all artistic forces to create the *Gesamtkunstwerk (the universal artwork)*.

MACHINISM

New phenomenology in Contemporary Art was a creation of Mobiles, developed by Alexander Calder. Mobiles are abstract sculptures, made of metal plates, plastics, wood, capable of movement with air breezes. It was the first time that movement has been added to plastic figures. Calder accomplished that in his creations which reached a dynamic equilibrium between their weight and levity. The equilibrium is the most important quality of his abstract sculptures: equilibrium are balances of statics and dynamics; statics in "mobiles" and dynamics in "stabiles" (AAVV; 2004). Kinetic researches started from concepts perceptively static (Optical art) and it has changed to dynamical perceptions by using infinite possibilities offered with the union of dynamic and luminous effects. Moholy-Nagy in 1922, started his researches on light, space and movement. He constructed his machine, the "Lichtrequisit" (Space-Light Modulator), introduced at the international exposition of Paris, in 1930, with describing it as "half-sculpture and half-machine construction". The construction was depicted by the author as:

"a superficial reflecting machine, in motion, on a circular base in which three cells create the movement in the given space. The rectangular metallic pieces move in irregular and undulatory way. The second part made of perforated metallic discs complete a vertical movement from above to the bottom, that frees one small black ball, which crosses this space. In the third cell, a spiral glass produces a virtual conical volume. The construction, moved by an engine, is fortified with hundreds of electrical light bulbs in various colors, connected and controlled by number of coils which create a complex light show". (Moholy-Nagy, 1947).

Munari in his Machinist's Manifesto in 1927 wrote:

"The artists are the only one to save our civilization from that danger. The artists need to learn about machines, to abandon their romantic brushes, palettes, canvas, frames and start to understand mechanical anatomy, the machine language, to understand the nature of machines by making them

function irregularly and using them to become artwork themselves, with their own means...."

He wrote that: "The Machine has to become Artwork"(Munari, Eco, 1962). The "Useless Machines" of Munari founded the precise need of "liberating abstract forms from the "staticity" of painting and suspend them in the air to live with us, in our environment" (Munari, Eco, 1962).

The machines of Munari anticipated Kinetic and Programmed Art with his concept invented for the exhibition of Olivetti group in Milan in 1962. Umberto Eco defines Kinetic art:

"It is the plastic art where Form movements of colors and plans unify in a mutational set. The goal of kinetic art isn't to obtain a fixed and definitive artwork". (Munari, Eco, 1962), and Eco (1962) introduced concepts of "open artwork" (opera aperta). According to Enzo Mari (1963), a researcher should experiment programming methods to "modulate" elements. It would be necessary to find a series of rules or programs where it is possible to create with great liberty of composition. This programming depends on the fact that the spectator sees only elements of artwork but not the Whole of artwork (Mari, 1963). He questions the concept of Artwork, conceiving that interactions between artwork and its creator through programming, indicate to the user its profound social meaning. Munari, in his complex researches, created "kinetic objects with variable structures" which have mechanisms/engines similar to small watches. Munari in his "Arrhythmic Machines" (1951-83) artworks realizes:

"researches of arrhythmic mechanics are linked to unknown energies, which are generated by mechanisms' functioning with regular rhythms"; with machine's irregular functioning in a sort of progressive humanization and anthropomorphic tendencies.

The "Arrhythmic Machines" need to be switched on by spectators. For Munari (1951), the machine has a human and emotional component because they are in their "identity crises" and questioned their own "existence" but reached equilibrium, where rules and randomness was accomplished.

"Rule is monotonic and randomness is uncertain. The combination between rules and randomness it's life, it's fantasy, it's equilibrium "....." The merry-go-round is kinetic and programmed artwork. Kinetic because it moves, rotates, but every rotation is a repetition and every repetition is a constant. The variants are people that goes on and off with changing composition of whole" (AAVV 2004).

Francis Picabia another important artist of Avantgarde movements, paints in his artworks the absurd machine, symbol of industrial society and wrote:

"..the genius of the Modern World is the Machine, and within machine, Art could find its vivid expression" (Camfield, 1979).

His intention wasn't to create absolutely new forms, but significant images of real light flows. In 1924, Archipenko realized a real machine that was conceived in a way that creates the illusion of pictorial subject's movement, analogously to slow-motion in the Cinema. Each side has several thin metal strips, overlapped as in a Venetian blind. Pieces of painted canvas, successively move in front of the spectators, due to an electrical engine. It obtains the real movement effect, and Archipenko underlines that the invention's essence and its origin shouldn't be given importance, but to the subject's becoming in its mutation (Popper, 1970).

THE CELIBATE MACHINES

Marcel Duchamp conceived one of the most enigmatic metaphors of artistic machine and its relations with human experience and artificial creation of technology: the celibate machines. They are based on visual anamorphosis, where mechanical, optical and technological implications relate to ambiguous biological forms: the interchange between human and machine, which as a cruel initiation links the machine to its metaphysical desire and subconsciousness. This is a description of his creative process:

"He took an ordinary and existential element and placed it with its utilitarian meaningdisappeared under a new title and a new point of view - had created a new concept of thesame object" (Marcel, 1959)

His artwork La Mariée mise à nu par ses célibataires, même (The Bride Stripped Bare by Her Bachelors, Even; The Large Glass), it's a prototype of his celibate machines.

METAMATICS

Jean Tinguely's experiments are mainly on motion and how to obtain it from mechanical means. His experiences realized a metaphorical development of mechanisms' symbolic value, where the mechanisms found their anthropomorphic semantic value and reach their auto-destruction. Significant artworks were his "Meta-matic" and "Cyclograveur", that with the mode of their automatic writing create a "ghirigoro"(dadaistic) signs, which resemble the naturality of children's drawings. Tinguely's meta-matic artworks in their complex movement, become "alive" with their motion and sound, always with elements of surprise, defined by the artist "functional usage of randomness". They intend to scare and surprise spectators of his artworks (Tinguely, 1982). One of his most important artwork is "Eureka" exposed at the Swiss National Exposition in Lausanne, in 1963.

CYBERNETIC MACHINES

Nicolas Schöffer uses different concepts of space, time, light, shadows and movement in his creation process with the theory of information. His creations use art languages as parameters of living mechanisms systems with rethinking to electronic models of Artificial Intelligence. Cybernetics is a natural order that maintains the equilibrium of elements' forces and in physiology, which is concerned with organism functioning, this biological principle is called homeostasis and it is applied to technological world.

The human brain has been used as a model by cybernetics pioneers in their first calculus and homeostatic machines. They became artificial organs that can receive, process and emit information within computers. Ross Ashby's cybernetic homeostatic model has been developed in the 1940s of the last Century, which is a regulatory system that permits to compare values and to adjust its internal equilibrium(Wiener,1988,reprinted). It simulates human organism's correspondent based on its complex mechanism that keep us alive with maintaining the exact level of sugar in the blood or with simulating the capacity of maintaining the constant body temperature of 37 Celsius degrees. These new relations between nature and culture have the necessity of various confrontation and regulation, complex activities of modern life.

Schöffer constructed a sculpture called CYSP1 in 1956, and the acronym means Cybernetics and Space Dynamics(*spatiodynamique*) more similar to a robot than to a sculpture. The Cysp1 project(Cybernetic and Space Dynamics) is a movable orthogonal construction that is guided with a computer hidden within a black steel and polychromatic aluminum roof. It possesses movements capacity which could be seen as a choreography.

With his sculptures sensible to light, sound and color he creates electronic ballet that evolve under impulses of numerous projectors (Schöffer, 1963). Schöffer's discovery of light's kinetic forces led to elaboration of "luminous-dynamism" in theory and practice, researches that had direct applications in 3D with mechanical, electromechanical and electronic devices. His "space dynamics" researches give new perspectives on the development of the motion concept in plastic art. He, in fact, had the idea to use cybernetic elements and electronic force not only to integrate music but to create interaction of used elements with external events. This integration between aesthetic machine and human body became a device for observing human neural cortex's responses to visual stimulation. Schöffer's project Luminotron placed the spectator inside a spherical environment named "Stimulation sphere" and his reactions are analyzed through various electronic devices and circuits of television cameras. (Schöffer, 1970) In 1973, in his cybernetic artwork Kyldex1 spectators were linked to electrocardiograms and electroencephalograms. Traces obtained from these medical instruments connected to human bodies were projected on monitors and interfered with sequences of visual spectacles. He transforms his artworks with modeling them on the Information and Communication theory and they became emitters of audiovisual signals. The spectators became receivers of those signals with their immersing into reaction paths (feedback) described by Cybernetics. This challenge is transforming spectators into actors where the artist invites the spectator to make a device "alive" to get actively involved, to participate.

In conclusion, the interpretation of the machine as a motor and a co-operator into human dynamics, anticipatively demonstrated the possibility for the development of actual virtual realities in the last century. Virtual realities give new possibilities for Society, Art and Science. The definition of renewed roles for the artwork, spectacle, author and the spectator is underlined as a necessity for creating new languages of meaning communication and interpreting a need for contemporaneity through the new virtual medium.

THE DYNAMIC SYSTEMS THEORY

This paragraph introduce a brief excursive on dynamic systems theories necessary to understand metaplastic systems. The general systems theory has been originally proposed by the Hungarian biologist Ludwig Von Bertalanffy in the 1948. Since Descartes, the *scientific* method has been developed under two linked assumptions. The system can be in fact, decomposed in its primary elements so that every member could be analyzed like an independent entity and the same members can be added in order to describe the totality of the system. Von Bertalanffy in 1951, extended his general systems theory including the biological systems.

An element in common to all the systems has been described by Kuhn (1974) and that element is the information content based on information unit, which is proportional to its quantity and inferred with the starting information.

The communication is the exchange of information, while the transactions are related to exchanges of matter and energy. All the organizational and social type of interactions are connected to communication and/or transactions. The Kuhn's model defines that the rule "moves" the system towards the equilibrium. The communication and the transaction supply therefore "a guide" for the system to the state of equilibrium.

"Cultures is communicated, learned patterns... and society is a collective of people havingin common a body and cultural process." (A. Kuhn, 1974).

When the society is seen as a system, the culture is seen like its pattern. The sociological analysis is therefore the study of "communicated, learned patterns common to relatively large groups (of people)." (A. Kuhn, 1974). Some dynamic systems definitions:

- **Element:** Any identifiable entity;
- **Pattern:** Any relation between two or more elements;
- **Object:** A pattern that changes in time;
- **Event:** A change of the pattern in time;
- **System:** Any pattern which elements are related between;
- **Active system:** Pattern where two or more elements interact;
- **Component:** Any element interacting in an active system;
- **Interaction:** A situation where the change in a member induces a change in another member;
- **Mutual interaction:** The situation in which the change in one member causes a change in some other member, which induces a related event in the first member;
- **Pattern system:** It is a pattern where two or more elements are interdependent. A situation where the change of a member induces a change in another one.

The abstract system or analytical system is a pattern system which elements consists in signs and/or concepts. The controlled exchange processes (cybernetics) maintains at least one variable of a system within specific limits or changes entire system toward reaching an internal equilibrium. The cybernetics is in fact, a discipline that ordinates systems. The input of a system it is defined as the "movement of the information" (for the abstract system) or with "matter-energy" (in material systems) from the external environment. The output is the movement of the information from the system to the surrounding environment. Both, input and output are defining the limits which every system represent. When all forces are balanced to the point that no more exchange is occurring, then the state of the system is defined in "static equilibrium" (Ashby, 1956).

DIFFERENT LOGIC SYSTEMS

The main issue of fuzzy logic is how to categorize the arguments. Cantor examined the modality of categorizing into *sets*. He defined the whole set related to an argument, *the universe of the speech*. For example, if we study the cat, the universe of the speech would be all the cats or all the mammals or all the living creatures. The important note is that the universe contains the variability. The complement of the set is anything which does not belong to the set. The studies of Cantor on set relations, have led him into the definition of *intersection* and *union*. The problem in the Cantor's set theory was the difficulty with the definition of the limits of every set. The limits were vague and imprecise. According to Peirce (1945), every object is comprised in the set by its inclusion in the continuum. At the same time, they become clearly part of the same set. In fact, the existing vagueness causes the determination of inclusion in the set to be very difficult. In 1920, Polish mathematician Jan Lukasiewicz proposed the idea that the simple true or false dichotomy, should contain a third possible logical value. With assuming the same Lukasiewicz's hypothesis, that defines any number as medium value is possible. Instead of simple true or false notion, the numerical value should be used to represent the degree of true-values.

The mathematician Max Black (1937) proposed the vagueness in matter of subjective probability. If as an example, 60% of the population believe that something is true then its value of truth it is of 0,6. The truth degree is always between 0 and 1, 0, 6 in this case. Lofti Zadeh theorized on the Black's researches that the inclusions to the set can be graded. Some elements belong completely to the set, while others only partially. Fuzzy properties are defined when the judgment and the context are used in order to assign property values. Zadeh indicates in the person's the ability to quantify set properties. The persons can easily assign a number between zero and one that represents the true degree assumption. The basic premises on which classic (Boolean) logic is founded:

- The existence of a set of conditions;
- The existence of universal truth and falsity.

The *set* is defined as a collection of distinct elements having specific common properties. More formally, the notion of *set* can be declared as a function that classifies all the objects pertaining to an abstract universe U in objects that have P property, and others that does not have it.

$F(Pu) = \{0,1\}; u \in U$

In this notation, 1 means that the object u, belongs to the universe U, and that it shares its property with other members, while with 0 this does not happen. In this case, 0 and 1 are the only values available for u. The difference between classic logic and the fuzzy logic regards the true and false notion. The classic logic has two values in order to define true and false, while in fuzzy logic it is possible to assume different degrees for the conditions of true or falsity: true, false, more true than, not so true, more or less true, the truest one, etc. The researchers in classic logic didn't believe in the concept of partial truth. They thought that the truth is an absolute value, without the possibility of partial true-values. Zadeh on the contrary, asserts that "as complexity rises, precise statements lose meaning and meaningful statements lose precision" (Lakoff,1987).This assumption is defined as a Law of the Incompatibility that helps into analyzing complex systems.

The problem with the acceptance of fuzzy logic is that it becomes natural to form categories of things. The linguist George Lakoff (1987) has worked with Zadeh on the description of the Hedges, which are generally used to modify fuzzy sets. The terms determines several categories like:

General Modifiers: (*very, quite, extremely*);
Truth Values: (*mostly true, mainly false*);
Probability: (*probable, not so probable*);
Quantifiers: (*most part, some, few*);
Possibility: (*impossible, quite possible*).

Some words, for example, *more* or *less* enlarges the set, others like *quite* concentrates and lessen the set. The hedges are vague and without an exact definition, they reflect the human thought. The human thought is *fuzzy*. The psychologist Eleonor Rosch (1978) has examined how the words relate to the fuzzy logic concepts. She has found that some words (prototypes) are better as examples for a class than others and classification of these words coincided with the intuitive meaning. "…..Every language is a vast pattern-system, different from other, in which are culturally ordained the forms and categories by which the personality not only communicates, but also analyzes nature, notices or neglects types of relationship and phenomena, channels his reasoning, and builds the house of his consciousness"(Whorf,1964). According to Bertalanffy (1983, reprinted):

"The linguistic pattern, described by Whorf, determines how an individual perceives and thinks about the world. This relativistic vision has its consistency with the general systems theory. Our culture and experience define our understanding of all systems. The fact that the general systems theory recognizes the relativity of the perception can serve intrinsically to explain the understanding of our role in the universe. This fact grants the possibility to us to examine and to understand the environment. The systemic vision offers a common method for the study of sociological and organizational pattern. It offers a defined dictionary in order to increase the interdisciplinary communication. The systems theory is a way of seeing things. It's a method consistent with a

surveying that should be applied to all the areas of social science."

This overview has been introduced to understand the Metaplastic metalanguage system theory.

THE METAPLASTIC LANGUAGE

"Here is learned and organized the movement inlogical relations". Klee

The conceptualization of the metaplastic theoretical model of virtual space leads to the definition of a fuzzy dynamic system constituted through a visual formal language *(Chomsky, 1959)* and it puts into effect with the composition rules of abstract languages. The *Metaplastic model of abstract language* is based on the theories of Paul Klee. The chromatic qualities of the objects define weights and measures according to conventions used to define the alphabet. The visual language is made with relations of meaning and events dynamics within the space. The movement is determined from the qualitative relations and quantitative elements between the environment and the user. These dynamic relations found the equilibrium of virtual space determining the representation meaning through its shapes, chromatic contrasts and the spatial relations of meaning.

Systemic coordination is accomplished according to cybernetic and artificial intelligence studies, with a fuzzy logic criteria. The expression of the interpretative system process, finally, finds its performance through the definition of a suitable plastic language with precise references to the poetics of Paul Klee. The example of abstract art languages for the definition of the visual language, require the constitution of the following primary elements of shapes and colors:

- *A formal grammar that contains construction rules*;
- *Plastic elements* will be recognized and defined in the same way of the words in spoken language, corresponding to the law of composition;
- The analytical part will specify its primary elements, designating different and complex parts to define;
- The synthetic part specifies possible laws for ordering the elements of the language;
- *Definition of a dictionary for all visual signs*. The formal grammar consists of a finite set of symbols (as alphabet letters of the written language)(Goodman,2003,p.116), a finite set of non-terminal symbols, a set of generation rules and an initial symbol.

Grammar defines the language, containing the final symbols deriving from the initial ones.

Language is definable as a fuzzy dynamic system model called "finite state machine" (FSM). It transforms sequences of sensory state input (stimuli) which produce feedback and sends it as state sequence (response) to the output of the system. The result is determined as a truth value that generates the activation of the decision making process. Transformations are completed within a dynamic cycle of internal states. New states are stored subsequently for use in the successive phases. The described model consists of the following parts:

Metaplastic language (FSM) = { Σ, Q, δ, s, F }
Where:

- Σ is a set of input events (Events) and output (Actions) of the system; (example: executed action, emotional change, color change);
- Q is a finite set of States, including the initial state; example: initial state, optimal state, excellent state, final state;
- δ are the functions of the Transition states. The function assumes the actual state from an input event and gives a new output value to the next state. Some states can be designated to be final states.

- s is a state that initializes the system;
- F is the set of final states.

The visual language grammatical definition is described with its necessary elements in the following way:

States: Space and abstract figures
Events: Proximity, haptic, vision, touch, time.
Transitions: Extended senses
sensory Events effectors
Rules: Composition design criteria;

More specific definition:

Abstract figures: Composed Elements.
Alphabet: Set of symbols (Figures).
Visual Syntax: The composition equilibrium Syntax
Semantics: Dynamic opposition Semantics.
Composition criteria: Quality and relations of the environment.

Visual syntax of the Metaplastic metalanguage includes these following characteristic elements:

- *The Abstract figure is the sign with the function of Subject*;
- *The Color and Sound, Form qualities, are the Adjectives*;
- *The action of Movement corresponds to the Verb of Metaplastic visual expression.*

ABSTRACT FIGURES

"We don't want form, but function" Klee

The figures qualitatively express its shape through chromatic modules with one or more colors. The formal image expressions uses plastic means of measure, weight and chromatic qualities to dispose and create opportune spatial relations. The color could be considered, by following Paul Klee's criteria, as quality indicator; secondly weight indicates a luminous intensity and it has also a measure of extension, because color is measurable. These formal criteria offer numerous possibilities of chromatic combination for the definition of all language elements. The gray point, infinitesimal figurative unit, expressed in the Klee figuration theory, grows activated from tension with its complementary figure. Every space element of

"...the inner forces of the things are polarized, are attempted, they are chosen and they are compensated on the plan of the concrete truth which damage place". (Klee)

The generative process defines plastic elements and the same virtual space, through the following constitution of plastic structures:

- *Essence*: inner space containing figures and its own information constituent;
- *Structure*: superstructure spaces composed with a variable number of chromatic qualities;
- *Appearance*: external space composed with a variable number of chromatic qualities.

Element possesses a delimited external surrounding space from a "spherical sensory Threshold" (dialogue system), that will be described later on, carries out interactive communication functions with the visitor, other figures and the surrounding environment. Klee theories describe with their criteria each of the three spaces characteristics to define the plastic element qualities of time measures and quantitative weights, which are subdivided in qualitative indications:
Measures

- *EXTENSIONS*. They indicate structural measures of the element;

- *REACTIVITY THRESHOLDS.* They indicate the measure of the element's sensorial limit;
- *SHAPE TYPOLOGY.* It indicates the topology of inclusions with the indication of the space volume's angular opening;

Weights

- *SHAPE EFFECTS.* It indicates the quantitative relations between weights of shape;
- *WEIGHTS OF SHAPE.* It indicates inserted construction weights of information in several parts of the element;

Quality

- *CHROMATIC QUALITY.* I indicate chromatic qualities of the element;
- *HERMETIC.* It indicates the element's different level of accessibility to proper information;
- *DEFINABLE.* It indicates the various degrees of element's qualitative definition;
- *CONSISTENCY.* It indicates the level of quantitative and qualitative consistency of the element;

Movement

The tensions generated from the relations on different level induce movements and reverse movements in elements that modify the inner and external space volumes' equilibrium.

- *SHAPE RHYTHM.* It indicates the definition of element's shape rhythm;
- *SHAPE PITCH.* It indicates the structural rhythm generated from the sequence of weights and measures.

THE FORM_RHYTHM

"Rhythm could be acquired with ours' three senses: firstly, hear it; secondly, see it; and thirdly, feel it in our muscles." (Klee, Teoria I, page 267)

Figurative elements in theirs structural equilibrium, statically and dynamically compose shape rhythms of time and space. The vertical and horizontal shapes prevail in static relations, while in the dynamic relation curvilinear shapes are evidenced. The possibility to unify the two generated space in the way that inner and external space could coincide in an organic polyphony full of contrasts. Rhythmical pattern repetitions are transformed from the form compositions to the space-time structural movement of the figures. The metaplastic Virtual World shows its own dynamic semantics through the visual formalization of the rings structure which are linked between them, compose grammatical rules of the visual language. (The rings' structure is included within the Red and Black visual semantics that will be explained later in the chapter). Groups of rings define the signs of language set, the selected token and the relations among them.

The settings vary with rotations of their axes. Different rings structure configuration define new kind of metaplastic visual language. The abstract plastic figure is composed with the following form elements:

- *Form_Rhythm structure*: is a structural equilibrium composed with dynamic relations of form_weights, distributed along figure's dimensional extensions;
- *Form Sequences*: are structural patterns for figure definition which identifies a basic unit of points within the form volume;
- *Form Variations*: are rhythmical balance relations between form weights. The form variations and sequences are the components of Rhytm_forms;
- *Form Aspect*: are figure's appearance qualities, determined with:

- *Color qualities*: indicates the chromatic qualities;
- *Consistency*: indicates the level of element's qualitative consistency;
- *Brightness*: indicates the light quality;
- *Hermeticity*: indicates the element's different level of accessibility to proper information.

The following formulations define the Form Rhythm development criteria and how its own complex form is computable:

*f*Form = Rhythm_form f (DE, W, D, S, P, V, C, B, Co, H);

where the following parameters are also figure states(memberships) of the FSM system:

Form parameters: Width(W) and Depth(D) = Form width and depth angles degree;
Form weights (P) = Form weight;
Form variations (V) = Rhythmical balance relations.
Structure parameter: *Sequence connections* (S) = Form weights links;

Appearance parameter grades: Chromatic quality(C)=Chromatic; Brightness(B)=Light quality; Consistency(Co)=Density; Definability(De)=Smoothness; Hermeticity(H)=Accessibility grade.

f Appearance = Form Appearance function;
Rule Light-Color-Sound$_i$ = (W+D)**f*Appareance (interp(C,B,Co));

The following function defines the complexity of the rhythm form as:

Rhythm Form Complexity (synthetic expression)
 = (W+D)* *f* Form (interp(P, V, S));

The Rhythm_Form Complexity expression written in full version:

$f = (W+D)*interp[(frap(gvariant, fsign(gvariant,P[i]))_{int}, frap(gvariant, fsign(gvariant,P[i]))_{ext})];$

Where:

frap = Relationship function; fsign = Form sign function;
gvariant = function Form variation; int = internal index; ext = external index.

The gvariant values below define the form complexity meaning:

- [0...0.1]: Ascendent / Descendent;
- [0.1...0.2]: Linear / Articulate;
- [0.2...0.3]: Rotational clockwise / anticlockwise;
- [0.3...0.4]: Acute / Obtuse;
- [0.4...0.5]: Incremental Less / More;
- [0.5...0.6]: Squared Large / Narrow;
- [0.6...0.7]: Smooth Less / More;
- [0.7...0.8]: Looped Less / More;
- [0.8...0.9]: Continue / Discontinue;
- [0.9...1.0]: Eccentricity Less/More.

If fsign function gets an input<0.5, gives a negative output value otherwise is a positive output value.

The frap numeric values below define relations modalities between functions:

- [0.0...0.2]: Identity;
- [0.2...0.3]: Add;
- [0.3...0.4]: Neg;
- [0.4...0.5]: Multipling;
- [0.5...0.6]: Fractional;
- [0.7...0.9]: Exponential.

VISUAL ALPHABET

The alphabet is a sign set with expressed meaning attribution in its three dimensions of quality, weight and measure. The relations between the different quantitative and qualitative gradation allow to compose meaningful form for the definition of the element property. New language is defined as an alphabet of signs and constructed through composition criteria, the following example (Figure 1) explain its application:

The Figure 1 describes an example of metaplastic language composed with six spaces, each of them characterized by its own set of properties:

A Form_Rhythm [Complexity .4, Variance .5, Chroma .35];
B Form_Rhythm [Complexity .2, Variance .8, Chroma .63];
C Form_Rhythm [Complexity .7, Variance .7, Chroma .85];
D Form_Rhythm [Complexity .1, Variance .15, Chroma .72];
E Form_Rhythm [Complexity .8, Variance .6, Chroma .18];
F Form_Rhythm [Complexity .9, Variance .9, Chroma .26].

METAPLASTIC ONTOLOGY

Ontology of the metaplastic virtual world includes notions and relations between:

- The "abstract figure" as the fundamental ancestor element of the model;
- The virtual space, entity subclass which is a fuzzy dynamic system(FSM) (Mura,2007).

Virtual space entity subclass contains other space elements (children) and communicate with them and the user within its sensorial threshold. This virtual space or entity subclass is an abstract cyberworld. The meta-plastic visual language contains a set of Abstract figures and each has a sign-phoneme of the visual alphabet. This virtual world is a fuzzy dynamic system (FSM) which communicates with its sub-elements and the user within his sensorial threshold. It interacts and solves specific actions within the applied visual metalanguage rule configurations.

The virtual user, entity subclass, is represented with his extended sensorial state values that the system output generates within interaction processes. Every "abstract figure" within the Metaplastic Virtual World is made with a set of input properties' values. Some of the properties' elements results are shared with other subentities. The system acquires knowledge during its processes

Figure 1. (Left) Rhythm_form structure; (Right) Example of signs' alphabet©2010 Gianluca Mura

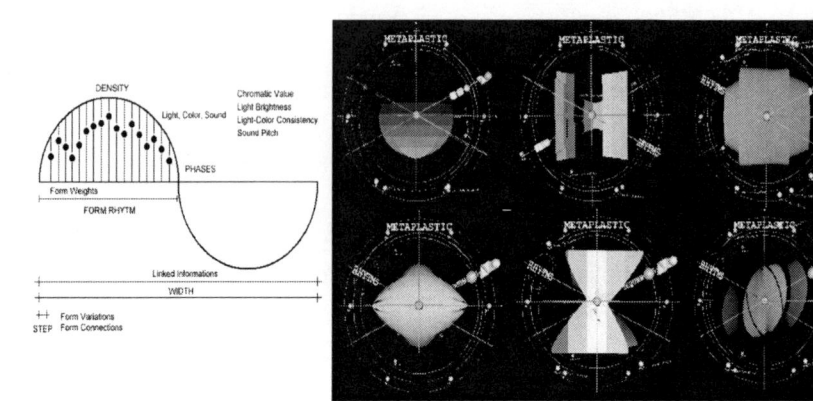

from every element. Simultaneously, parts of the ontological knowledge base is shared between entities. The elements' structure are described in term of a group and individual roles hierarchy.

"The visual language is made of related meaning and events dynamics within the space. The movement is determined from the qualitative relations and quantitative elements between the environment and the user. These dynamic relations establish the equilibrium of the virtual space determining the meaning representation through its shapes and spatial relations" (Mura,2007).

Figure 2 summarize the ontological hierarchy of metaplastic virtual worlds.

THREE SYSTEMS

The Metaplastic Virtual World is an automata or "semi-machine" composed through its Form-Rhythm Complexity structure. The dynamic

Figure 2. Metaplastic Virtual World Ontology. ©2007 IEEE (G.Mura, The Metaplastic Virtual Worlds,CYBERWORLDS 2007).Used with permission

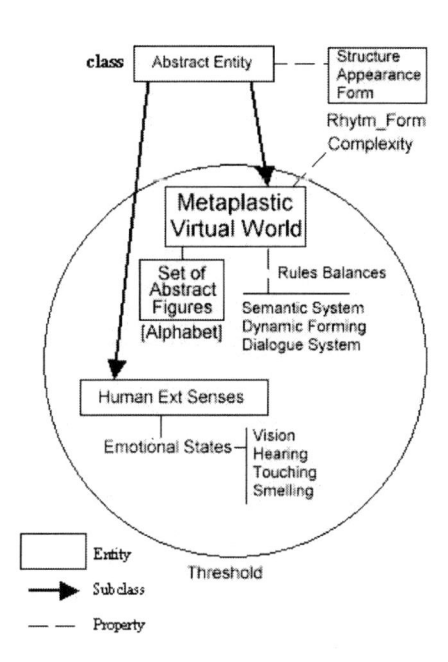

equilibrium relations of three subsystems fulfill different functions and interactions between them, with direct definition of artwork's behaviors and meanings through the metaplastic visual language codifications. Three metaplastic subsystems are:

- *The Dialog System* that fulfill its fundamental function for the management of relations between the human and the machine;
- *The Dynamic Forming* system compose the dynamic rules of visual syntax;
- *The Red and Black Semantic* system constructs meaning through its visual codification.

The following paragraphs describe every part of the metaplastic system with its peculiarity and applied methodology.

EXTENDED SENSES AND THE DIALOGUE SYSTEM

The virtual environment dynamics, as it has been previously described, realize the expressions of human psychology, with this statement:"Psychology is the physics of virtual reality"(Woolley, 1993, p.21). The Dialogue System has a fundamental function of relations management between human user and machine. The movement which generates shape principles of the environment is also at the same time creator of human sensorial perceptions. The images are perceived with the movement of the oculomotor muscles, that lead to the simultaneous composition of the shapes, dynamic relations and the their objective antithesis. Klee(Edschmid, 1920) describes the process with these words:

"All the ways meet in the eye and they lead, converted in shape from the crossing point to the synthesis of external vision and inner contemplation."

From the aforesaid, the crossing point originates completely different artifact from the real image of an object, without contradicting it from the view of the whole. The human perceptive process transposed in artificial space is identified with this Paul Klee (in Edschmid, 1920) words:

"...orients into three dimensions space and its current position inside the space is a base of concepts evaluation."

The self is the center of space and simultaneously the center of virtual environment. In fact, every direction of movement in the illusory space is in relation with the natural sense of orientation which characterizes itself with the same center of the environment. The last development in psychological fields introduce many structural models of human behavior applied also in cybernetics.

However, there are still remarkable difficulties in the research and complete understanding of the human brain's functioning complexity. In lack of certainty in matter, exists several hypotheses of systemic models. Some of these models were introduced for their importance regarding this chapter. They are: homeostatic system model (John H. Milsum, 1966); Human Systems interacting in social systems (A. Kuhn, 1974); Human Behavior determination model (Miller, J.G., 1978); Emotional Cognitive Structural system for productive or creative thought(Gray,2007). From comparative studies of several listed models, a famous psychologist and neurophysiologist Charles A. Fink (1979) has elaborated a successful experimental model applied to the artificial environment.

The inadequacy of mathematical models for the description of objects and immaterial processes, has motivated the development of this dynamic model in analogy to the methodologies used in economic systems. The results of these analyses concur to compose, on reasonable bases, the models for human behavior. The dynamic system developed by Fink from 1975 to 1999, becomes important for cognitive models, because it describes the human being as a whole made of different stimuli. Analyses and decision-making are included in a subprocesses, structures or interconnections, with complex cycles of feedback, and a set of possible responses.

The decision on a specific methodology in order "to convert" human sensorial system and the virtual reality is of fundamental importance for metaplastic theory. It has been developed a suitable cognitive system, called Dialogue System (Figure 3). Within the human-machine interaction processes, human emotional states (objective tasks, worlds, cognitive awareness, emotional awareness, physiological awareness) are induced by the artificial context ("Sensing"), becomes analyzed by the system ("Thinking") through the sequence of feedback stimulus-to-decision making-to-response.

Every stage of the process is compared with previous experiences. The importance of the response is analyzed and if judged positively, it produces a consequent action. In the opposite case, results are stored in the system and made available for successive appraisals. The final phase is produced from a human action towards the environment ("Acting") that at the same time induces in itself the emotional consciousness ("Feeling").

The dialogue system acts within all the interaction processes of virtual reality model developed in this study. It is a specific dynamic system methodology for the interpretation of sensorial interaction processes between users and the system.

Some essential definitions are brought back for clarity:

- *Dialogue System*: dynamic system for the interpretation of sensorial interaction processes between human and machine;
- *Extended Senses*: sensorial motor channels of transmission;

The Metaplastic Virtual World Theory

- *Sense Threshold*: synthetic responsive membrane for the interpretation of sensorial interaction processes between users and the system;

The thresholds "perceive" from the physical laws and system quantities rather than physiological. Tables of analyses in the following pages, introduce different perception modalities between the human organism and the artificial environment. Figures of language are expressed from analysis on the necessary structures that should have proper qualitative and quantitative information. The critical attention to several parts follows the indications of the Figuration Theory (Gestaltung), which suggested relation between the parts and the whole.

The map in Figure 3 indicates the complexity of the human sensorial combinations of relations with the virtual environment. It is useful to visualize the changes which occur into the user's emotional behavior when it is placed in the virtual space.

Perception and Cognition

The process of perception, based on Gestalt psychology, includes actions of the observer in relation to observed objects and emotions. The perceptual knowledge is learned through previously acquired "common sense" of the form. The cognitive process uses spatial meanings to create proper shape image with effects of proximity, similarity, orientation and symmetry (Figure 4).

A-B-C. Proximity (A) Elements unify in groups. The observer perceive the group as three vertical lines or dots. The group (B) is perceived as three horizontal lines or dots. The dots in (C) are equally spaced and do not suggest an orientation. (D) Similarity Elements have forms that are apparently 'similar'. The observer sees separate white diagonal lines and black diagonal lines rather than vertical or horizontal lines of black and white dots. (E) Orientation Elements form groups with similar orientations. The observer sees it later when he looks at relationship with its frame, which also influences our interpretation of a shape. (F) Closure Elements unified together give the appearance of closed shapes (Figure 4).

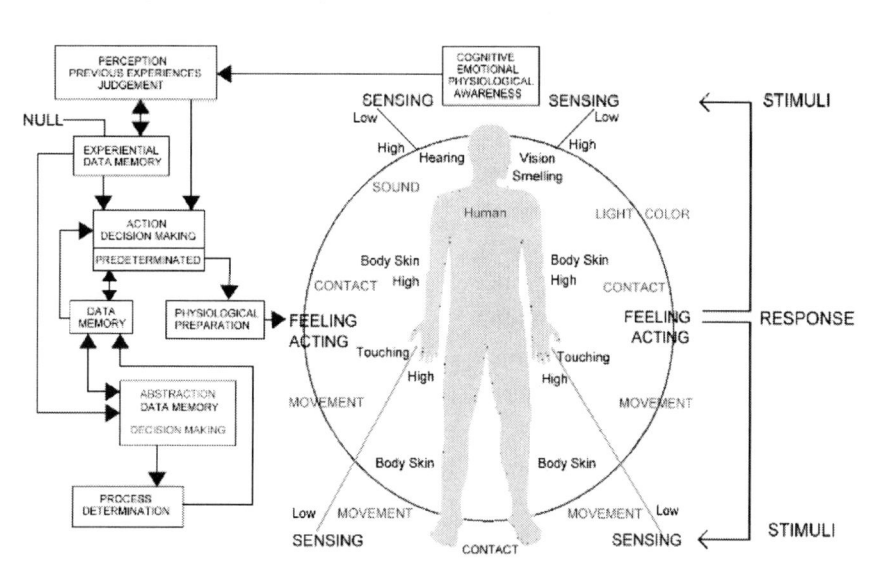

Figure 3. Extended dialog system sensory model. ©2010 Gianluca Mura

Figure 4. (Left) Visual perception figures. (Right) Synaesthetic interaction process schema. ©2010 Gianluca Mura

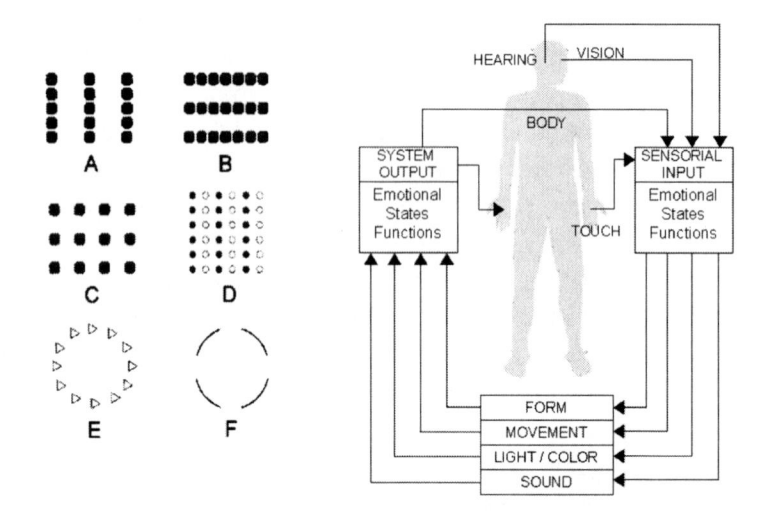

HUMAN SENSORIAL EXPERIENCE (SYNAESTHESIA)

The sensorial results coincide with the composition of a human behavior model developed from cybernetic studies (Carver, 1998, reprinted). The human perceptions' decision-making processes are divided in sub-processes, structures or interconnections, with complex cycles of feedback, and a set of possible sensorial responses. On the basis of our artificial cognitive system called *Dialogue System* (Mura, 2007), the human emotional state changes the *human-machine* interaction processes. This fuzzy functions relations produce a synaesthetic feedback results to the human interactor. The synaesthesia is a particular emotional-feeling state that involves mutual influence between sensorial representations. Certain qualities of feeling, may be linked to other qualities of imagination and, as result, the system gives a high-level human feedback quality of synaesthetic perceptions which creates a sensorial path. The intensity of a feeling state may be linked to the brightness of an figure; the color may influence the level of feeling; some kind of movements may "feel" with the interactor. The quality of the synaesthetic effects depends on fuzzy functions connections or overlaps of our sensorial systems. In traditional psychophysics, the term "fuzzy function" was defined as:

"Any modeling involving a representational system and either an input channel or an output channel in which the input or output channel involved is a different modality from the representational system which it is being used" (Grinder, 1975).

In this way, fuzzy functions are typically characterized by terms such as see-feel or hear-feel circuits. In the extended Dialog System this concerns mean a particular application of the fuzzy-state machine.

*Rule Light-Color-Sound$_1$ = (W+D) * fAppearance (interp(C, B, Co));*

User Emotional Interaction Evaluation

According to the theory of P.T. Young (1967, pg.32-40) about the emotional transition, it is depending on three main factors such as sign value, intensity and duration:

- Sense weight sign. We can find two phases in our activities during the interaction with the object giving as signal result the sign + and – respectively.
- Emotional intensity. Emotional process is affected by the intensity of input signal and the intensity can be drawn on the sense weight values with + and -.
- Emotional duration. The signal varies and is evoked depending on the duration of the interaction process with the objects. The user's internal emotional state variation is supposed to be caused by input signal and inclosed into a fuzzy inference mechanism. The user's emotional level is evaluated through a simple system composed within five fuzzy inference rules shown with the list below. The related five Sense's defuzzified values can be used as an expression of of the user's stimulation and its new emotional state caused by external environmental inputs. The emotional states values could be defined as:

    - [0.8...1.0]: Highly involved;
    - [0.7...0.8]: Happiness;
    - [0.5...0.7]: Interested;
    - [0.2...0.4]: Mildly interested;
    - [0.0...0.2]: Relaxed.

Emotional evaluation fuzzy rules:

Rule 1: if sign value -and intensity -100 then w = 0.00;
Rule 2: if sign value +/- and intensity 0 then w = 0.20;
Rule 3: if sign value +and intensity 50 then w = 0.50;
Rule 4: if sign value +and intensity 100 then w = 0.75;
Rule 5: if sign value +and intensity 150 then w = 0.85.

The defuzzified w value is used in the further function to obtain final E emotional state for different output channels results (vision,hearing,body,touch).

Emotional state $E_i = f\,S_i(w)t * (1 - f(\mathrm{interp}(S_i(w)\,\mathrm{vision}, S_i(w)\,\mathrm{hearing})));$

where: E = emotional state; S = Senses function; w = Sense weight [0..1]; t = action time;

i = dialogue state.

THE DYNAMIC FORMING SYNTAX

"The Artwork is forming through its becoming."
Paul Klee

The space dynamics is an essential function to ascribe meaning and expression in the plastic language. The movement in the several elements relations assumes fundamental importance, as it emphasizes Paul Klee in (Spiller, 1959, p.84):

"...contrive of the parts to the total function in a figured liberate polyphonic organism. The movement rhythm is given from successfully increasing and decreasing amount and quality of the employed energy. Direction and development of the movement, depend on density and expansion of the linear tensions and the centers... The motor organism is arranged to act (simultaneously) of the organs as a whole, quietly enlivened or lively soothed; but it will be complete only when to the movements will be added and the movement against or in an infinite movement."

This declaration is the fundamental principle for the structured genesis in the virtual space. Every element pertaining to the artificial environment finds in this dynamics, its own origin and its measured action in cause-effect relations. The generating element is the symbol of reversed treble clef as it appears in the painting *Gleich Unendlich*, and it has equal genesis to infinite symbol.

"The eight horizontal - treble clef and sign of infinite - traced on a divisionist background, represents the plan, musical and mathematical, of one structured genesis of the shape."(Damisch, 1984)

Semiotic explanation in two levels: plastic level, of forms, colors and forces; and iconic level of denomination and figuration, Klee in (Edschmid,1920) says that:

"...the painting cannot explain but from the inside of a minimal definition of the cause-effect relation, considered becoming plastic from the point to the surface, because the same movement in which the active function of the line has got worn out, the surface becomes active, determining the cyclical resumption of the cause-effect relation in a definition of the space that is already three-dimensional, architectonic or sculptured."

The equilibrium generate surfaces, with alternate accentuation on external and inner spaces, active and passive, light and dense through reciprocal penetration, spatial joints, transparent and with polyphonic intersections.

The interactions of energetic flows generated from the encounter between remote senses and senses caused by the tensions of the movements and inverse movement of the plastic elements, which generate in the space a continuous energetic cycle. This is Klee's "continuous movement". It is:

"the eight sign, a twofold circle, that is an intercrossed circle and bipartite, whose motorcenter "dominates" the two "contra-opposed" cycles "(Damisch,1984)

It is the mathematical sign of the infinite. It is a tensive cycle value that consists in an alternation of the condensation and relaxation states, expansion and concentration of the space.

It leads all the Shapes in its flow to the space's rhythm of particles, through the states of formation, genesis and becoming. The cyclical phases of the Dynamic Forming (Figure 5) consist in an alternation of:

- Phase 1: *CONDENSATION*;
- Phase 2: *RELAXATION*;
- Phase 3: *EXPANSION*;
- Phase 4: *CONCENTRATION*.

The Dynamic Forming assumes a fundamental role which stabilizes the fulcrum of plastic space expressions. The movement is a fundamental element of virtual environment, with its dynamics expression, transformations, shape mutations. The same principles on the motor processes animate the conceptions of creation of Paul Klee, with studied and detailed norms. The antithesis of the static-dynamic values and the qualitative expressions are reassumed primarily in the opposition to all. This contrast, it is therefore possible for every comparison on the concepts and perceptible dynamic values about shapes and the environment.

Figure 5. Dynamic forming balance schema. ©2010 Gianluca Mura

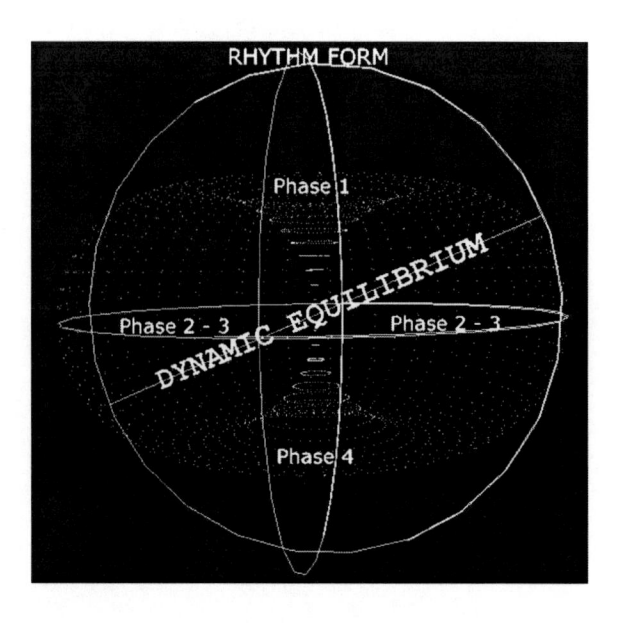

The theory of the equilibrium develops every plurimotoric order of its plastic means and of theirs meaning in perennial mutation. The color emerges as a qualitative element in the dynamic development of the space through its strong tensions and in its close relations to the movement.

The balance function is computed between x e y figure state values, for each dialogue state, and is defined as:

State gradient = truth function(x);
Balance $f(x,y) = y*(\text{gradient } x(0) + \text{gradient } x(0.5) - \text{gradient}(0)/0.5))$;

In conclusion, the visual syntax rule for each dialogue state is:

Dynamic Syntax rules(dialogue$_i$){
    state=min(gradient, 1.0);
Inference = Senses(weights) * State$_k$;
Rhythm Balance = $\Sigma_{k=1}$(Inference(state$_k$), Inference(state$_{k+1}$)); }

Where: State$_i$ = Abstract Figure membership$_i$; dialogue$_j$ = dialogue state; i,j= state;

f=trigger function; k=item shown;

THE RED AND THE BLACK SEMANTIC SYSTEM

The discipline of Semiotics in plastic art was introduced by Greimas in 1984. It studies conceptual analysis and semantic description of Space. These researches use represented plastic images and they were considered as a language because they are independently significant from eventually different meaning of the figures. In opposition to indicated thesis, it is still disseminated the idea that abstract languages cannot contain signs, because they have to relate to a meaning, otherwise they wouldn't be considered signs. The logical reasoning conduces, therefore to apparent conclusion that abstract languages don't relate to anything, because they don't have any signs values. These concepts has been introduced and explained in the book "L'estétique de l'abstraction" of Charles-Pierre Bru (1955) that underlines:

"The abstractions are nor schemes, nor signs, because they don't relate nor to concept of things nor to ideas; they don't relate to anything else rather than themselves".

The correspondence of the sign between Poetry, Linguistics and Plastic Art is confirmed by Saussurre in (De Mauro, 1967) that linguistic sign unifies a concept to visual acoustics. In the same way, linguist Jacobson confirm this hypothesis with reminding the example of experimentation done by Picasso in abstract art as a semiotic model (Jakobson, 1973, pg. 133).

Levi-Strauss disagreed with this hypothesis and in his opinion:

"With this analogies, (colors and music) we could become victims of an illusion. Colors exists in nature but musical sound doesn't, there are only rumors [...] Painting and Music are not placed at same levels. Painting founded in Nature has its matter: colors are primary elements and vocabulary are secondary, derived level, even within small linguistic differences: blue night, blue peaflow [...] In other words, in painting there are no colors if there are no beings or colored objects; and it's only because of abstraction that colors could be isolated from their natural sights and treated as parts of a new system". (Levi-Strauss, 1958)

The only possible solution was contained in this phrase: "It's only because of Abstraction" (Roque, 2004). This is, in fact, the main statement within researches on abstract languages of the last Century. Critical analyses are still insufficient for plastic signs to be a semantic element of an autonomous language. Further analysis has been conducted by Nelson Goodman (2003) which defined several forms of Symbolization, one particular form called Denotation. For Levy-Strauss, this is the only way to define abstract art.

For Goodman, abstract art is a category to which belongs an image referred to the object that it represents. He defines Representations and individuate other Categories of Expression and of Exemplification:

"what is exemplified, it is abstracted"..."the Representations consider Objects and Events,while the Expressions consider Feelings or other related properties."(Goodman, 2003).

These conclusions are fundamental:

"...That's how a White spot could express Emotions as Joy, Emotions as Force, Plenitude, orRichness. It's a fundamental concept for Abstract Art, which obtains its theoreticalrecognition, with this concept; abstract art besides other form of Symbolization, itacquires Meaning" (Roque,2004).

Goodman writes:

"The importance of Denotation (Representation or Description) of Expression and ofExemplification ("formal" or "decorative"), varies with Art, with Artist, with Artwork.Sometimes one aspect dominates the other two, until their virtual exclusion. For comparisonexample Debussy's La Mer, Bach's Goldbergh Variations, Charles Ives's Fourth Symphony;or water color painting of Dürer, a painting of Jackson Pollock, a lithography of Soulages. In other cases two aspects are important,or all three aspects, Functional or Contrapuntal"(Goodman, 2003 and Roque, 2004).

The semantics indicates the scientific study of the meaning. In this context, because of numerous existing semantic theories, we will limit the problem in considering our system "the organization of diversities in the unit" of meaning. Greimas, linguistic theorist, developed three fundamental concepts to consider for the reasoning process: the "spatialization", the "temporalization" and the "actorialization". Greimas and Courtes (1982) explained:

"the spatialization(in the first place...), involves the space localization procedures so that itserves to apply on speech-enunciated space organization more or less independent, useful asa frame for the presented narration material and its concatenation of events ".

The space, the shapes and colors constitute with evoking and contrasts, for Klee, the expressive plan of a deep and complex sense: "Here is learned and organized the movement in logical relations". The Klee painting *Le Rouge et le noir* is shaped as a demonstration on the possibility of existence of pure, abstract signifying that do not need figures from the real world. The sense creation indicates that every act of signifying demands three fundamental elements: the expression, the reference and the content. We should introduced the analysis of Felix Thürlemann (1982), for the explanation of the expressed hypothesis. In the considered context, it should be assumed that every pictorial prose can be analyzed regarding these three complementary dimensions: chromatic, eidetic and topological. Every dimension describes a discursive function, and each of it has an expression plan (category level) and a plan of content (the valuation level). The metadiscourse from The Red and the Black semantic system (Figure 6) is used within chromatic dimension which constitutes the *explicandum*, and the eidetic and topological dimensions are considered as *explicans*. The *explicandum* of pictorial metadiscourse of The Red and the Black is the content of the chromatic dimension, read at the valuation level. Thürlemann (1982)proceeded in his description of the chromatic articulation based on the categories of the radicals, the saturation, the value and the matter.He also analyzes its reflection on the logical relations that exist be-

The Metaplastic Virtual World Theory

tween the eleven chromatic radicals, and its "red" and the "black" opposition extension dualities: chromatic and achromatic, color and "non-color". The analysis concluded in recognizing a signifying apprehension procedure through a transcoding operation, that according to Greimas considers the only way of creating meaning. Thürlemann therefore indicates the content valuation of eidetic and topological contrasts, synthesized in a more general content opposition between dynamism of the Red and statics of the Black. From these conclusions, to give "order to the movement", the structural outline of the painting can therefore be considered as *Le rouge et le noir* (The Red and the Black) semantic model (Figure 6).

SEMANTIC APPRAISAL SCHEMA

We proceed with some simple formulations used in the operations of plastic elements transcoding. Levi-Strauss (1990) frequently uses the mathematical formula of the similarity for the comparison of terms. The use of this formula is motivated with the possibility to obtain an effective "semi-symbolic" (for the concept of "Semisymbolic Semiosis", Greimas, Courtes (1979) and Greimas, Courtes (1982) on term "semisymbolic") coding between the plastic categories of the content and categories.

Law of similarity A: B:: x: y

To codify with the "The Red and The Black" and the previous formula it becomes in the following way: *RED: BLACK:: x: y*

The Klee poetics of the movement has a definition of "static" and "dynamic" as terms of appraisal of the semantic schema: *RED: BLACK:: "dynamic": "static"*

This example follows the application of variable eidetic categories in opposition (in this case "vary" and "uniform") of meaning in "static" and "dynamic" reference:

R: N:: {Irregular, eccentric, modulated} vary:
{regular, not eccentric, homogenous}
uniform:: "dynamic": "static"

In conclusion, the application of previous schema to the algebraic logical relations of the semiotic square, produces a spatial guideline adapted to obtain the dynamic appraisal of meaning from plastic elements, as shown in the following schema of Figure 6.

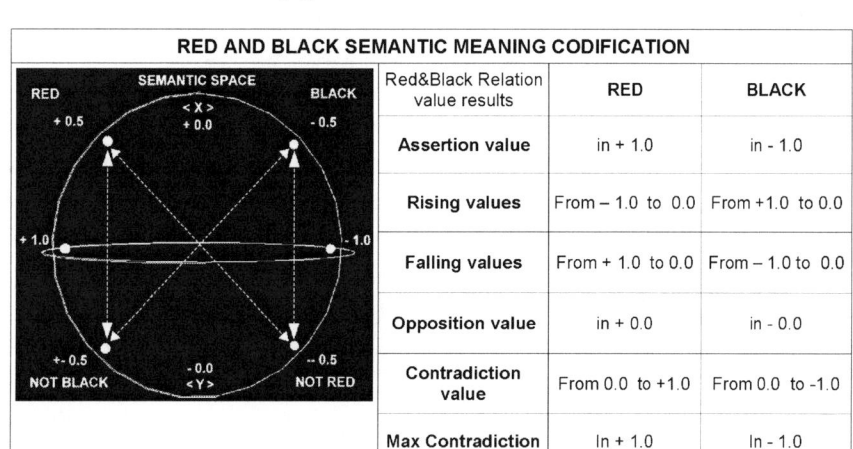

Figure 6. Red and black semantic configurations. ©2010 Gianluca Mura

RED AND BLACK SEMANTIC MEANING CODIFICATION		
Red&Black Relation value results	RED	BLACK
Assertion value	in + 1.0	in - 1.0
Rising values	From – 1.0 to 0.0	From +1.0 to 0.0
Falling values	From + 1.0 to 0.0	From – 1.0 to 0.0
Opposition value	in + 0.0	in - 0.0
Contradiction value	From 0.0 to +1.0	From 0.0 to -1.0
Max Contradiction	In + 1.0	In - 1.0

Figure 7. The Metaplastic virtual world system.©2010 Gianluca Mura

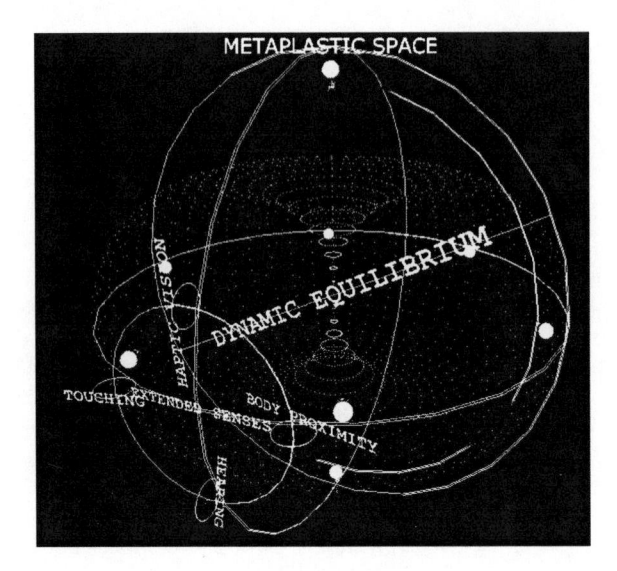

The different configurations of the virtual world space give meanings within its own dynamic "Red and Black" meta-codification system of references. The following formulas define how visual semantic meaning process are computed:

$State_i$ = Abstract Figure membership $_i$;
RB f(x) = 1.0 – truth function(x);

The visual semantic rule for each dialogue state is:

$Red\&Black\ Semantic$ rule $_i$ (dialogue $_i$) = $\sum_k$ (RB (state$_i$))*(1.0/nRB);

Rules for Black and Red are defined separately. The whole system processes are defined by:

VW_{ij}^{new} = f(Red&Black Semantic$_i$ (Dynamic Syntax$_j$ (Δproximity$_{ij}$)))+VW_{ij}^{old} ;
Proximity f(Sensing,Feeling,Acting)= f (E(User. position) – Threshold);

Where: RB=Red&Black weight; nRB=number of RB weight$_i$; x= state value; i,k= state; f=trigger function; k = item shown. E = Emotional states function(see synaesthesia paragraph);

dialogue$_i$=dialogue state; VW = Metaplastic Virtual World.

THE METAPLASTIC VIRTUAL WORLD PROJECT EXAMPLES

This paragraph describes two application examples of the metaplastic virtual worlds.

Example 1. Combinatorial Metaplastic Language

The first project illustrates a construction game of metaplastic language based on a combinatorial principle (Figure 8). It is underlined that all operations made within this project should be assumed with time constant T and the exclusion of dialog system's external interactive dynamics and relations.

General composition criteria in this example, is the formal development of virtual environments through combinations of two distinct figures groups formal complexities. For that reason, it should be defined two subsets with four elements each. The first group is composed of circular primitive morphologies, while the second one contains quadrangular elements of diverse complexities. In the table below, every element of two groups is indicated regarding its proper characteristics and related to other element of both groups. Every figure is described with his Form, Contour and Movement through permutations of some parameters within proper Form_Rhythm function:

New Figure k = f_{FORM} (interp(W,D, $P_{\sum_n A+B}$, $V_{A+B}, S_{A+B}, C_{A+B}, Br_{A+B},$ Co,H,De));
Metaplastic Language = Set of New Figures k;

The Metaplastic Virtual World Theory

Figure 8. Metaplastic language combinatorial set. ©2010 Gianluca Mura

				Set B [Rectangular figures]			
				B1	**B2**	**B3**	**B4**
A resulting new figure example				$P=[0.2,0.12,0.33]$ $V=0.2$ $C=0.3$ $S=0.15$ $Br=0.8$	$P=[0.7,0.43,0.9]$ $V=0.45$ $C=0.5$ $S=0.20$ $Br=0.6$	$P=[0.75,0.8,0.66]$ $V=0.5$ $C=0.8$ $S=0.7$ $Br=0.40$	$P=[0.42,0.3,0.8]$ $V=0.9$ $C=0.4$ $S=0.3$ $Br=0.8$
Set A [figures Sferiche]	**A1**		$P=[.2,.2,.2]$ $V=0.2$ $C=0.3$ $S=0.2$ $Br=0.8$	$P=_{interp}P_{\Sigma n}$ A1+B1 $V=_{interp}V_{A1+B1}$ $S=_{interp}S_{A1+B1}$ $C=_{interp}C_{A1+B1}$ $Br=_{interp}Br_{A1+B1}$	$P=_{interp}P_{\Sigma n}$ A1+B2 $V=_{interp}V_{A1+B2}$ $S=_{interp}S_{A1+B2}$ $C=_{interp}C_{A1+B2}$ $Br=_{interp}Br_{A1+B2}$	$P=_{interp}P_{\Sigma n}$ A1+B3 $V=_{interp}V_{A1+B3}$ $S=_{interp}S_{A1+B3}$ $C=_{interp}C_{A1+B3}$ $Br=_{interp}Br_{A1+B3}$	$P=_{interp}P_{\Sigma n}$ A1+B4 $V=_{interp}V_{A1+B4}$ $S=_{interp}S_{A1+B4}$ $C=_{interp}C_{A1+B4}$ $Br=_{interp}Br_{A1+B4}$
	A2		$P=[.2,.3,.6]$ $V=0.2$ $C=0.7$ $S=0.2$ $Br=0.8$	$P=_{interp}P_{\Sigma n}$ A2+B1 $V=_{interp}V_{A2+B1}$ $S=_{interp}S_{A2+B1}$ $C=_{interp}C_{A2+B1}$ $Br=_{interp}Br_{A2+B1}$	$P=_{interp}P_{\Sigma n}$ A2+B2 $V=_{interp}V_{A2+B2}$ $S=_{interp}S_{A2+B2}$ $C=_{interp}C_{A2+B2}$ $Br=_{interp}Br_{A2+B2}$	$P=_{interp}P_{\Sigma n}$ A2+B3 $V=_{interp}V_{A2+B3}$ $S=_{interp}S_{A2+B3}$ $C=_{interp}C_{A2+B3}$ $Br=_{interp}Br_{A2+B3}$	$P=_{interp}P_{\Sigma n}$ A2+B4 $V=_{interp}V_{A2+B4}$ $S=_{interp}S_{A2+B4}$ $C=_{interp}C_{A2+B4}$ $Br=_{interp}Br_{A2+B4}$
	A3		$P=[.3,.7,.4]$ $V=0.3$ $C=0.5$ $S=0.44$ $Br=0.6$	$P=_{interp}P_{\Sigma n}$ A3+B1 $V=_{interp}V_{A3+B1}$ $S=_{interp}S_{A3+B1}$ $C=_{interp}C_{A3+B1}$ $Br=_{interp}Br_{A3+B1}$	$P=_{interp}P_{\Sigma n}$ A3+B2 $V=_{interp}V_{A3+B2}$ $S=_{interp}S_{A3+B2}$ $C=_{interp}C_{A3+B2}$ $Br=_{interp}Br_{A3+B2}$	$P=_{interp}P_{\Sigma n}$ A3+B3 $V=_{interp}V_{A3+B3}$ $S=_{interp}S_{A3+B3}$ $C=_{interp}C_{A3+B3}$ $Br=_{interp}Br_{A3+B3}$	$P=_{interp}P_{\Sigma n}$ A3+B4 $V=_{interp}V_{A3+B4}$ $S=_{interp}S_{A3+B4}$ $C=_{interp}C_{A3+B4}$ $Br=_{interp}Br_{A3+B4}$
	A4		$P=[.2,.8,.2]$ $V=0.4$ $C=0.9$ $S=0.4$ $Br=0.6$	$P=_{interp}P_{\Sigma n}$ A4+B1 $V=_{interp}V_{A4+B1}$ $S=_{interp}S_{A4+B1}$ $C=_{interp}C_{A4+B1}$ $Br=_{interp}Br_{A4+B1}$	$P=_{interp}P_{\Sigma n}$ A4+B2 $V=_{interp}V_{A4+B2}$ $S=_{interp}S_{A4+B2}$ $C=_{interp}C_{A4+B2}$ $Br=_{interp}Br_{A4+B2}$	$P=_{interp}P_{\Sigma n}$ A4+B3 $V=_{interp}V_{A4+B3}$ $S=_{interp}S_{A4+B3}$ $C=_{interp}C_{A4+B3}$ $Br=_{interp}Br_{A4+B3}$	$P=_{interp}P_{\Sigma n}$ A4+B4 $V=_{interp}V_{A4+B4}$ $S=_{interp}S_{A4+B4}$ $C=_{interp}C_{A4+B4}$ $Br=_{interp}Br_{A4+B4}$

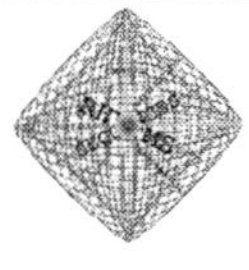

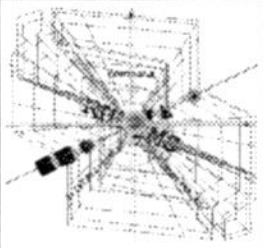

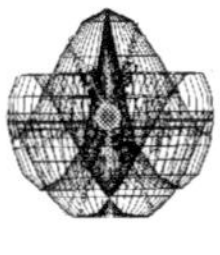

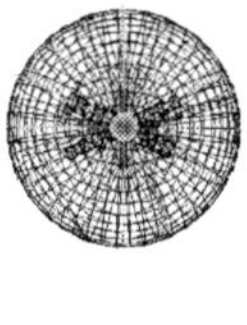

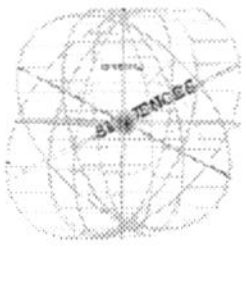

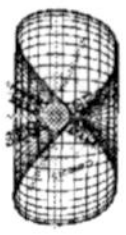

where:

f_{FORM} = Rhythm_Form function; A = Set of Figures A; B = Set of Figures B; P = Form weights; V = Form Variance; S = Form sequence; C = Chromatic quality; Br = Brightness; Co = Consistency; De = definability; n,k = item; interp = linear interpolation function.

This example demonstrates that all obtainable combinations with two initial groups of four elements could generate metaplastic languages up to a maximum of 4096 different virtual spaces.

Example 2. A Virtual Game Project

This project is a basic model for a complex metaplastic virtual world. It describes a mythological story on the research of the Truth. The virtual environment is placed into an oneiric and abstract dimension of two opposite worlds, where one is the Higher world and the other is the Lower world. The human (visitor) could go into and freely interact within these two worlds in searching for valuable treasures of information contained in each of the two worlds. The visitor has to correctly evaluate the given signs within virtual environments to identify the place typology and discover possible treasure locations. The choice of the "positive" treasure would reveal the Wisdom to Human visitor, while false notions would reveal the "negative" treasure which will lead him out of these two worlds. The visitor could interrupt in any moment his exploration, by leaving the oneiric dimension. This example of an online metaplastic semantic and philosophical virtual game shown below, clearly explains the methodology application through these process phases:

1. Composition Design Criteria

The construction of storytelling needs the creation of figurative structures and behaving properties of virtual environments. It is achieved through the following methodologies:

The surrounding rings structure define the metaplastic cyberworld composition criteria within its own configurations. The fuzzy visual settings of the "Form_Rhytm" generate the virtual environment below with the following parameters:

Rhytm_Form rule $RF_i = (W+D) * fFORM(interp(P, V, S))$;

Light-Color-Sound rule $LCS_i = (W+D) * fAppareance(interp(C, B, Co))$;

The game consists in finding two treasures which contain information. The visitor could be an animated character or an avatar realized through a representation as a "third person" typology or in another representation modality not used by the Dialog system configurations. In this virtual world example, it has been chosen the remote eye and remote hand modality as a representation of human user.

Where: FORM=Rhythm_Form function; W= Form width; D=Form depth; P=Form weight;

V=Rhythmical variations; S=Sequence connections; C=Chromatic qualities; B=Brightness;

Table 1. Schema of the virtual world game configuration.©2010 Gianluca Mura

Feedback [emotional changes]		System Dialog States								
		Visitor Sensing			Visitor Feeling			Visitor Acting		
Senses weights	Visual depth	Highly involved (+1.0)			Highly involved (+1.0)			Highly involved (+1.0)		
	Hearing intensity	Highly involved (+1.0)			Highly involved (+1.0)			Highly involved (+1.0)		
	Touching	Not involved (0.0)			Medium involved (0.5)			Highly involved (1.0)		
	Interactive Threshold zone	External			Sensing zone			Feeling zone		
		World I	World II	Objects	World I	World II	Objects	World I	World II	Objects
Form-struct	1. Form_Rhytm Complexity	Smooth	Looped	Linear	Smooth	Looped	Linear	Smooth	Looped	Linear
	2. Form Variation	Mean .5	High 1	Low .2	Mean .4	High 1	Low .2	Mean .5	High 1	Low .2
Appearance	3. Form Sequences	Low .2	High 1	Low .2	Low .3	High .8	Low .3	Low .2	High 1	Low .2
	4. Chromatic qualities	High 1	Low .2	Mean .5	High 1	Low .2	Mean .5	High 1	High 1	High 1
	5. Brightness	High 1	Low .3	Low .2	High 1	High 1	High 1	High .8	High 1	High 1

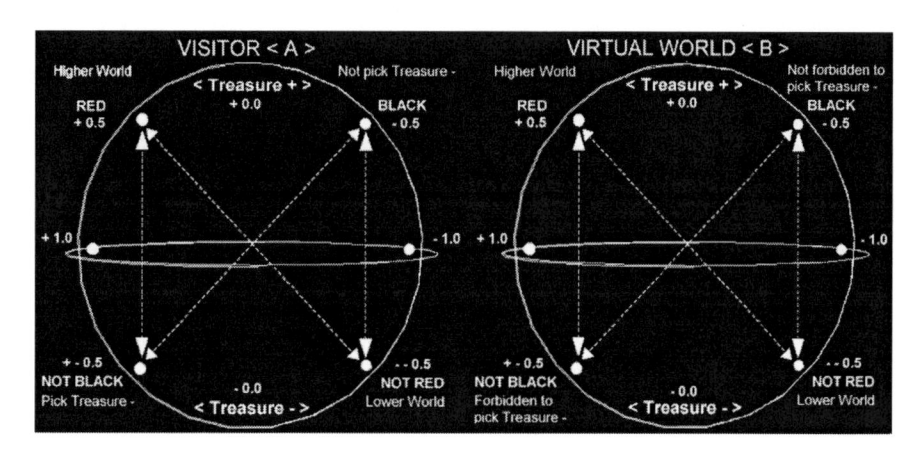

Figure 9. Visitor and the virtual environment narrative semantic schema ©2010 Gianluca Mura

Co=Consistency; LCS=LightColorSound function.

2. Setting the Dynamic System Rules

The narrative structure of the virtual world is constituted with four semantic functions in sequence and demonstrated with semantic schemes of "The Red and Black"(Figure 9) as follows:

1. The Wisdom Treasure has been given to the virtual environment.→
2. It is forbidden to have the false treasure.
3. Visitor's action to act/not to act the infraction of having the false treasure.→
4. The Price/Punishment as consequence of having the Treasure True/False of Information: Wisdom or Exit from virtual world.

The Rhythm Form complexity determines dynamic relations of visual syntax between virtual worlds' elements and they directly define the system behavior. The engine executes the resulting configurations of the cyberspace through the dynamic rules between each of the visual components within a simple manipulation of its visual elements. The formalizations of metaplastic language are synthetically explained below with all the interaction processes made among the user and the virtual system.

Visual inference = $Ss(w) * State_k$;
Balance$[i] = \Sigma_i (f Ss(w_i) * f(interp(RB_k, Rb_{k+1})))$;
Dynamic Syntax $[D_i] = \Sigma_i (f D(E(w_i))i * f Balance[D_j])$;

3. Semantic Meaning Results

The dynamic relations of visual syntax, for each step of the interaction dialogue system define the semantic meaning through the Red and Black meta-codification signs of information.

$RB\ i = \sum_i (1-truth(RB_j))*(1.0/nRB)$;
$RB\ Semantic_i = \Sigma_i (1-truth(RBi, i+1)*(1/nRB))$;

The whole system processes are defined by:

$VW_{i,j}^{new} = f(RB\ Semantic_i\ (Dynamic\ Syntax_j\ (\Delta proximity_{ij}))) + VW_{ij}^{old}$;
Proximity $f(Px) = f(E(User.position) - Threshold)$;

Where: VW = Metaplastic Virtual World; f = trigger function; i,j = state; k = item shown;.

E = Emotional states function(see synaesthesia paragraph); D_i = dialogue state; w = weight;

Table 2. Game rules configuration. ©2010 Gianluca Mura

"Searching for Wisdom" Virtual World Game						
Virtual World						
MAN	B1	B2	B3	B4	B5	B6
A1	Man in Higher World	Man in Higher World	Man don't pick treasure - but it is not forbidden	Contradiction! Man in Lower and Higher World	Contradiction! Man in Higher World with negative treasure	Man in Higher World and it is forbidden to pick treasure -
A2	Man in Higher World with Treasure +	Man in Higher World with Treasure +	Man in Higher World with Treasure + but it is not forbidden to pick the treasure -	Contradiction! Man with Treasure + in Lower and Higher World	Impossible! Man with treasure + and with Treasure - in Lower World	Man in Higher World with Treasure + and it is forbidden to pick treasure -
A3	Man in Higher World I'm not picking the Treasure -	with Treasure + I'm not picking the Treasure -	I'm not picking the Treasure - but it is not forbidden	I'm not picking the Treasure - Man in Lower World	I'm not picking the Treasure - but I have Treasure -	I'm not picking the Treasure - It is forbidden
A4	Contradiction! Man in Higher and Lower World	Impossible! Man in Lower World with Treasure +	Danger! Man in Lower World with no restrictions to pick the treasure -	Man in Lower World	Man in Lower World with Treasure -	Man in Lower World with restriction to pick the treasure -
A5	Impossible! Man in Lower World with Treasure + and Treasure -	Impossible! Man in Lower World with Treasure +	Man in Lower World with Treasure - but it is not forbidden	Negative! Man in Lower World with Treasure -	Negative! Man in Lower World with Treasure -	Negative! Man in Lower World with Treasure – and it is forbidden
A6	Impossible! Man pick the treasure – and he is in Lower World	Opposite! Man pick the Treasure – and has the Treasure +	Man pick the treasure- and it is not forbidden	Man pick the Treasure- and is in Lower World	Negative! Man in Lower World pick the Treasure -	Negative! Man in Lower World with Treasure – and it is forbidden
	VISITOR WINS				VISITOR LOSES	

SS=Senses function; RB = RB semantic weight; RB Semantic = Red&Black semantic function.

The metaplastic virtual world during its dynamic processes gives their results within the following running phases (Sensing, Feeling, Acting):

if (Px f(User_enter to the Space)> 0) then VW[D sensing] = VW.rules[Sensing interactions];
if (Px f(User exchange information)= 0) then VW[D feeling] = VW.rules[Feeling interactions];
if (Px f(User compose the information)< 0) then VW[D acting] = VW.rules[Acting interactions].

The storytelling follows a procedure dynamics through different states (Sensing, Feeling, Acting) of interaction processes. The possible narrative cases evolve with the following situations list:

The situations different narrative combinations give following results (Table 2):

- The visitor wins in A1-B2, A2-B1, A2-B2, A2-B6 and A3-B2;

The Metaplastic Virtual World Theory

- The visitor is in a dangerous situation in A1-B3 and A4-B3;
- Opposite situations occur in A6-B2 and A2-B5;
- Contradictory situations occur in A1-B4, A4-B1, A6-B3 and A3-B6;
- The impossible situations that give null results are A4-B2, A5-B1, A5-B2, A6-B1 and A6-B2;
- Absurd situations that give null results are A6-B3 and A5-B3;
- The infraction rules occur in A5-B6 and A6-B6;
- The visitor loses in A3-B5, A4-B5, A5-B4, A5-B5, A5-B6, A6-B4, A6-B5 and A6-B6;
- All others events give a neutral virtual world condition.

The system runs until the visitor wins with finding the right treasure or loses with picking the wrong treasure or simply by deciding to leave the virtual world.

THE METAPLASTIC CONSTRUCTOR

The project (Figure 10, *Metaplastic Installation at the Museum of Modern Art in Toluca*) is related to an abstract art language methodology applied to the virtual world concept. The Constructor is

Figure 10. (Above) 4 frames of the Metaplastic Virtual World virtual game ©2010 Gianluca Mura (Below) The Meta-Plastic installations in the Hall of the Museum of Modern Art, Toluca, Mexico and online immersive shared installation. ©2009 IEEE (Gianluca Mura, Cyberworld Cybernetic Art Model for Shared Communications, CYBERWORLDS 2009). Used with permission

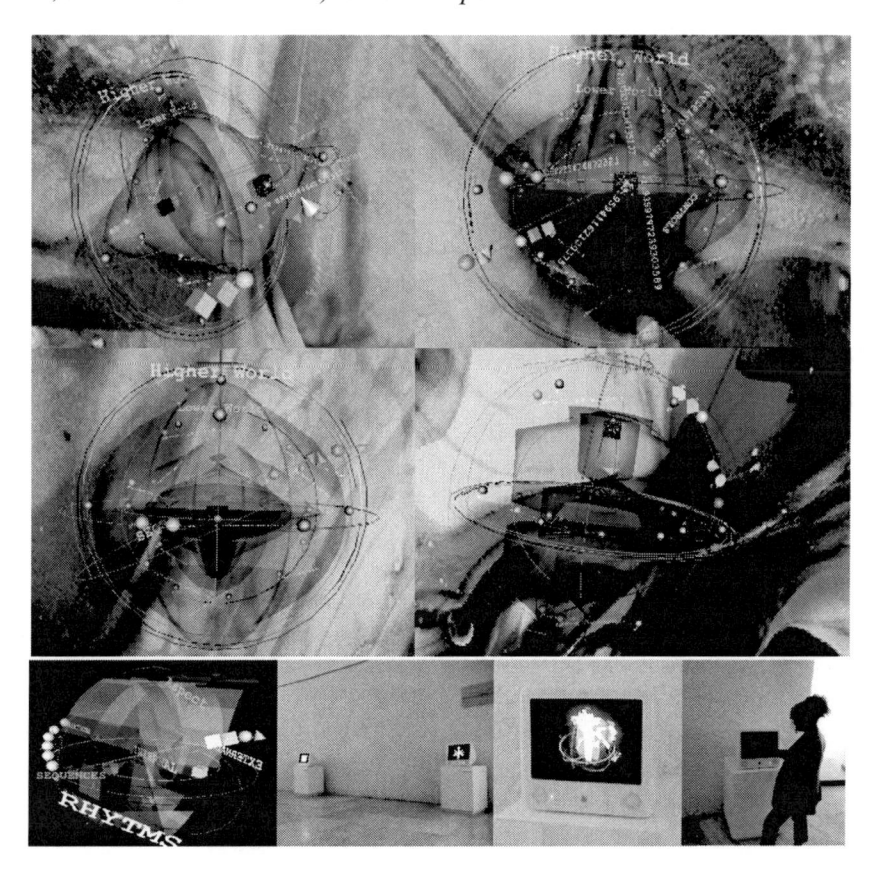

a conceptual 3d visual engine which builds the virtual space through formal elements expressed with many balances of light, chromatic and sound values. The construction principles of this metaphysical worlds are based on the *Gestaltung* theories of Paul Klee. The inter-actor (user) modifies the visual elements relation rules by moving the structure rings through its own feedback process. As results, the user builds its own concept of virtual space. The resulting repertoire of these virtual spaces, define a metaplastic abstract art language where each space with its configurations become a visual sign of this language.

CONCLUSION

This chapter has demonstrated the implementation of a "Red and Black" fuzzy methodology semantic space and a becoming of new typology of cyberspace: the Metaplastic virtual world. The main reason of this theory is to create and communicate conceptual meaning by using artistic machine as a new type of virtual metaphors. The system innovates the virtual world concept through a creation of union between art concept and fuzzy logic, into achieving new digital aesthetics and semantics concepts.

In conclusion, the semantic system of metaplastic virtual worlds opens a possibility for a profound analysis of social issues that emerge into social communities and networks on the Web. The metaplastic vision already contains its Klee's original sense and social ethics indicated by Stendhal's homonymous literature masterpiece. The metaplastic virtual world is proposed as a conceptual tool of Contemporary society and indicate the opportunity for social content evaluation on the Net by underlining problems and contradictions through the Red and Black semantics. Within actual globalized panorama the tensions of sustainability, economy and development, this semantic model offer a new possibility of social topics representation and explanation from virtuality to reality.

The further future applications of this theory, its evolutions and new applications, will include digital environments for different usage (web3d environments, interactive games, virtual museums, digital archives, interface systems, mobile applications, virtual art platform and installations).

ACKNOWLEDGMENT

This chapter is based on the doctoral dissertation of Gianluca Mura,"Un modello di spazio virtuale del Design" developed at the Politecnico di Milano University, Italy.

I would like to thank Dr. Nataša Duboković for her collaboration in writing this chapter.

REFERENCES

Artsmachine Media Lab. (2010), from http://www.artsmachine.com

Ashby, R. (1956). *Introduction to Cybernetics*. London, England: Chapman & Hall.

Balla, G., & Depero, F. (1958). Ricostruzione Futurista dell'Universo. In *Archivi del Futurismo*. Roma: Manifesto.

Bellasi, P., Corà, B., Fiz, A., Hayek, M., & Magnagnagno, G. (2004). *Tinguely e Munari. Opere in Azione*. Milano, Italy: Mazzotta Editore.

Bertalanffy, V. L. (1983). *Teoria generale dei sistemi*. Milano, Italy: Mondadori.

Black, M. (1937). Vagueness: An exercise in logical analysis in *Science, n. 4, pp 427-455.*

Boccioni, U. (1971). *Manifesto tecnico della scultura futurista*. Milano, Italy: Feltrinelli.

Bru, C. (1955). *L'estetique de l'abstraction*. Paris, France: Presses Universitaires de France.

Camfield, W. (1979). *Francis Picabia his art, life and times*. Princeton, NJ, USA: Princeton Univ. Press.

Carver, C. S., & Scheier, M. (1998). *On the self-regulation of behavior*. London: Cambridge University Press.

Cherchi, P. (1978). *Paul Klee teorico*. Bari, Italy: De Donato.

Chomsky, N.(1959). Three models for the description of language in Information and Control. *On certain formal properties of grammars, n.2*, IRE Transactions on Information Theory.

Corrain, L. (2004). *Semiotiche della pittura*. Roma, Italy: Ed. Meltemi.

Damisch, H. (1984). *Fenêtre jaune cadmium: ou les dessous de la peinture*. Paris, France: Gallimard.

De Mauro, T. (Ed.). (1967). *Corso di linguistica generale*. Roma-Bari, Italy: Laterza.

Dorfles, G. (2002). *Il divenire delle Arti. Ricognizione nei linguaggi artistic*. Bologna, Italy: Bompiani.

Duchamp, M. (1994). *Duchamp du signe*. Paris, France: Flammarion.

Eco, U. (1962). *Opera Aperta*. Milano, Italy: Bompiani Editore.

Eco, U., & Munari, B. (1962). *Arte Programmata (catalogo Olivetti)*. Milano, Italy: Olivetti.

Edschmid, K. (1920). *Paul Klee Schopferische Konfession in (Vol. XIII)*. Berlin: Tribune der Kunst und Zeit.

Fink, C. A. (1979). *Searching for the most powerful behavioral theory: The whole of the Behavior Systems Research Institute and the Behavioral Model Analysis Center*. Falls Church, Va: Fink.

Goodman, N. (2003). *I Linguaggi dell'Arte. L'esperirenza estetica: rappresentazione e simboli, trad. it* (Saggiatore, I., Ed.). Milano, Italy.

Gray, W. D. (Ed.). (2007). *Integrated Models of Cognitive Systems*. USA: Oxford University Press.

Greimas, A. J., & Courtés, J. (1979). *Sémiotique. Dictionnaire raisonné de la théorie du langage*. Paris, France: Hachette.

Greimas, A. J., & Courtés, J. (1982). *Semiotics and Language: An Analytical Dictionary*. Bloomington, Indiana, USA: Indiana University Press.

Grinder, J., & Bandler, R. (1975). *The Structure of Magic II: A Book About Communication and Change*. Palo Alto, CA: Science & Behavior Books.

Hang, H. P., & Alessi, N. E. (1999). *Presence as an Emotional Experience in Medicine meet Virtual Reality. The Convergence of Physical and Information Technologies options for a new era in HealthCare* (pp. 148–153). Amsterdam: IOS Press.

Holtzman, S. R. (1995). *Digital Mantras*. Cambridge, MA, USA: MIT Press.

Jakobson, R. (1973). *Questions de poetique, Paris, France: Seuil, Kuhn, A. (1974) The Logic of Social systems*. San Francisco, USA: Jossey-Bass.

Kosko, B., *Il Fuzzy pensiero*, trad. it, Milano, Italy: Baldini&Castoldi

Kuhn, A. (1974). *The logic of social systems*. San Francisco, USA: Jossey-Bass

Lakoff, G. (1987). *Women, fire and dangerous things: what categories reveal about the mind*. Chicago, IL, USA: University of Chicago Press.

Laurel, B. (1990). *The Art of Human-Computer Interface Design*. Reading, Massachusetts, USA: Addison-Wesley.

Lerat, P. (1917). *Les Recherches Futuristes*. Paris, France: Imprimerie Levé.

Lévy, P. (1992). *Le tecnologie dell'intelligenza, trad. it*. Ed.A/Traverso.

Lévy, P. (1997). In Cortina, R., Ed.). Milano, Italy: Il Virtuale.

Lévy-Strauss, C. (1958). *Anthropologie structurale 1*. Paris, France: Plon.

Lukasiewicz, J., & Borkowski, L. (1970). *Selected works*. London, England: North Holland.

Marcel, J. (1959). *Histoire de la peinture surréaliste*. Paris, France: Seuil.

Mari, E. (1963). *Arte Programmata*. Milano, Italy: Olivetti Group.

Mari, E. (1964). *Libertà nell'ordine*. Milano, Italy: Feltrinelli.

Marsciani, F., & Zinna, A. (1991) *Elementi di Semiotica Generativa. Processi e sistemi della significazione*, Progetto Leonardo, Bologna, Italy: Ed. Esculapio

Miller, J. G. (1978). *Living systems*, New York, Usa: McGraw-Hill

Milsum, J. H. (1966). *Biological Control Systems Analysis*. New York, USA: McGraw Hill.

Moholy-Nagy, L., & Kemény, A. (1922). *The Constructive Dynamic System of Forces: Manifesto of Kinetic Sculpture*. Berlin, Germany: Der Sturm.

Moholy,-Nagy, L. (1947) *Vision in Motion*, Chicago, US: Paul Theobald

Moles, A. A. (1990) *Art et Ordinateur*, Paris, France: Ed. Blusson

Moszynska, A. (2003). *L'Art abstrait*. Paris, France: Thames&Hudson.

Mura, G. (2005) *Virtual Space Model for Industrial Design*, Unpublished doctoral dissertation, Politecnico di Milano University, Italy

Mura, G. (2006) Conceptual artwork model for virtual environments in *JCIS journal* (ISSN 1553-9105), Vol.3, n.2, pp.461-465, BinaryInfoPress

Mura, G. (2006). *The red and black semantics: a fuzzy language in Visual Computer, n.23* (pp. 359–368). Berlin, Germany: Springer.

Mura, G. (2007) The Metaplastic Virtual Space in Wyeld, T.G., Kenderdine, S., Docherty, M. (Eds) *Virtual Systems and Multimedia* (pp. 166-178), Berlin, Germany: Springer

Mura, G. (2008) The meta-plastic cyberspace: a network of semantic virtual worlds in *ICIWI 2008* WWW/Internet International Conference, Freiburg, IADIS

Mura, G. (2008) The Metaplastic Constructor in *CAC 2 Computer Art*, Paris: ed. Europia

Mura, G. (2009). *Cyberworld Cybernetic Art Model for Shared Communications in Cyber-Worlds 2009, Bradford University*. UK: IEEE.

Newman, W., & Sproul, R. F. (1987). *Computer Graphics principles*. New York, USA: McGraw Hill.

Parker, S. P. (Ed.). (2002). *McGraw-Hill Dictionary of Scientific & Technical Terms*. New York, NJ, USA: McGraw-Hill.

Pierce, C., & Weiss, P. (1935). *Collected papers of Charles Sanders Peirce, Cambridge*. Ma, USA: Harvard University Press.

Popper, F. (1970). *L'arte cinetica*. Torino, Italy: Einaudi.

Popper, K. R., & Eccles, J. C. (1977). *The self and its brain*. New York, USA: Springer International.

Ralston, A., & Reilly, E. D. (1983). *Encyclopedia of Computer Science and Engineering*. New York, USA: Van Nostrand Reinhold.

Reilly, E. D. (2004). *Concise Encyclopedia of Computer Science*. Hoboken, NJ, USA: Wiley&Sons.

Roque, G. (2004). *Che cos'è l'Arte astratta?* Roma, Italy: Donizelli.

Rosch, E. (1978) Principles of categorization, in Rosch E. and Lloyd B.B.(eds) *Cognition and categorization*, (pp.27-48):New York,NJ,USA: Erlbaum, Hillsdale

Schöffer, N. (1963). *Schöffer*. Neuchatel, Switzerland: Editions du Griffon.

Spiller, J. (1959). *Teoria della forma e della figurazione*. Milano, Italy: Feltrinelli.

Thürlemann, F. (1982) *Paul Klee. Analyse sémiotique de trois peintures.* Lausanne,Suisse: L'Age d'Homme

Tinguely, J. (1982). *Catalogue Raisonné, Sculptures and Reliefs 1954-1968.* Zurich, Switzerland: Edition Galerie Bruno Bischofberger.

Whorf, B. L., & Carroll, J. B. (1964). *Language,Thought and Reality:selected writings*. Cambidge, MA, USA: MIT Press.

Wiener, N. (1948). *Cybernetics or Control and Communication in the Animal and the Machine*. New York, Cambridge, MA,USA: MIT Press.

Wiener, N. (1988). *The Human Use of Human Beings:Cybernetics and Society.* Cambridge, MA, USA: Da Capo Press.

Woolley, B. (1993). *Virtual Worlds*. London, England: Blackwell Publishers.

Young,P.T.(1967) Affective Arousal in *American psychologist*, n.22, pp.32-40

Zadeh, L. (1969). *Biological application of the theory of fuzzy sets and systems in the proceedings of an international symposium on BioCybernetics of the central nervous system* (pp. 199–206). Boston: Little Brown.

KEY TERMS AND DEFINITIONS

Dialogue System: Artificial cognitive system of the Metaplastic virtual world.

Form_Rhythm: The structural equilibrium composed with dynamic relations of form_weights to describe the metaplastic virtual world.

FSM: A fuzzy dynamic system model called "finite state machine"(FSM).

Fuzzy Logic: A form of algebra employing a range of values from "true" to "false" that is used in decision-making with imprecise data, as in artificial intelligence systems. In mathematics, a form of logic based on the concept of a fuzzy set. Membership in fuzzy sets is expressed in degrees of truth—i.e., as a continuum of values ranging from 0 to 1. In a narrow sense, the term fuzzy logic refers to a system of approximate reasoning, but its widest meaning is usually identified with a mathematical theory of classes with unclear, or "fuzzy," boundaries. Control systems based on fuzzy logic are used in many consumer electronic devices in order to make fine adjustments to changes in the environment. Fuzzy logic concepts and techniques have also been profitably used in linguistics, the behavioral sciences, the diagnosis of certain diseases, and even stock market analysis (Wiley & Sons Encyclopedia, 2004).

Metaplastic Language: Conceptual models based on abstract art languages rules applied to digital symbolic systems.

Ontology:(Philosophy/Logic) The set of entities presupposed by a theory (Wiley & Sons Encyclopedia: 2004).

Red and Black Semantics: The Metaplastic virtual world semantic model.

Synaesthesia: (psychology) The condition in which a sensory experience normally associated with one sensory system occurs when another sensory system is stimulated.(McGraw-Hill Dictionary of Scientific & Technical Terms, 2002).

Virtual Reality Archetypes: Plastic arts movements artworks that preceded virtual reality (Duchamp, Avant-Garde movements, Futurists movement, Picabia, Moholy-Nagy, Tinguely, Schöffer, Munari and others).

Chapter 3
The Tangiality of Digital Media

Paul Catanese
Columbia College Chicago, USA

Joan Truckenbrod
School of the Art Institute of Chicago, USA

ABSTRACT

This chapter introduces the concept of Tangiality, a defining characteristic of digital studio practice. It proceeds by illustrating how Tangiality is present in the work of contemporary interdisciplinary, hybrid media and multi-modal artists by providing several case studies regarding how Tangiality manifests in artworks. We conclude this chapter with a detailed discussion of how the emergence of Tangiality demonstrates that artists engage with post-digital materiality.

INTRODUCTION

Paradoxical contradictions of virtuality and materiality have undergone a contemporary transformation; the concept of *Tangiality* addresses a radical shift in our sensory perceptions as they are evolving to absorb, incorporate, and adopt the immateriality of the virtual. This evolution is catalyzed by the intervention of materiality with digital media. As our sensory experience of virtuality embodies the emotional, physiological and the kinesthetic, digital materials too are imbued with resonant, visceral, kinetics. This shift has precipitated hybrids: digital artworks that manifest in myriad forms that contradict their origins as choreography, visual artifacts and substance. *Tangiality* addresses the inherent, ongoing re-examination of the idea of material and its use in contemporary art practice. Concomitantly, digital media embodies a multitude of dimensions that have been explored and expanded on, such as: transparency, opacity, immersion, immediacy, translation, malleability, novelty, play, illusion, interaction, multimodality, narrativity, hypermedia and virtuality. Artists give palpability to experiences with virtual materiality, or *Tangiality*, a defining characteristic of digital studio practice.

The term: *Tangiality* was introduced by Dr. Slavko Milekic in his paper *"Towards Tangible Virtualities"* presented at the 2002 *Museums and*

DOI: 10.4018/978-1-60960-077-8.ch003

the Web conference in Boston. In his paper, Dr. Milekic outlined *Tangiality* as a proposed domain between physicality and virtuality where *"interactions with virtual data produce tangible sensations"* (Milekic, 2002). While Dr. Milekic's primary focus applies *Tangiality* to a conversation about the potential for additional dimensions of feedback for human / computer interface devices, the domain between the physical and virtual is an intriguing threshold in the context of artists' materials.

Tangiality in digital media is embedded in digital raw materials; the gesture and performance of studio practice infiltrates and transforms the entropic, itinerant, nomadic, transitory code. Interrogating the construct of materiality inherent in the practice of digital media precipitates a theory of art practice that does not strictly represent media, fine or performing arts, but instead migrates among these classifications using the capabilities of the digital realm to embrace the potential of a hybrid art practice. Artists whose work exemplify this approach include *Wafaa Bilal,* whose durational performative installation work allows the public to remotely interface and interact with him; *Dr. Angela Geary*, who has developed input processing systems for virtualizing the engraving process for printmaking; and *The Resonance Project* which conducts performance as research creating dance with real-time 3D movement scanning for tele-presence choreography. In addition, artists whose work illustrates additional aspects of *Tangiality* that we will be discussing in this chapter include *Ann Hamilton, Catherine Richards, Thecla Schiphorst, Koosil-ja Hwang, Scott Killdall* and *Victoria Scott*, the artist collective *Blast Theory,* as well as examples from our individual art practices. In the artists' work we will be examining, the virtuality of the screen erupts as intervention with the materiality of paper, video projection, performance, and objects or space in installation. These artists negotiate the tension between an impulse to create by hand and the ingrained framework of the Cartesian system that pervades construction processes of computed form, by employing novel conceptual strategies that privilege the physicality of experience. As the hand reaches through the liminal space of the screen, the physicality of these connections is critical to the hybrid practice of digital artists, the process of constructing meaning and the relationship between perception and the body.

Case Study: Dr. Angela Geary

Dr. Angela Geary has conducted extensive research on the affects of computer technology in the material realm of traditional processes in drawing and printmaking, and as it transforms into digitally driven processes. Despite pervasive virtuality, the sustained importance of physicality and tacit intelligence for artists is clear. Using digital interfaces in the creative process finds the digital embedded in the artworks. Digital processes leave distinctive

Figure 1. In this image, an artist uses haptic computer interaction to work with a carvable, virtual 3D block for printmaking. © 2006 Angela Geary. Used with permission

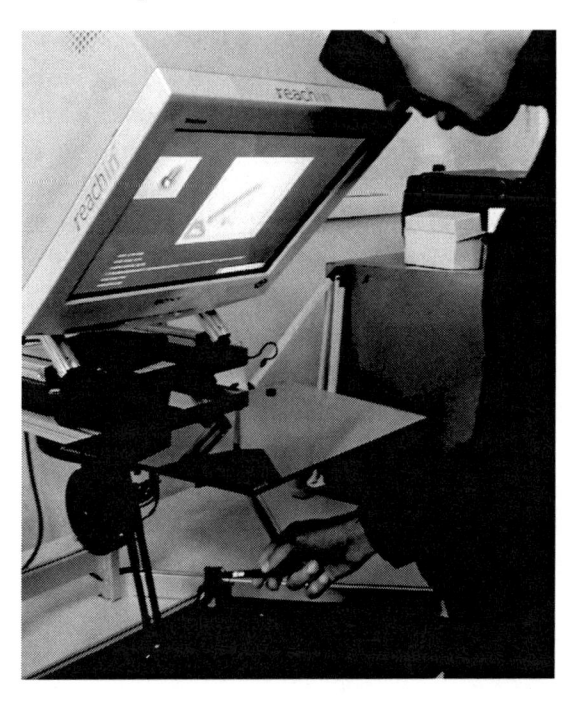

traces, which are physically manifest yet materially vulnerable. Digital physicality presents an opportunity to invigorate digital practice with the human creative excitement that arises through tangible engagement and material outcomes.

Dr. Geary's research focuses on artworks "becoming digitally physical" (Geary, 2009). The Virtual Haptic Interface for Printmaking (VHIP) was one project in which she used touch-based interactive, haptic, devices to explore this potential. The focus of this research involved how the human experience of physically making art, and understanding the accompanying tacit skills, could be served by emerging tangible interaction technologies. Haptic devices, such as *Sensable's Phantom*™, seemed to offer the potential to extend traditional mediums in fine art practice into the virtual context, providing a route back to materiality for artists. Initially it appeared that the use of haptics could mitigate the difficulties of traditional techniques, such as engraving a steel plate, while keeping the authenticity and boundaries of the physical experience of drawing intact through virtual force-feedback. Undoing marks on a steel plate is difficult, for example, so the intermediary digital platform allows consideration of the actual mark making until there is satisfaction with the virtual artifact. Thus there is a delay in the commitment to the making of marks in the physical production process. This availability of creating and previewing artwork in multiple versions in the intermediate digital stage, prior to producing a perfect plate through digital manufacturing does not appear to contribute to enhanced creativity. Geary postulates that the opposite in fact is true. In the arts and sciences alike, restriction can be the fuel of creativity. This is confirmed by studies in psychology (Weiner, 2003) that show that the mind retreats from open-ended choices and reverts to repeating known patterns and options. Problem solving promotes exploration, and some artists embrace restrictions or deliberately impose limiting rules to exploit this. Hence, human creativity does not necessarily thrive on the apparently boundless possibilities enabled by digital and virtual technologies. Creative intent can be dulled when there is no boundary to push against. Digital filters and other manipulation devices can lead to predictable and uninspired outcomes even when approached from different starting points. Noticeably original or atypical work more often emerges from practitioners, such as Frank Gehry, who prefer physical over virtual methods in the germinal stages of the design process.

The potential of haptic interaction has not developed in the way that many expected in the mid 1990s, when accessible interface technologies began to emerge. In Geary's research, the VHIP project, they found, in practice, haptic interaction seemed to add little value or perceived benefit in the context they were exploring and even seemed detrimental to the users engagement. To quote Geary: *"As technology advanced it has become clear that the real problem is not the technology, but with the human"* (Geary 2009). Sensory processing in relation to haptic feedback is not yet completely understood as cognitive psychologists continue to explore this field. However, it is a well-documented psychological phenomenon that humans ignore or filter out sensory information that is repetitive or not currently useful and foreground more relevant stimuli (Weiner, 2003). This process known as sensory gating or filtering is iterative and automatic. Repetitive sounds or ambient odors, for example, soon fade into the background of our awareness.

Studying applied haptic interaction research, Geary speculated that if such a technology is going to find gainful application, then it really must add perceptual value to the application it is engaging with and be sympathetic to human selective attention and sensory filtering. McLuhan (1969) stated long before haptic devices existed that *"tactility is not a sense but an interplay of all senses"*. Professor Geary found that in the experience using haptic devices in the creative process, artists need an engagement with material

The Tangiality of Digital Media

that embodies an interplay that triggers an excitement. So while her research practice clarified that multi-sensory interfaces, particularly (haptic) touch feedback systems, are an unexpectedly poor vehicle for the physical expression of creativity in the virtual realm, this realization has invigorated her examination of how digital and traditional processes interoperate. Professor Geary's print-making research aims to fuse new methods with traditional techniques, to expand the conventions of the medium. As all research should do, this discovery, though unexpected, led to new inquiries. In her artwork *Faux Pano,* her approach was to disrupt the conventional optical data input step of digital photography by becoming the "camera", using a human generated drawing of a scene in the "photographic frame". Then the digital workflow was re-engaged by subjecting the drawings, once digitized, to a panorama-generating algorithm. The software attempts to correct the analogue discrepancies in perspective and tonal continuity, introducing new artifacts that accentuate the hand disruption. According to Geary, the physicality of analogue making or processing seems to be stubbornly persistent in fine art practice, despite the widespread availability, convenience and advanced sophistication of digital tools and processes. (Geary 2009) Disrupting, combining and applying physically material methods – both human and machine controlled - presents new opportunities for extending or reinventing traditional media in the digital age. The intersection of physical materials and digital control seem to be a particularly rich area of interest and offers an avenue for artists who seek a tangible engagement or outcome in relation to their work. Her *Disrupted Drawing* project is continuing to investigate human made disruptions to digital procedures and current work is examining the exploitation of human cognitive and perceptual limitations as a capture device in this context.

Case Study: Catherine Richards

The artist Catherine Richards investigates proprioceptive experiences; examining how physical stimuli are produced, and perceived within the body. Her projects have included creating illusions of the physical body or bodily presence, and inversely precipitating an experience of disembodiment of the physical body. In her 1997 installation *Charged Hearts* she investigates the nature of the object, and how it changes in new media and information environments. The objects in Charged Hearts are nostalgic containers in the form of the glass heart and the world (terrella). The glass terrella is a model of the natural wireless electromagnetic dynamo that surrounds the earth: the northern lights. The aurora works in a similar way as an artificial electromagnetic system: the cathode ray tube, the historical basis of TV and computer screens. The *real* object in *Charged Hearts* is an electromagnetic field that creates a fluctuating ambivalence between the material and the virtual. The hearts and the terrella are containers for these electrons. They are windows that frame the activity. When the viewer picks up the heart, a shadowy heart forms in phosphorescent gases. When someone else holds the second object, they see a shadowy heart in the phosphorescent gases also as a result of the excited electrons. Hearts are charged both literally and figuratively. In this artwork, the invisible becomes palpable; the electromagnetic spectrum that surrounds the viewer and precipitates sensations is experienced physically. Richards postulates that new technologies are continually absorbed into familiar *material* objects and environments. Thus digital technology becomes a parasite as physical objects become carriers or hosts – a remainder of materiality. Assimilated by objects, the boundaries between object or body and technology become porous.

Similarly, in an earlier work from Richards entitled *Virtual Body* in 1993, the body becomes a nostalgic material object. The artist succeeds at creating an optical illusion that dislocates the

Figure 2. Documentation from "Virtual Body" by Catherine Richards. In this image, clockwise from top-left: Partial view of installation with the artist; detail view of viewfinder on top of cabinet; detail view of illuminated miniature room seen through side opening of cabinet; virtual body collage of images. ©Catherine Richards. Used with the permission of the CARCC, 2010

senses. In this installation a wooden cabinet stands on a platform in the middle of the gallery. As spectators enter they see a miniature glass room on top of the cabinet. The walls are semi-transparent and illuminated, with light glowing from a patterned floor. On approaching they see that the images on a monitor portray a Rococo room.

The viewer steps to the platform to take a closer look and sees a peephole in the top or ceiling. As they look inside they see that the Rococo ceiling is a reflection – a momentary reflection that is interrupted by the image of the floor on the monitor. Next the floor image is overtaken by the ceiling, destabilizing the image of the space in the miniature room. On one glass wall there is a hole in which to put one's hand to enter the room. As the viewer inserts their hand, the spectator is *sensed* and the glass walls of the room suddenly become opaque and lose their transparency for anyone standing on the outside. The viewer looking down at the room's floor, the monitor, seeing their hand spread across the floor. The floor pattern then begins to scroll and the viewer experiences a sense of motion – a body illusion. One's hand appears to be infinitely traveling away from the body. Then the arm begins to take the body with it. It is as if miniature space is folded into infinite space, as if stillness is folded into motion. The body loses all references: inside/outside, giant/miniature, spectator/object, part/whole. "*The virtual environment becomes material, and our own part in this environment is tangible*," asserts Richards in her artist statement (Richards, 2009). Working with scientists from Brandeis University in Boston and University of Alberta in Edmonton Professor Richard's research into the mechanisms used to perceive our bodies, and the precarious boundaries between the physical body and the one experienced in virtual reality, led to her project *Spectral Bodies*. In exploring the body and how a sense of physical presence is created, there is a parallel study of individuals who experience the loss of the sense of body. Case studies demonstrate people that experience some ambiguity about the presence of or boundaries of the body, particularly arms or legs. In these examples people intellectually understand the form of the body, yet in reality the experience of the world through the body is compromised or disabled to varying degrees. An example is when the brain knows the hand is touching a ball, but the normal physical sensations are not experienced. Using VR, researchers have been able to create *illusions* allowing a person to virtually experience the compromised or *invisible* part of the body. Experiments were conducted with virtual representations and simulations of arms and hands. To begin this process a person's body is simulated in virtual reality, re-representing the body as a virtual body. Then as an image of the person's actual arm is mapped onto the virtual body, a physiological sense erupted, transforming the virtual body into the person's actual experience of the physical body. The virtual image became physical. In Richards words "*a kind of fiction made physical*" (Richards, 2009). The virtual image took on the *costume* of the real arm, simultaneously as the person experienced physiological sensation triggered from the virtual representation. The Tangiality of the body exists simultaneously in the physical world and the virtual world.

Case Study: Thecla Schiphorst

Artist Thecla Schiphorst has worked at the threshold of the physical and virtual in a number of her projects, blurring the boundaries into an emerging territory. Her installation titled *Bodymaps: Artifacts of Touch,* creates a physical space activated by a virtual body, mediated by technology that viewers are enticed to "touch". The site is a material construction in which the virtual body is experienced by the viewers' propriocetively. The viewer experiences the position and reactive movement of *the body* perceptually through vision and touch. The physical presence of the viewer reveals, activates and appears to arouse the representation of the body. Perceptually the viewer *feels* a connection with the body, even though the body is not physically present.

The viewer perceives a sensory, palpable, tangible response to this virtual body. The effect is erotic and sensual, yet disturbing.

In this work, beneath the white velvet surface of a large table, resides a body writhing in response to the viewer. The body is a projected image on the surface of the white velvet, but appearing deep beyond the surface. It is alternately revealed and

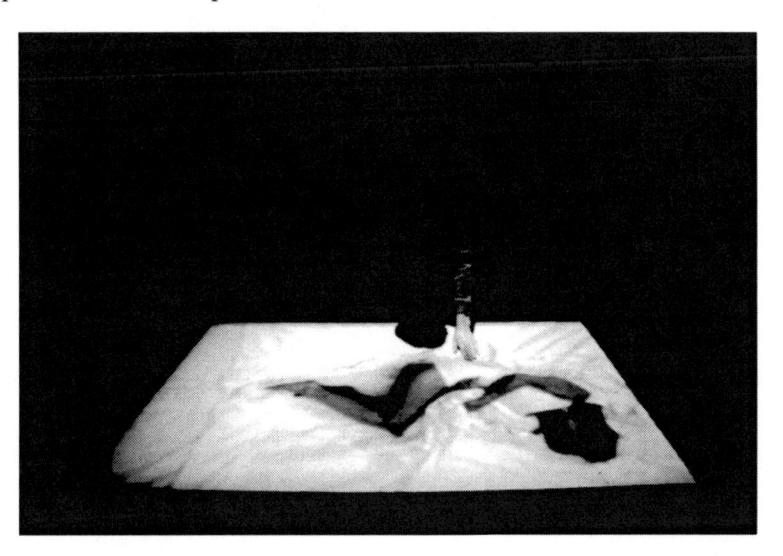

Figure 3. Documentation of installation "Bodymaps: Artifacts of Touch" by Thecla Schiphorst. © 1996-1998 Thecla Schiphorst. Used with permission

concealed as it twists and squirms in relation to the viewer's behavior. In the words of the artist, this surface *"yearns for contact and touch"* (Schiphorst, 2009). Embedded in the surface are electromagnetic field sensors that operate much like Theramins and force sensitive resistors. Together they detect proximity, touch, pressure and the amount of force applied to the surface. These physically oriented inputs are then processed using a complex algorithm; the response of the body is unpredictable, giving an experience of "liveness", of physicality to the body, even though it is immaterial or virtual. In the acclaimed 2002 book, *Understanding Virtual Reality*, authors William R. Sherman and Alan Craig identify the four defining elements of Virtual Reality consisting of: virtual world, immersion (physical and mental), sensory feedback, interactivity (individual & collaborative). Of these, the definition they offer for virtual world is of key interest, in which they state:

A virtual world is the content of a given medium. It may exist solely in the mind of its originator or be broadcast in such a way that it can be shared with others. A virtual world can exist without being displayed in a virtual reality system... much like play or film scripts exist independently or specific instances of their performance... When we view that world via a system that brings those objects and interactions to us in a physically immersive interactive presentation we are experiencing it via virtual reality. (Sherman & Craig, 2002)

The question then is: what if the artist does not choose to "broadcast" virtual world into the realm of virtual reality? What if artists are concerned with the notion of virtual world as a characteristic of their work, but choose to manifest art forms that are not wholly digital, immersive, or interactive; works that inhabit materials which reside among the virtual and the actual? How does one begin to discuss work that maintains relations with a virtual world as a space of discourse, but which erupts in a virtualized sensorial realm? *Tangiality* in digital media is embedded with unique entropic material residue. Perceptually we intuit *Tangiality* as residing in the material world – *but does it?*

Case Study: Wafaa Bilal

Wafaa Bilal's installation *Domestic Tension* is a new media artwork that definitively illustrates an immaterial artwork embedded with physicality; this work resides at that precarious threshold embracing both the physical and the virtual, exemplifying *Tangiality* in digital artworks. Bilal is an American citizen who was born and raised in Iraq, and who came to the US as a teenage refugee from the Gulf War, leaving his parents and siblings in Iraq. The terror and destruction of the war in Iraq is painfully intense for Bilal, amplified by his separation from his family, country and heritage. Horrific images of the death of soldiers and civilians in Iraq, together with destruction of cultural sites such as the Art Museum, and historic, archeological sites in Iraq, create jarring waves of emotion. These emotions were further intensified in the grief and turmoil that he and his family experienced after the death of his brother in the unspecific violence of the Iraqi war. And yet, he observes others seeing and experiencing this through the frame of the TV screen positioning this war in the realm of the unreal – at best removed, distanced from the viewer, or as a video game totally removing any threads of reality. Through this mythology of virtual games and online interaction, Bilal chose to engage the audience – both physically and online – in the experience of the constant tension and anxiousness of being shot at, being an unsuspecting target in the 24 hour each day war.

Professor Bilal created a life size *video game* in the gallery, embodying for him and participants worldwide, the provocative experience of the ceaseless and unpredictable violence of this war from the perspective of an Iraqi. He fused people in the gallery, with people worldwide online,

Figure 4. Installation details from "Domestic Tension" by Wafaa Bilal.. © 2007 Wafaa Bilal. Used with permission

together. In this *game* he made himself the target of a robotically controlled paintball gun, symbolic of the war. It was conceived of as a project that would be largely immaterial. The space was minimal, consisting simply of a room, a bed, a chair, a lamp, a computer, a paintball gun, and a camera. The space itself served, and was designed to serve, much more as platform than object of encounter. Bilal lived for thirty days in this small room, with his every movement recorded by a webcam attached to the top of a paintball gun. Viewers, anywhere in the world at any time of day, could log on to a website, remotely control, aim and shoot the paintball gun at him, or could choose to direct the gun away. Participants could also interact with each other through space provided in an online chat-room. His intention was that the interaction with the viewer would begin with the space provided, but that the encounter—that push and pull between work and participant—would quickly expand outward, grow richer in dialogue, and foster its own aura.

As this durational installation project unfolded, it gained an intense amount of attention and participants; responses were strongly opinionated and varied. Ultimately, the project took on a life of its own as it reached across political and cultural boundaries. In just over the course of a month, *Domestic Tension* mushroomed into an entity whose reaches were as deep as they were previously unknown as it rapidly exploded into a heated dialogue among the viewers and participants. From this tumult, a group of programmers emerged that periodically subverted the installation's paintball control software via the Internet, modifying the parameters for viewers shooting paintballs at Wafaa, intermittently preventing others from attacking. Other participants, however, continued to deluge the system with instructions to fire paintballs whenever they could. The competition to gain virtual control of the paintball gun became fierce: viewers organized themselves into communities to stop the shooting. Continually vigilant, one group would constantly re-aim the paintball gun, hoping that any shots that came through would at least miss their mark. The central issue of "shooting an Iraqi" quickly became evident, even though some participants continued to view it as an online game. The experience of this work was filtered through individual impressions of war, video games, online interaction and participatory media.

The engagement with the audience was intense. Participants experienced the threat of shooting and killing in a powerful manner – with many people attempting to protect Wafaa though different strategies online. Threads of material presence were intimately woven with virtual participation through the *Tangiality* of this project. For Wafaa, this experience was on the edge of unbearable. While his friends continually brought food and supplies to him and he was able to exit his "room" to use the bathroom in the gallery, the anxiety of being a *target* 24 hours a day for a month took a toll on his emotional state, as he experienced intensely the radical physical and emotional demands of the war on his family in his home country. He endured the entire month of this constant shooting and dialogue. With *Domestic Tension*, Bilal *"sought to override the heavy reliance of experience on the object"* (Bilal & Lydersen, 2008). Certainly, the space of the project itself allowed for the encounter, encouraged interaction, and furthered the progression of conversation. But the feedback created with *Domestic Tension*, the interweaving of Bilal's work and responses to it, removed weight from the physical object, once its initial statement was made. The participation itself became the *object* of the project. From the initial encounter, both his work, and the work of others contributing to the project entered into a participatory cycle. Each side became mutually constitutive: served not only to inform, but to author. The *Tangiality* of the *space-as-object* itself became re-contextualized, directed by the very opinions it helped to shape.

The artist has stated that the inspiration for the project occurred when he heard a news broadcast of a U.S. soldier in Colorado who was dropping bombs remotely, over Iraq. The successful lack of even momentary interaction between this U.S. soldier and the Iraqis being bombed is what allows such actions to continue. As he considered this, in light of his own loss of family and friends, it seemed that a project creating just this sort of removed presence, or virtual presence was a project that would spark emotion and exchange between people in a global space emanating from his room in the gallery. This artwork is discussed in detail in the book titled "Shoot an Iraqi: Art, Life and Resistance Under the Gun" published by Wafaa Bilal.

Sensory perception, by definition, originates from the Latin meaning to seize, to understand. Our long understanding is that our senses are the conduit between our bodies and the world around us, integrating the input from the senses to construct this realm. Marshall McLuhan expresses the importance of our ability to think, feel and express kinesthetically, in a way that brings hearing, seeing, tasting and touch, and today virtuality, together. In his book, *The Gutenberg Galaxy* (McLuhan, 1962), he discusses a segmentation of sensory perceptions, due to the invention of print and the printing press. In print, words became divorced from their related modes of expression, including voice, gesture, dance, song and animated behaviors such as rituals or storytelling. He says that when a sense becomes locked in a technology, it becomes separated from the other senses. This portion of one's self closes, as if it were locked in steel. Prior to such separation, there is a complete interplay among sensory experience. Joan Truckenbrod calls this *Integrated Creativity* in an article published in "Leonardo Music Journal" (Truckenbrod, 1992). The ratio or balance of sensory perception changes when any one sense, bodily or mental function, is externalized into technological form. Is this ratio or balance among the perceptions and faculties, which exists when they are not embedded in or "outered" in material technologies critical for a digital studio practice? Does virtuality and/or virtual reality fragment sensory experiences or is it possible that artworks engaging virtuality have the potential for re-establishing this interplay of the senses, the kinesthetic?

Case Study: The Resonance Project

The Resonance Project emerged as a collaborative project between UC Berkeley's Center for Information Technology Research in the Interest of Society (CITRIS), the University of Urbana-Champaign Computer Science Department (UIUC), the Department of Theater, Dance, and Performance Studies at UC Berkeley, and the Dance Department and Inter-media Program at Mills College.

Both CITRIS and UIUC have established laboratories that utilize multi-camera tele-immersion systems that are able to capture three-dimensional data in real time. In the context of this collaboration, the computer scientists, dancers and choreographers utilized performance as a research practice, experimenting with the capabilities of the tele-immersion systems to instantiate a new type of space for exploring motion, bodies, space and time. Specifically, the collaboration embarked on a series of experiments in the tele-immersion laboratories that have been used to capture the movement and gestures of full dancer's bodies. As they are digitally captured, the data of bodies and movements are compressed and transmitted to remote locations. Practically, this has meant streaming the data of a performance experiment from one laboratory to another as well as streaming from one or both of the laboratories to a theatre space. There have been many variations developed that intermix pre-recorded three-dimensional data with live three-dimensional data; projections of body-as-data and live dancers on stage; disembodied and re-embodied combinations that shine light on the gulf between the physically and the digitally remote. Curiously, the capture of gesture and movement is not a perfect representation; body surfaces are implied by individual voxels that float independently of one another in three-dimensional space as opposed

Figure 5. Lisa Wymore in the Tele-Immersion system UC Berkeley Electrical Engineering and Computer Science / Center for Information Technology Research in the Interest of Society. © 2009 Lisa Wymore. Used with permission

to being visually constructed in a smooth manner. Bodies digitally scanned and translated to point cloud data in real time, then compressed for streaming become grains of sand that shift, erode, collapse, combine, merge and evaporate as though under the force of some unseen digital wind.

The visual poetics result both from the gesture and movements of the dancers as well as the occlusions of the body from the view of the camera that occur which cause parts of the dancers to suddenly appear and disappear. Though these occlusions arise because of artifacts of the technology, they also provide an unusual and captivating new space for choreography where, as Lisa Wymore, one of the principle choreographers commented, the "*body can literally dissipate and atomize or fragment… on a virtual stage, bodies can even pass through one another*" (Shreve, 2009). This physical impossibility of co-location provides choreographers with a compelling and contemporary new space for dance to occur. In reference to one of the performance experiments entitled *The Reception* that was presented during the 2006 Humanities, Arts, Science and Technology Advanced Collaboratory (HASTAC), another choreographer involved in the collaboration, Katherine Mezur, observes that:

The lab dancers wearing pale and royal blue tops and pants, appeared to pulsate on the screen, their image appearing very clearly on some nights and on others, they were more like liquid images, in softer focus. And when they moved an arm or a leg in slow arcing movements of the 'carving sequence' the digital debris of their curving pale glowing arms streamed away, making their movements grow and echo out into space and time. (Mezur, 2009)

The tele-immersive performance experiments of *The Resonance Project* transform gesture; and while digitally merged, co-located bodies remain physically remote. Regarding this phenomenon, Mezur remarks "*our visuality has changed our*

Figure 6. Clockwise from the top left: Portable Tele-Immersion System; Dancers Lisa Wymore and Renata Sheppard performing in Panorama: Multi-Media Happening; Dancers Lisa Wymore and Renata Sheppard pictured working with technologists Gregorij Kurillo and Ramanarayan Vasudevan at the portable Tele-Immersion System; Sheldon Smith looking at images created by the Tele-Immersion system. © 2009 Lisa Wymore. Used with permission

bodies, *our bodies have actually morphed: we use more of our perception, we use seeing deeply and broadly"* (Mezur, 2009). This virtualizing of the body itself into a digital raw material reframes the percept; provides palpability to virtual objects, *Tangiality*.

Case Study: Koosil-Ja Hwang

For several years, dancer and choreographer Koosil-ja Hwang has pioneered a new performance method she refers to as *Live Processing*. This method expands the practice of live cinema remixing to include gesture and movement by incorporating dancers as though they have networked bodies, which are directly connected to media. Regarding her work, *Dance Without Bodies*, Koosil-ja remarks that this approach allows her and her dancers to *"grow nervous systems outside of our skin and extend them to the video monitors and to the other machines like a microphone and bass guitar. It is like becoming a new body"* (Hwang, 2009). In this work, the live cinema remix becomes a digital raw material for shaping gesture and movement; choreography unfolds in real-time, in full view of the audience. Live video takes shape on monitors or projection screens on stage; filled with images of gestures and movements that are remixed in real time by Koosil-ja via custom interaction software for live cinema performance.

Images of gesture and movement are culled from video documentation of studio based improvised dance movements, 3D animation data extracted from Machinima, and short clips from a diverse range of films including *Onibaba*, *Scarface* and the anime cult-epic *Akira*. These images are then combined and remixed while dancers onstage perform to the movements presented to them on screen, though there is often a slippage between what is occurring on screen and the embodiment of potential shapes and gestures. Hwang refers to this tension when she remarks *"our bodies become slippery and hollow to optimize this function for processing the information somatically"* (Hwang, 2009). The audience is presented with both the

Figure 7. Documentation from "Blocks of Continuality/Body, Image and Algorithm". Dancers Melissa Guerrero (front) and Elise Knundson (far) perform; 3D work in the center by Peter Blanco and on right Olorunseun Shogunro. © 2009 Koosil-Ja Hwang. Used with permission

onstage and backstage and division between these two realms is not distinguished. Similarly, the territory of the body is merged with the algorithm, which itself is not always moving in a clearly identifiable manner. While this performance method certainly incorporates aspects of improvisation and collaboration, rather than seeking repeatability, it extends beyond them through reveling in the slippage that the body algorithm introduces to the overall system, which manifests as embodied movement. The digital raw material is not activated without the bodies of the dancers; it is the live act of performance that makes the witness-able threshold between dancer and media palpable. This palpability indicates how digital studio practice reshapes the perceptible; it disregards borders of physical and virtual, resulting in artwork that manifests across boundaries where virtual coordinates, physical locations and cultural practice interrupt and inform one another. While the domain of *Tangiality* provides a membrane between physical and virtual; functionally, it is a performative space where liveness and human gesture catalyze aesthetics and form. In Florian Cramer's 2005 publication, *Words made Flesh*, he addresses the notion that code gives rise to material form as a pre-existing condition into which digital computing emerged. The invocation of code as magic is one of many threads examined in terms of the historic, literary and poetic – but in the end, his conclusion is direct. Cramer offers that:

Computation [is] rich with contradictions... [and] underneath those contradictions... lies an obsession with... the phantasm that words become flesh. It remains a phantasm, because... the execution [of code] fails to match the boundless speculative expectations invested into it (Cramer, 2005)

Yet, if code is meant to refer to instructions interpreted not just by machines, but those that involve human effort, we encounter a broad spectrum of forms directly manifest through their use: choreography, architecture, cooking, drawing – or if one prefers forms overtly lensed through digital computing, we might speak of mixed reality gaming, robotic performance and rapid prototyping.

Case Study: Blast Theory

The artist collective, *Blast Theory*, is well known for creating thought provoking, pervasive, augmented reality game experiences that utilize multiple forms of digital media in innovative narratives that conflate physical experience, virtual worlds, social interaction and live performance. Viewers of their works are both participants and audience members, equipped with multiple entry points into the artworks where physical and electronic locations become a site for narrative and play. Significantly, their works unfold simultaneously in physical urban locations as well as online environments in real time. The permutations of how physical and virtual, player and audience member, narrative and play modulate each other are not always straightforward; part of the experience of these works is embedded in how media is both a point of orientation and disorientation. Real time actions in physical locations initiate prompts for online users in virtual versions of the same locations and viceversa. Often, players are presented with tasks or scenarios that invite exploration and invention; in some of their works (*Rider Spoke*, *Uncle Roy All Around You*), participants are provided affordances that allow them to engage one another. In others, such as the groundbreaking: *Can You See Me Now?* – winner of the 2003 Golden Nica for Interactive Art at Prix Ars Electronica – members of Blast Theory inhabit the physical environment, while players inhabit the online world. There is a playful relationship between the two as online participants try to use virtual avatars to track and tag players (whose physical locations are transferred into electronic locations) running through the real world as though it were a game board. In much of their work, mobile devices provide points of entry to the narrative, prompts, play, and are tools for orienting oneself, but the devices may also reveal location or other information to the game world or other players. The slippage between multiple simultaneous realities, or mixed-reality, demonstrates ingenuity with digital raw materials that is arresting and complex. These mixed-reality experiences, which rely on mediated liveness demonstrate *Tangiality* in the context of site-specific performances.

Case Study: Scott Kildall & Victoria Scott

Digital raw materials and their infiltration of physicality are present in the work of artists Scott Kildall and Victoria Scott. Perhaps this is the case most strikingly in their 2007 installation, *No Matter,* which is comprised of "imaginary objects" that first exist in cultural imagination that were coaxed to erupt into physical form by way of the virtual environment Second Life. For this project, the artists developed a taxonomy of "imaginary objects" culled from mythology, literature, music, film, and other diverse sources such as history, thought experiments and alchemy.

The artists then populate these categories with a cross section of dozens of cultural artifacts that are recognizable, clever, and evocative including: the Holy Grail, Excalibur, the Philosopher's Stone, the Tell-Tale Heart, the Maltese Falcon, Rosebud, Schrodinger's Cat, Time Machines, the Red Herring and the Book of Love. Of course, all of the examples embody the notion of the "imaginary object" – as in not actually existing, yet at the same time are wholly embedded within culture. Once this list was established, the artists proceeded to contract individuals who have experience with digital-object crafting skills within the virtual environment, Second Life, to go about the process of developing virtual versions of these imaginary objects. Some of the objects (Kryptonite, Pandora's Box) are embedded with interactions that avatars in Second Life can be affected by; others (the Tree that Fell in the Forest, Can of Worms) merely can be observed. All of the "imaginary objects", however, inhabit a vir-

Figure 8. No Matter: Trojan Horse, photograph of paper model. © 2009 Scott Kildall & Victoria Scott. Used with permission

tual exhibition in Second Life, yet the manifestation of the artwork does not conclude there. Once established as virtual objects, the geometry, textures and materials of these imaginary objects, instantiated in a virtual world, were then extracted from Second Life and reconstructed as paper replicas in physical space. Ultimately, the installation and the artwork exists both in the context of the virtual world and physical world simultaneously, while cleverly disrupting the notion of real and imaginary. An installation such as this, which does not simply cross back and forth across a threshold, but makes us aware of its artifice and potential could be said to display *Tangiality*. Imbued in that potential is an alluring artists' material. During an artist's talk at the 2004 Interactive Media Forum at Miami University in Oxford, Ohio, prior to the premiere of her performance: *The End of the Moon*, Laurie Anderson spoke about the unpredictable nature of walking through the world as a form of artwork: a continuous unfolding real-time experience. She commented how her walks were suffused with internal dialogues that provide texture and counterpoint to actions that occur in real-time, even if they also occur in virtual space. This notion of whether a work is meant to translate or recompose liveness has been discussed in many contexts. In the book *Software Studies,* edited by Matthew Fuller, an essay by Michael Murtaugh entitled *Interaction* discusses the simulation of liveness. Murtaugh states "*a result of liveness is that an interactive system is one that supports a sense of playing or performing with the system*" (Murtaugh, 2008) – and in the 2008 edition of Philip Auslander's book *Liveness: Performance in a Mediatized Culture,* Auslander observes that:

We are at a point at which liveness can no longer be defined in terms of ... the presence of living human beings... the emerging definition of liveness may be built primarily around... the extent [to which]... virtual entities respond to us in real time; they feel live to us and this may be the kind of liveness we now value. (Auslander, 2008)

Murtaugh's definition of liveness in the context of interactivity, Auslander's redefinition of values for liveness and Anderson's desire to frame liveness share a common feature: the production and transmission of affective experience in real time. The implications of liveness inherent in the work of art collectives like *Blast Theory*, whose alternate reality games fuse virtual data and live performance; or the work of *Scott Kildall & Victoria Scott* whose mixed reality experiments with virtual labor, catalyze the exchange of virtualized senses and material traces, residue or objects in the physical world, additionally suggest *Tangiality* functions as framework, rather than threshold.

TANGIALITY AND THE EMERGENCE OF POST-DIGITAL MATERIALITY

As virtual worlds are imagined and constructed, the hand moves through the liminal surface of the screen. Is there an ambiguity of raw materials verses digital raw materials? The DIY and the Crafting movement have exploded in contemporary culture and in digital culture. In his book *FAB, The coming Revolution on Your Desktop – from Personal computers to Personal Fabrication,* Gershenfeld discusses a class at MIT engaged in digital fabrication in which students are inspired by traditional arts and crafts to create physical material forms through the digital or the virtual. Gershenfeld's students are envisioning in the material world, and then using digital technology, they shift towards physical crafting. The hand becomes digital – connecting computation with fabrication – injecting materiality into computation. The boundaries between the hand and the machine seem to dissolve, and in fact are irrelevant to these students. The process of rapid prototyping, and laser cutting CAD files represent form.

This fluctuating state of perception is probed in an article titled *Skin; Textile; Film* in the journal *Textile, The Journal of Cloth and Culture*, (Dormor, 2008). This article discusses the relationship between perception and the body as a reflexive shift between haptic and the scopic. What is experienced predominately is haptic, while what is seen, like the screen, is primarily scopic. In this context Ms. Dormor discusses an installation from 1990 entitled *linings* by Ann Hamilton. In a small room, constructed inside a larger room, the walls are covered with glass slides which themselves are covering water-stained writing. The writing under the glass slides on the wall is obscured, being smeared by water. A monitor embedded in one of the walls displays a video that of an open, overflowing mouth filled with what appears to be water. In Hamilton's 2006 monograph *An Inventory of Objects*, the description of this movement "*where the water seems to be flowing either in or out of the mouth*" is indicative of Hamilton's "*persistent exploration of thresholds, particularly the ambiguity of what is inside and what is outside, and the slippage of borders*" (Hamilton & Simon, 2006). In the space of the installation, the walls have a watery sense to them, creating a pool-like appearance over the surface of the walls and floor, reflecting the video on the surfaces around the room. These surfaces create a haptic visuality: the eye becomes an organ of touch, as the wet surface of the eye interacts with what is seen, roams over its surface, bringing skin and film, scopic and haptic into one place as an intertwining relationship. *Virtuality* resides within this dichotomy - the realm of the visible, the revealed, simultaneously with the concealed, the experienced yet invisible. In digital artworks there is a collaboration of these dimensions of experience – the absence and the presence.

This phenomena parallels lived experience in the Australian Aboriginal culture where the abstract nature of Aboriginal painting reveals the powerful presence of their ancestors in their everyday life. This artwork opens a portal to the ancestral realm, analogous to the internet allowing access, and creating a passage to other realms. Transforming this manifestation of the ancestral presence from painting to new media, Aboriginal artists were trained and given the use of video camera to create animated images, as described in the book *Shimmering Screens, Making Media in an Aboriginal Community*. One of the first images the artists captured was the undulating luminescence of water that glimmered on the screen, amazingly similar to their paintings. (Deger, 2008) At the heart of the Aboriginal endeavor are the experiential dimensions of recognizing the emergence *into visibility* of the underlying spirituality in their culture. In a similar manner, our sensory experience has evolved to include experiences of virtuality. It is not separate; rather there is a unity of the visible world and the invisible world. We give a palpability to these experiences; they have a virtual materiality, a *Tangiality*.

We create an experience of this presence through its traces. In fact memories deposit residue, traces of activities through material deposits. In the book *Memory Work, Archaeologies of Material Practices*, (Mills & Walker 2008), material traces provide insight into understanding the interaction of human and materials. Materiality is a window or portal to understanding the connections between people through time, and across space as people inhabit their worlds in very different ways. The physicality of these connections is critical to the process of constructing meaning. This extends into virtual worlds currently being constructed, leaving material traces or objects in the physical world in the construction process.

Joan Truckenbrod's studio practice embodies this fluctuating state of digital and physical materials. Multimedia installations catalyze processes of transformation created by a fusion of video projection and objects. A virtual dialogue erupts as the video residing in objects, precipitates a virtual realm with a material encasement. Video projection creates a virtual architecture that immerses the viewer in visual narrative and sound. In this architecture, video constructs space like that created by the flickering of firelight in a fireplace, injecting narrative. A costume is imposed on the object causing the social context of the object to becomes elastic, fluctuating in its meaning. The light and shadow of the video activates the spirit of the environment, while the narrative activates the object. This juxtaposition of video with object is provocative as the narrative of the video is conflictive with the inherent meaning of the object. Video cultivates a skin that is virtual like that in Second Life.

This crossover of experience of creating in a virtual space, has parallels to physical materials, in fact embodies some of the same physiological experiences as working with physical materials. As Artists we no longer differentiate between sculpting with clay or sculpting with light. There is an intuitive flow from envisioning physical form to constructing digital form, then mirroring back into physical form through rapid prototyping, 3-D printing and other digital fabrication techniques. Virtual processes are embedded in the hand, while the hand participates physically in the virtual. This circular process was used to create an outdoor sculpture in stainless steel and dichroic ecoresin for a public art commission in Green Bay Wisconsin.

In Paul Catanese's work, gestures made by hand are recorded, amplified, translated and finally are re-performed during the carving of relief printmaking blocks by way of industrial CNC machinery. Creating work that incorporates the mathematical precision of machine control, soft-bodied human gestures and irreversible ink marks on paper occupy a space for image making that is not strictly virtual or physical. And though designers and engineers utilize computer-aided drafting programs to interface with CNC machinery, the end result is incredibly precise, mechanical and industrial. In a 2008 article for the Media-N Journal of the New Media Caucus he states:

The visual grammar of the drawings that I want to achieve is more immediate and organic: I write my own drawing software to control CNC machinery that is custom tailored to my marks. And while shaping code is a component of my art practice, gesture drives the eruption of materiality. (Catanese, 2008)

The physical object is powerful - tension erupts between our impulses as artists to create by hand, and the drive to computer form. Physical objects are critical in the creative practice of digital artists, paralleling Sherry Turkle's examination of the role of physical objects in the creative life of scientists – in her book "Falling for Science: Objects in Mind". She postulates that *"neither physical nor digital objects can be taken out of the equation"* (Turkle, 2008) of creative endeavors.

Object passions or obsessions, bring us to the same enthusiasm for "what-is", that computation inspires for the "what-ifs". We envision in the digital the waves and echos of our explorations as to "what kind of sand is best for building castles, or the stubborn complexity of soap bubbles, or studying the details of light bent by a prism. (Turkle 2008)

CONCLUSION

The concept of *Tangiality* creates threads of materiality in digital studio practice sited at the threshold of the virtual and the physical. As the unifying, continuous filament weaving together these seemingly contradictory domains, *Tangiality* resides in the work of contemporary interdisciplinary, hybrid and multi-modal media artists. Perhaps most importantly, *Tangiality* indicates a rupture in the evolution of our senses; a radical shift in our sensory perceptions, as they are evolving to absorb, incorporate, and adopt the paradoxical materiality of the virtual.

Furthermore, the emergence of *Tangiality* reveals that artists are engaging with post-digital materiality. *Physical materials* are nomadic, moving through contradictory domains, interpreted differently by the hand, the eye, or the body. The phenomena of material has become transitory. It is itinerant, as it migrates between the virtual and the physical worlds. How do we know where it is sited? As artists engage with digital and physical material simultaneously, there are differences in their charged relation to the body but not in a bounded or segregated manner. The concept / construct of *Tangiality,* inhabiting the digital convergence of materiality suggests the emergence of a parallel indigenous realm that artists traverse at the confluence of the virtual and the physical.

Material precipitates *dimension* – in both the virtual and physical realms. Parallels in architecture are dialogues established between architecture and its occupants. Scandinavian architect Sverre Fehn uses the example of a column; is it slender enough that you can encircle it with your arms, or does it take two people to encircle it? Thus a narrative erupts with the materials. In this instance there is a contrast between human and architectural scale. Fehn states, *"this is a dimension you must submit to, and out from this confrontation, you build up an extraordinary world because you have something to work against."* (Fjeld 2009) This is also the relationship between the physical and virtual realms. They seem counter to one another – in scale and materiality. And yet it is this *confrontation*, these factors working against one another – the immaterial and the physical that precipitates the constructing of "extraordinary worlds".

REFERENCES

Auslander, P. (2008). *1999). Liveness: Performance in a Mediatized Culture*. New York, NY: Routledge.

Bilal, W., & Lydersen, K. (2008), *Shoot an Iraqi, Art, Life and Resistance Under the Gun.* City Lights Publisher, San Franscico, California. from: http://www.wafaabilal.com

Catanese, P. (2008). *The Small Magnetism: Interrogating the Nature of Collection.* Appeared in *Imaging in the Digital* Fall 2008 edition of *Media-N Journal of the New Media Caucus*, Retrieved December 1, 2009, from: http://www.newmediacaucus.org/journal/issues.php?f=papers&time=2008winter&page=catanese

Cramer, F. (2005). *Words Made Flesh: Code, Culture, Imagination*. Piet Zwart Institute: Willem De Kooning Academy Hogeschool Rotteradam, Retrieved December 1, 2009, from: http://pzwart.wdka.hro.nl/mdr/research/fcramer/wordsmadeflesh/

Deger, J. (2006). *Shimmering Screens, Making Media in an Aboriginal Community*. Minneapolis: University of Minnesota Press.

Dormor, C. (2008). *Skin, Textile, Film, Textile. The Journal of Cloth and Culture,* November, 2008.

Fjeld, P. O. (2009). *Sverre Fehn*. Monacelli Press.

Geary, A. (2009, February). *Digital Physicality - Exploring Hybrid Practices in Drawing and Printmaking,* Paper presented at the College Art Association conference, Los Angeles, CA.

Gershenfeld, N. (2005). *FAB: The coming Revolution on Your Desktop, From Personal Computers to Personal Fabrication*. New York: Basic Books.

Hamilton, A., & Simon, J. (2006). *An Inventory of Objects*. New York, New York: Gregory R. Miller & Company.

Hwang, K. (2009). Artist Statement from Artist's personal website. Retrieved December 1, 2009, from http://www.dancekk.com/dance2/welcome.html

McLuhan, M. (1962). *The Gutenberg Galaxy*. Toronto, CA: University of Toronto Press.

McLuhan, M. (1969). *Counterblast*. London, UK: Rapp and Whiting.

Mezur, K. (2009, February). *Invisible Intimacies and Cold Burn: Haptic Migrations in 3D Tele-Immersion Choreography*. Paper presented at the College Art Association conference, Los Angeles, CA.

Milekic, S. (2002, April). *Towards Tangible Virtualities:Tangialities*. Paper presented at the meeting of Archives & Museum Informatics: Museums and the Web conference, Boston, MA.

Mills, B., & Walker, Wm. (2008). *Memory Work, Archaeologies of Material Practices*. Santa Fe, New Mexico: School for Advanced Research Press.

Munster, A. (2006). *Materializing New Media, Embodiment in Information Aesthetics*. Dartmouth College Press, University Press of New England.

Murtaugh, M. (2008). *Interaction* in M. Fuller (Ed.), *Software Studies: A Lexicon* (pp. 143-148). Cambridge, MA: MIT Press.

Richards, C. (2009). *Artist Statement* Retrieved December 30, 2009 from: http://www.catherinerichards.ca

Schiphorst, T. (2009). *Artist Statement* Retrieved January 15, 2010 from: www.siat.sfu.ca/faculty/thecla-shiphorst/

Sherman, W. R., & Craig, A. (2002). *Understanding Virtual Reality: Interface Application, and Design*. San Francisco, CA: Morgan Kaufmann Publishers.

Shreve, J. (2009). *All the Labs a Stage*: Publicity release on the website for the Center for Information Technology Research in the Interest of Society (CITRIS), Retrieved November 14, 2009, from: http://ucberkeley.citris-uc.org/publications/article/all_lab%2526%2523039%3Bs_stage

Truckenbrod, J. (1992). Integrated Creativity, Transcending the Boundaries of Visual Art, Music & Literature. *Leonardo Music Journal*, *2*(1), 89–95. doi:10.2307/1513214

Truckenbrod, J. (2008). *Digital Raw Materials*, in *Imaging in the Digital* Fall 2008 edition of *Media-N Journal of the New Media Caucus*, http://www.newmediacaucus.org/journal/issues.php?f=papers&time=2008_winter&page=truckenbrod

Turkle, S. (2008). *Falling for Science, Objects in Mind*. Boston: MIT Press.

Weiner, I. B., Freedheim, D. K., Schinka, J. A., Gallagher, M., Nelson, R. J., & Velicer, W. F. (2003). *Handbook of Psychology*. Hoboken, NJ: John Wiley and Sons. doi:10.1002/0471264385

ADDITIONAL READING

Banes, S., & Lepecki, A. (Eds.). (2006). *The Senses in Performance*. New York, NY: Routledge.

Barker, J. M. (2009). *The Tactile Eye: Touch and the Cinematic Experience*. Berkeley, CA: University of California Press.

Birringer, J. (2008). *Performance, Technology and Science*. New York, NY: PAJ Publications.

Botler, J. D., & Grusin, R. (2000). *Remediation: Understanding New Media*. Cambridge, MA: MIT Press.

Broadhurst, S. (Ed.). (2007). *Digital Practices: Aesthetic and Neuroaesthetic Approaches to Performance and Technology*. New York, NY: Palgrave Macmillan.

Broadhurst, S., & Machon, J. (Eds.). (2006). *Performance and Technology: Practices of Virtual Embodiment and Interactivity*. New York, NY: Palgrave Macmillan.

Chapple, F., & Kattenbelt, C. (Eds.). (2006). *Intermediality in Theatre and Performance*. Amsterdam, Netherlands: Rodopi.

Dixon, S. (2007). *Digital Performance: A History of New Media in Theater, Dance, Performance Art, and Installation*. Cambridge, MA: MIT Press.

Eleey, P. (2009). *Getting Beyond Ourselves: Interview with Olaf Blanke, in exhibition catalog for The Quick and the Dead curated by Peter Eleey at Walker Art Center. Minneapolis, MN*. New York, NY: Distributed Art Publishers.

Hansen, M. B. A. (Ed.). (2006). *Bodies in Code: Interfaces with Digital Media*. New York, NY: Routledge.

Hayles, N. K. (1996). Embodied Virtuality: or how to put bodies back into the picture. In Moser, M. A., & Macleod, D. (Eds.), *Immersed in Technology: art and virtual environments* (pp. 1–28). Cambridge, MA: MIT Press.

Ihde, D. (2001). *Bodies in Technology*. Minneapolis, MN: University of Minnesota Press.

Jones, C. A. (Ed.). (2006). *Sensorium: Embodied Experience, Technology, and Contemporary Art*. Cambridge, MA: MIT Press.

Kozel, S. (2008). *Closer: Performance, Technologies, Phenomenology*. Cambridge, MA: MIT Press.

Massumi, B. (2002). *Parables for the Virtual: Movement, Affect, Sensation*. Durham, NC: Duke University Press.

Massumi, B., Mertins, D., Supybroek, L., Marres, M., & Hubler, C. (2007). *Interact or Die: There is Drama in the Networks*, Rotterdam, Netherlands: NAi Publishers.

Munster, A. (2006). *Materializing New Media*. NH, Lebanon: Dartmouth University Press.

Pallasmaa, J. (2005). *The Eyes of the Skin: Architecture and the Senses*. Chichester, UK: Wiley-Academy.

Patterson, M. (2007). *The Senses of Touch: Haptics, Affects and Technologies*. Oxford, UK: Berg, Oxford International Publishers Ltd.

Paul, C. (Ed.). (2008). *New Media in the White Cube and Beyond: Curatorial Models for Digital Art*. Berkeley, CA: University of California Press.

Penny, S. (1996, April). *From A to D and Back Again: the Emerging Aesthetics of Interactive Art* from the *Leonardo Electronic Almanac*, Retrieved on December 1, 2009 from: http://sophia.smith.edu/course/csc106/readings/Penny_interaction.pdf

Shanken, E. A. (2009). *Art and Electronic Media*. London, UK: Phaidon Press.

Truckenbrod, J. (2000). Torn Touch: Interactive Installation. *Leonardo, 33*(4), 265–266. doi:10.1162/leon.2000.33.4.265

Zielinski, S. (2006). *Deep Time of the Media: Toward an Archaeology of Hearing and Seeing by Technical Means*. Cambridge, MA: MIT Press.

Chapter 4
Pervasive Virtual Worlds

Everardo Reyes-García
Monterrey Tech at Toluca, Mexico

ABSTRACT

Digital media have become pervasive. Most of ordinary tasks in everyday life, from leisure to work, are mediated through electronic devices and their respective digital content. The variety of tasks and their integration into computers and portable devices allow us to think about media as a 'society of media'; a collective of media elements that exchange information and act upon those exchanges. In a society of media, the world is constituted by the hyperspace created by digital media and human uses. Within this context, an interesting role of digital media is that they 'virtualize' human senses and capabilities. Historically, the notion of 'virtuality' has been associated to seminal media concepts such as virtual worlds, which imply the use of interactive computer graphics imagery (CGI) to perform simulation and immersion. Another concept is virtual reality (VR), which has fostered prominent developments including generated environments, multimodal interaction, 3D modeling, and digital animation.

INTRODUCTION

In this chapter, we aim to study the notion of virtual worlds from a cultural and social standpoint. Our intention is to extend the notion of virtual worlds beyond its classical supports and devices. We claim the pervasiveness of digital media has constructed a world that is mixed, extended and mediated through media. In this world, the society has developed contemporary processes that produce a virtualization of individuals. Our perspective inspires us to present three projects that were conducted recently. First, an artistic installation named TR • 33 which proposes to users a physical environment that is enhanced with interactive projection and multimedia hand-held devices. The second is an interactive piece of audible and visual figures that reside in a Web environment. This work was called Beatanz. The third is a contemplative artwork that uses dynamic and animated spheres

DOI: 10.4018/978-1-60960-077-8.ch004

as triggering figures to motivate virtualization. The purpose of discussing works from the field of art is mainly to explore innovation sources for media developments. In the last part of this chapter, we advance some conclusions related to virtual worlds. We also introduce a model for studying processes of creation and development in new media, grounded on four elements: information technology, media, art and, society.

DIGITAL MEDIA AND THE SHAPE OF WORLDS

We owe to Canadian theorist Marshall McLuhan one of the most well-know definitions of media: "media are extensions of man" (McLuhan, 1964). The importance of this observation relies in the fact that we humans have created artificial objects and intellectual technologies in order to expand our innate capabilities. In this sense, the car extends our legs; the glasses extend our eyes; the clothes extend our skin; the house extends our body. From this simple identification, we must add that those objects not only extend us, they shape us. The car introduced a new perception of time; glasses a new vision; clothes a new opportunity to adapt or explore other environments; the house a new way to fashion collectivities and cities. But when we talk about media we are dealing with a more abstract issue. Media are composed basically of symbolic content and thus they entail production and interpretation. Media primarily extend our thinking and senses; they are more an intellectual technology. As it could be anticipated, intellectual technologies reflect on the shape of objects, places and ways of being. For example, McLuhan observes behavioral distinctions between oral and written-based communities which have affected the shape of cultures. The first ones cultivate social participation and the latter motivate personal and intersubjective schemes. Both levels of participation are further categorized as 'hot' and 'cold' media. The TV would be cold –let's remember that back on 50's TV sets were installed on living rooms and watched as family time– while a book would be hot because it allows less participation. In any case, what is interesting to observe is that media are powerful vectors for extending social schemes and introducing different ways of being.

In the contemporary context, where media have become digital and new types of media have emerged, the distinction between hot and cold becomes fuzzy. One medium such as the Web is both participatory and intersubjective. Although this statement can be discussed with further detail, two fundamental properties of digital media (and new media) may be regarded: they are dynamic and 'zoomable'. First, dynamic because they constantly change and their nature is composed of several other media. They may be arranged, rearranged, remixed in order to represent thinking or senses. Second, they are 'zoomable' in respect to their use. When using new technologies or media, we experience a constant passage from interface to content. Because media are nested and mixed, we zoom in or zoom out within a medium, from its surface (the interface) to another (the content of interfaces), and vice versa.

Dynamic and zoomable characteristics of media have become a common performance, an almost imperceptible kind of interaction. People learn to use new media, explore emerging innovations, and enhance one medium with another. For us, the way people aggregate new media and gather extensions of man is a particular mode of constructing communities, societies, environments, cities, and worlds. A fundamental characteristic of these places is that they all support and embody people interactions. In recent years, examples of virtual worlds depict multiuser electronic environments where 3D avatar models explore and act on digital generated spaces. However, virtual worlds may also be associated to other examples such as social networks, which mostly rely on Web and networking technologies, as well as to artistic installations –often located in public spaces– and hand-held devices –multimedia artifacts carried by users–. Virtual, mixed and extended worlds function in the extent that 'virtuality' is a state of being. As Pierre Lévy (Lévy, 1994) has noted, the virtual state is often confused with the inexistent, the immaterial, and the unreal. But drawing upon philosophical concepts, mainly from French philosopher Gilles Deleuze (Deleuze, 1969), a closer look to virtuality allows for a redefinition. The virtual is not a state that opposes itself to the real; the virtual state of being is also real and it extends senses and capacities of individuals. According to this, the real and the virtual are different and deserve a deeper study.

This notion of virtuality is helpful to understand digital media as content adapted to suit the multiplicity of technological supports and devices. Moreover, we can think of media from their influence on shaping virtual ways of being. An important part of virtualizing human senses and capacities resides on graphical representation of body and actions. In the case of avatars, for example, they are designed accordingly to represent features and adaptation to environments. However, avatars are only actual solutions to a creative process that goes from fuzzy to cultural objects, a passage that Gilbert Simondon called 'concretization' (Simondon, 1989). In their primitive state, objects are limited to function and signify to the designer only; their structural functions are not yet well defined. Once the process is in mature stages, the object is confronted to other users (communities or collectivities) and may be redesigned to better cope with functions and intentions. Virtual objects are a special type of media because they benefit of all digitization and communication possibilities. Objects may be designed to communicate an operational function of the system and, at the same time, they communicate an allegory of the system, the world, and the society of media. At this point, we must say that virtual worlds do not have to represent reality in a verisimilar fashion, such as art is no longer about appearance but is concerned with apparition (Ascott, 1994). Virtual worlds may also be composed of abstract and complex forms produced by the interaction of heterogeneous elements; not only pure individuals but machines as well; not only physically existent things but also virtual.

We have discussed elsewhere (Reyes & Zreik, 2008) that digital media become processes, so what is important is not always the content but the processes and actions. Consider contemporary software to manipulate digital media, they all share common tasks and procedures, be it a video, an audio, a text or an image. The question we ask is therefore: is it possible that virtualized individuals enter this continuum of processes? In other words, in what extent man could be media?

VIRTUALIZING WAYS OF BEING

New and digital media are the result of a series of evolution in both technologies and artifacts with the intention to connect or disconnect people. This means that media fulfill a social function, either they transmit some information or they provoke some reactions in people.

In his seminal book, *Qu'est-ce que le virtuel?* (Lévy, 1994), Pierre Lévy examines the notion of virtualization from an original point of view. The Canada-based philosopher agrees with other thinkers that 'the virtual' state is not the opposite of 'the real' state. Furthermore, he also traces the process of virtualization from 'the actual' to 'the virtual'. The virtual state is better understood in comparison to the other three states of being. Let 'the real' be the first state of being. Lévy recalls the real as characterized by an occurrence of existence, by being here and now. The second state is 'the potential' or 'the possible'; it may be seen as similar to 'the real' but not yet manifested. The potential is latent; it will exist without any deep transformation in its nature. These two states have in common that they refer to 'substances', not to 'events'. Moreover, they can be separated by an axis. To the left is the potential, an axis of latent; not yet materialized. From this pole, a potential

Figure 1. States of being according to Lévy. (© 2009, Everardo Reyes. Used with permission.)

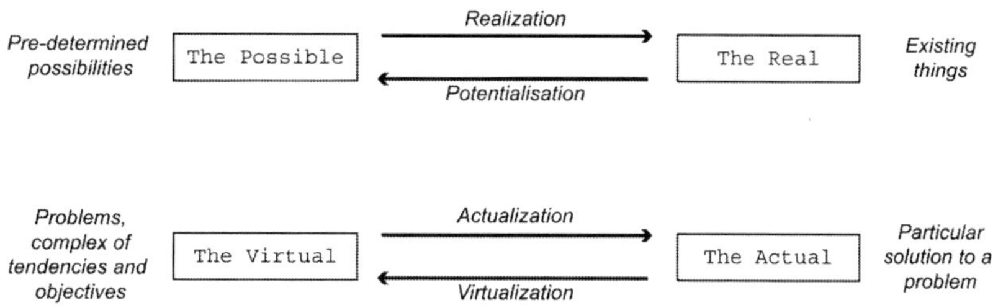

```
Pre-determined            Realization                        Existing
possibilities   The Possible  ──────────→   The Real         things
                              ←──────────
                            Potentialisation

Problems,                  Actualization                     Particular
complex of      The Virtual  ──────────→   The Actual        solution to a
tendencies and               ←──────────                     problem
objectives                 Virtualization
```

thing enters a 'process of realization' towards 'the real', which is on the axis of manifest, this process implies the use and expense of resources. Accordingly, the inverse process, from the real to the potential would be called 'potentialisation' and is characterized by the production of resources. To summarize, the potential is an assembly of predetermined possibilities, while the real are persistent and resisting things. The potential insists; the real subsists. The other two states follow a similar order. On the left axis, the latent, we find 'the virtual', while on the right side there is 'the actual'. These states are related to 'events', not 'substances'. The actual is characterized by a solution to a particular problem; it *arrives* here and now. When there is a creative relaunch to solve a problem, the actual enters a process of virtualization. In the virtual, the state *exists* as series of problems, nodes of tendencies, forces and objectives. The virtual needs to be actualized, which means to solve a problem. Thus, the virtual is not opposed to the real but to the actual. The actual state is a solution. It innovates a new form out of dynamic configurations of forces and goals. The virtual is a mode of being that is fertile and powerful. It fosters the process of creation, opens future paths. We relate the process of virtualization as similar to the process of inventive creation. When people create new things or actions, these events exist as virtual, sometimes they only exist for the person who invented them. To exemplify with a banal case, consider the use of stones. Primitive humans abstracted the function of hammering from using a stone. But is the stone the only thing that allows hammering? Is hammering useful for only one thing? Moreover, how to communicate the abstract function of a stone to other members of the collective when no stone is around? We can see that functions and processes go beyond a particular object. When we have identified the possibilities to extend our senses and capacities, we have learned to virtualize our world. At this point, it is interesting to note that virtualizing actions is a matter of problem solving. Indeed, we can think about the universe of objects that surround us as the result of solving problems in a determined way. What if hammers would have been shaped as gloves?

In the evolution of techniques and technologies, Gilbert Simondon has studied the passage from an abstract object to a concrete object. Inventors discover or produce new things to facilitate or make something more efficient. Then, once inventions enter a socialization process, they are either accepted or rejected by the community. Perhaps, some objects are rejected at a particular time, only to be revisited at another time. In order for an object to be considered as concrete, it must accomplish an evolution in both functionality and aesthetic directions. The technical object is considered as such when it coexists with humans in the sense that they cooperate, they benefit from each other in equal manner. But users are producers of inventive properties too. Most of objects that live with us may foster new uses. Michel De Certeau (De Certeau, 1990) has identified four modes of appropriation of objects. In the first case, users extend the ideal function of an object.

Imagine cell phones, which nowadays are more used to write than to talk. In the second case, users adapt the object to fulfill new uses. For this matter, take vintage TV cases that have been adapted as digital media centers; users modify the arrangement of internal parts. The third case extends functions of a given object: hammers not only accomplish their primary function but some models have magnetic field to attract nails. And finally, users may confer a completely different function to objects. A giant clay-made hammer would be part of a contemporary artistic installation. Simondon concludes that users needs tend to diversify continuously. There is no time when all required human functions have been satisfied.

To illustrate how virtualization is embraced in technological objects, allow us to discuss the iPhone device. It certainly is an existing object; it occupies a matter and form in the space and responds to the need of communicating. It realizes the act of speech and writing, such as any other cellphone device. It potentializes itself when we charge its battery, when we apply some energy to its use. But the most interesting aspect of iPhone is the fact that it supports virtualization. Because most of its primary functions have been designed as software, i.e. virtualized, users are not forced to interact with given hardware-made keyboards; it opens possibilities. In this respect, Apple and third-party developers have been creating applications that extend the function of the artifact. It could be used, as any other computer-based media, for working, leisure, or teaching/learning. But it also allows to experiment with artistic and experimental practices to convey emotions. Finally, because it has networking capabilities, it is possible for users to exchange information and media connections with other remote media.

Digital media and content developers have well understood that cultural data is different from computer data. When bytes are no longer perceived as such it is because humans have added some signification. Written documents become letters, reports, or poems. In the same regard, images transform into photographs, visualizations, or icons; video becomes film and audio becomes songs, speech messages, or alert sounds. Producers or new media content have abstracted the function of stones, they have learned to convey functions where no stones are around. Artists understand that to create a pictorial artwork there is no further need of pencils or brushes. Picture materials and surfaces are everywhere; any object or action carries in itself the power of expression. Artists no longer learn how to use Photoshop or Illustrator, they learn how to virtualize senses, how to provoke expression from computer code, 3D generated environments, or projections on walls, roofs, or soft surfaces. If to achieve an artwork there is need to learn how to program in a computer language or how to model CGI 3D objects, it is just a matter of technique, of using real and actual tools whose use might introduce new problems to do things.

Following Michel Serres (Serres, 1996), we agree when he says that information technologies demand us to be intelligent. In an era where humanity has generated the hugest amount of information and where technologies assist us to discover it, we have virtualized the very essence of being human. Our memory could have diminished in respect of dictionary models, but it has certainly increased in respect of using tools and being receptive to emerging ones. We are now extensive users of Wikipedia, social bookmarking, and massive storage of media assets. Users become intelligent when they learn how to connect such information and how to live and take the most out of them. The new models for fostering problems would consider –but are not limited to– the pervasive character of media and the approach of users with media hardware and software. With the virtual, given physical spaces are no longer required. Because media are pervasive and the virtual is latent, all places are subject to explore new functions. As Lévy suggests, the dynamic of virtualization consist on discovering the main problem, then to orient entities toward this issue and finally to redefine the actual starting point as a response to a particular question. The actualization goes from a problem to a solution; it is the materialization of the virtual. The virtualization goes from a given solution to another problem.

THE OPPORTUNITY FOR ART AND MEDIA ART

In this part of the chapter, we concentrate on the following question: What is the opportunity of innovation/virtualization in the era of digital media and pervasive computing? Pioneer digital artist Roy Ascott has already drawn an interesting departing point. Body and mind are changing radically; we are acquiring new faculties and new understandings of human presence. "We are all interface. We are computer-mediated and computer-enhanced. These ways of conceptualizing and perceiving reality involve more than simply some sort of quantitative change in how we see, think, and act in the world. They constitute a qualitative change in our being, a whole new faculty, the post-biological faculty of 'cyberception'" (Ascott, 1994:320).

For Ascott, the notion of 'cyberception' involves a convergence of conceptual and perceptual processes where network technologies couple with human senses and capabilities. It is perception mediated through digital media and technologies.

We understand digital artworks as a twofold effort. On one hand, they take advantage of digital media as a vehicle to express ideas and sensations. On the other hand, they introduce innovative ways to use and manipulate media. It is like an abstract state of objects that require the participation of users to complete their meaning. However, digital artworks are not always concerned with achieving Simondon's concreteness of objects. Artists tend to confront users with unexpected functions and actions of digital objects; there is no pursue of solving usability issues or improving schemes of efficiency. They try, instead, to trouble, to encourage problems. As such, media art could be seen as an act of virtualization. To approach the question of what virtual modes of being are carried by artworks, we can advance three. First, pervasiveness is about being here and now in several places simultaneously. People are surrounded by computing artifacts, each one of these with its own graphical interface; maybe there are differences at the level of graphic design but they share similar functions. While we are physically in some place, we can be acting in a remote place. Accordingly, while we act in that remote place, we can do similar tasks as those of the physical space.

Second, we zoom from media to media. Imagine the process of performing any task. At a given time we might be blinded by the content itself, i.e. forgetting the employed medium. But we change actions, select different buttons or keys from our devices; we even wander from one digital space to another. We exit some media to zoom to another. This process has been previously described as 'remediation' (Bolter & Grusin, 1999), but we must add that zoomable media concerns a desire of virtualization. To zoom from one part to another means that we have recognized a specific need to be actualized; that we have understood the capabilities of media to enhance us; that we are making sense and desire to solve a problem in a different manner.

Third, media are no longer about content but about processes and actions. All digital media share common characteristics: they are bytes and as such, they can be programmed and manipulated by computers. Interactive images behave similarly to video and texts and audio. It is the process that has been virtualized and it applies similarly to all media. From this standpoint, the computer would be a meta-medium, a medium that creates new media and that abstract processes to create or manipulate them (Manovich, 2008). To exemplify some explorations of virtual worlds produced with artistic intentions, we discuss three recently artworks developed at Monterrey Tech in Toluca, Mexico.

TR • 33

The project TR • 33 aims at responding to the premise: In a digital world, is it possible to get inspiration from natural objects in order to explore new ways of interactivity with technical objects? TR • 33 is an artistic installation that proposes to users an innovative form to interact with multimedia hand-held devices. The natural world is used as a metaphor to stylize a fractal generated tree that is retro-projected in the installation. Users experience the action of "engraving their trace in an object", in this case, the tree's trunk or branches. It seems that the ubiquity of portable multimedia devices has converted their use and dependence into something natural, almost imperative for social and cultural interaction. But what happens when the traces of all users are assembled into a same discourse? We are now entering the domain of what we call 'collective cinema', a mode of creation that demands social participation to embrace its content. Three interactive phases were explored. First, the installation was mounted at a public space. Users discovered it inside their regular environment and thus it arouses some expectation and curiosity. The installation is 6m. long, 2.4 m. height and 3 m. width. Users are attracted to come into the space by means of the fractal tree projection, once inside they notice hand-held devices hanged and stored in leaf-like cases. When users grab leafs, Blackberries are in video mode and only the button to start recording is available to activate.

In a second phase, a day after video recording, the installation is used as sound environment. For this matter, every one of the six Blackberries reproduces different sounds, all of them synchronized to convey the sensation of listening to music or orchestration. It would be like listening posts hanged in a forest. Indeed, some users evoked the

Figure 2. The TR • 33 installation. (© 2009, Everardo Reyes. Used with permission.)

sentiment of listening to the sound of trees. This phase is transitory to the last one. The third phase is contemplative. It shows random montage of movie clips recorded from the first phase. These clips are projected as superposed sequences to the fractal tree image, as if they would be leafs. Meanwhile hanged Blackberries still act as sound emitters for the installation. During our sessions, we could observe some interesting actions performed by users. The total amount of recorded clips was 95, involving subjective sightseeing of objects and persons, from daylight to early evening. We noticed that recurrent users started playing roles in front of devices; some others came back with other users to be recorded. As the daylight went off, the interaction changed and users started recording full messages to save the nature or dedicating love messages to distant persons. This mode of interaction with multimedia hand-held devices refreshes the idea of virtualization. Users stop focusing on technical possibilities and start thinking of them as extensions of memory. Every time a user understands that her message will be transmitted to other receivers in different contexts and time, they explore the idea of augmenting communication.

BEATANZ

The second project to discuss is an interactive artwork made out of visual and audio signs. Beatanz is set up to propose three different interfaces related between them, yet independent of one another. On one side, two stations show a similar interface to control and trigger white squares. Once activated, squares start changing colors and producing a particular looping sound. On the other side, spectators observe users manipulations, but they are the only ones that see a third interface, which combines both interfaces into a dynamic generated one.

Beatanz looks for intuitive interaction between users, without speech negotiations to obtain a

Figure 3. The Beatanz interface. (© 2009, Everardo Reyes. Used with permission.)

particular result. While users interact with squares they do not see each other and they do not see what spectators are seeing. Moreover, they do not see what the other user is manipulating. After users start noticing that sounds are similar to beats and drums, users start playing blind. During its exhibition, we noticed that after some time of manipulation, users communicate intuitively in order to achieve a rhythmic sequence. The dynamic interface, only seen by the audience, modifies itself accordingly to activated squares from both user stations. This interface may be considered as visualization of musical tones but above all, as the collective participation of users.

As it can be noted, Beatanz demands for collective interaction. If there were only one user, the artwork would have poor semiosis activity. Of course, a single user may activate visual and audio signs and discover the whole meaning after manipulating each interface separately, but the goal is to foster intuition in users and to captivate spectators through contemplativeness.

Beatanz seems to have been perceived as an artifact that looks for equilibrium. Watching users to approach and to discover functionalities leads us to depict a tendency in making order from chaos; making sense where it seems there is none. Technically speaking, Beatanz was created using ActionScript 2.0, within the Adobe Flash environment. There are no standard HCI rules, i.e. no color or white square changes the cursor shape; no instructions nor manual are given. To facilitate the communication between interfaces, we created a local database to store variables of states. Anytime a square is activated, it automatically sends its current state to the server, which is then accessed by the third interface (that which is only visible for spectators). As it might be guessed, the latter interface is a combination of states that dynamically reacts to users.

COLOR ATTRACTION

The last artwork that we introduce in this chapter is more contemplative than interactive. Color Attraction is a single piece of artwork that displays 17 spheres of three different colors and several sizes. The spheres start at a random position and then simulate to paste to the border of the application -in other words, to the app frame. Spheres slowly follow each other, one by one. After a couple of hours they all gather accordingly to their color. The intention is to simulate that spheres attract

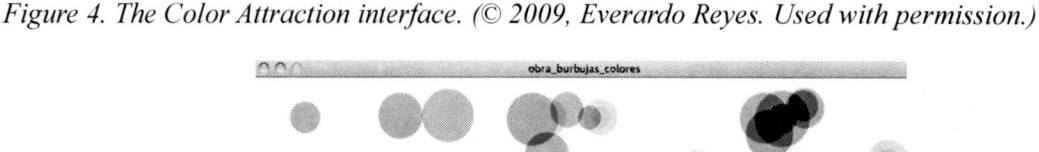

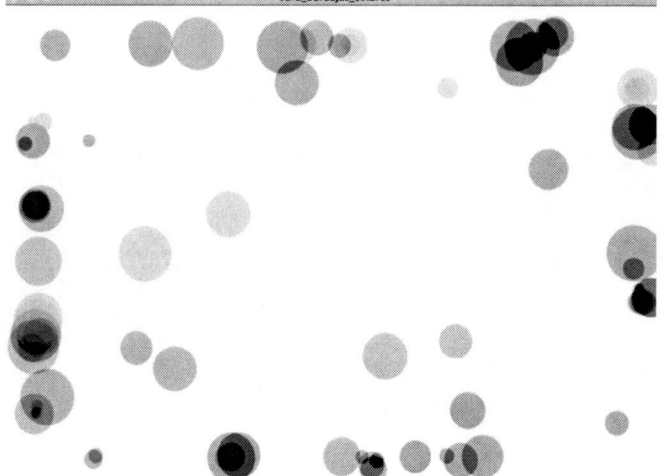

Figure 4. The Color Attraction interface. (© 2009, Everardo Reyes. Used with permission.)

each other for the same color. This artwork was produced with the Processing environment. Color Attraction is designed to run indefinitely. When all spheres have finally formed color groups, the remaining sphere restarts the process all over again, i.e. the last sphere will refuse to group and will attract the grouped spheres to gather into another place of the border frame. This piece of art has been exhibited publicly in scholar contexts. A simple session can run for hours. Spectators usually contemplate it for minutes, mentioning that spheres evoke sentiments of membership; of oblivion; of absence. It is interesting to note that simple shapes with simple behavior may provoke unlimited semiosis in people. How do users remember a past travel or the desire to go on vacation from the slow movement of spheres? We believe that it is because we identify ourselves with the process of movement, not with the media that conveys the action. The virtualization process is then un-personified; generalized not to iconic images of users but to process of displacement.

A FOUR-ELEMENT MODEL FOR NEW MEDIA

We have discussed elsewhere a model for studying processes of creation and development in new media. This model would contain at least four elements: information technology (IT), media, social, and art (Reyes, 2009). Its graphical representation can be expressed as Figure 5.

The first element, IT, refers to artifacts, hardware and software, specially designed and employed to process information. The attention is centered on those technical possibilities that machines perform better than humans. An example

Figure 5. A four-element model. (© 2009, Everardo Reyes. Used with permission.)

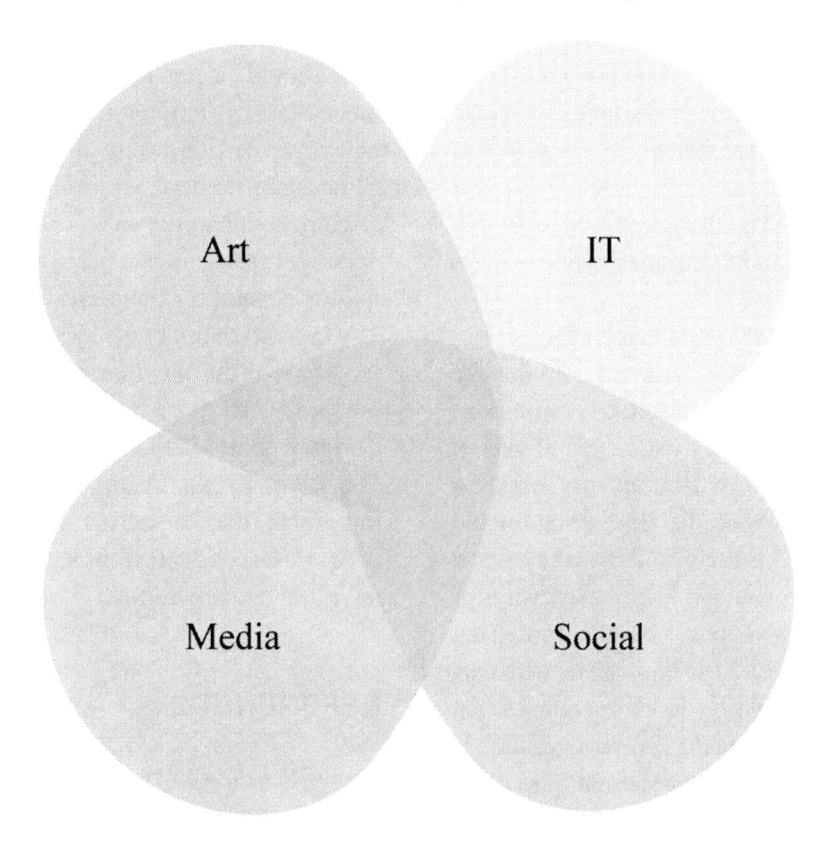

could be related to calculating multiple mathematical operations and storing massive amounts of data.

The second dimension, media, contains those technologies that, to put it in terms of the media theorist Marshall McLuhan, extend body and mind. The idea is to consider data as cultural forms mediated through its communicative status. In this respect, data becomes media when it is assembled to transmit a message in a human-understandable manner. The communicative status of information depends on a series of systems of signs that produce meaning, e.g., language, graphic symbolism, social modes of interaction, cultural codes, etc.

The third dimension, social, focuses on arrangements of actors that participate to gather a collective. Following social scientist Bruno Latour, the importance of studying a collective –not a society– is not to categorize it accordingly to given schemas but to regard its dynamism and heterogeneity of elements (Latour, 2005). At first glance, social interaction in new media systems is based upon users' activity. The size of a Web 2.0 application is determined by the quantity of users registered and by the operations performed on a lapse of time. At second glance, consider Net Art. Some artworks exist independently of users while others cannot function without participation, and others are tightly coupled between users and systems.

The last dimension, art, is centered on stimulating actions and feelings in users. Perhaps the clearest examples come from Net Art productions that provoke users and defy classic paradigms of user-centered design. Media elements might be stylized in volume, symbolic representation and composition within a surface. Artworks offer a prolific field to explore and inspire new ways of innovation. The term 'innovation' is preferred instead of 'production'. The latter deals with perceptive processes: users 'produce' meaning when they perceive objects. On the contrary, creation is related to innovation in the sense that it involves structuring new universes out of forms and objects.

CONCLUSION

In this chapter we have approached the question of virtual from its philosophical investigation, mainly by the thinking of Gilles Deleuze and Pierre Lévy. We have related the virtual –as state of being– to social and cultural schemes of interaction that allow us to foresee a virtualization of individuals. In the era of digital media and computer-mediated communication, we argue that virtual worlds are a combination of both physical and electronic environments that put special attention in processes and actions, rather than at objects or perceptual signs.

A simple definition of world would be an environment where individuals or collectivities perform tasks. It is a world precisely because there is a multiplicity of actors that can interact with each other. But what characterizes a virtual world is the space where individuals project, defy, and confront their mode of existence. It is a space where actors do not focus on objects but on their meaning; on their abstract function and logic. This world could be materialized on physical environments through innovation, creative practices, or new ways of using things. However, when we pass to the digital domain, actors virtualize themselves. Digital media reflect on new ways of being when actors comprehend that processes such as writing do not depend on materials but on knowledge. The last part of our essay introduced three artistic projects with the intention to exemplify how digital media is used to evoke senses and expressions. Through common devices and defying graphical interfaces, users enter unexpected worlds. Within this frame, it is the action of making sense that invites users to start thinking on how to extend their human capabilities.

REFERENCES

Ascott, R. (1994). "The Architecture of Cyberception" in Shanken, E. (ed.) *Telematic Embrace*, Los Angeles, CA: Unversity of California Press, 2003.

Bolter & Grusin. (1999). *Remediation: Understanding New Media.* Cambridge, MA: MIT Press.

De Certeau, M. (1990). *L'Invention du Quotidien. 1. Arts de Faire.* Paris: Gallimard.

Deleuze, G. (1969). *Logique du Sens.* Paris: Minuit.

Hansen, M. (2006). *Bodies in Code: Interfaces with Digital Media.* London: Routledge.

Latour, B. (2005). *Reassembling the Social: An Introduction to Actor-Network Theory.* Oxford: Oxford University Press.

Lévy, P. (1994). *Qu'est-ce que le Virtuel.* Paris: La Découverte.

Manovich, L. (2008). "Understanding MetaMedia" Opening Keynote at *Computer Art Congress 2008 [CAC.2].* Toluca, Mexico. March 26, 2008.

McLuhan, M. (1964). *Understanding Media.* New York: McGraw Hill.

Reyes, E. (2009). "Hypermedia as Media" in *20th. ACM Conference on Hypertext and Hypermedia.* New York: ACM Press.

Reyes, E., & Zreik, K. (2008). "A Social Hyperdimension of Media Art" in Zreik & Reyes (eds.) *Proceedings of the Second International Congress Computer Art Congress 2008 [CAC.2].* Paris: Europia.

Serres, M. (1996). *Atlas.* Paris: Flammarion.

Simondon, G. (1989). *Du Mode d'Existence des Objets Techniques.* Paris: Aubier.

KEY TERMS AND DEFINITIONS

Virtuality: Deleuze studied virtuality as an state of being. The main characteristic of this state is that it exists, along with reality. In this manner, what is real can also be virtual. Later, Pierre Lévy extended this notion by opposing virtuality to the actual. In this respect, virtuality is a creative manner to solve problems which will be actualized (materialized) before relaunching another problem again.

Virtualized Self: The activities and representation of man through media.

Zoomable Media: {roperties of media to imbricate other media and the action to focus at one level or another by humans.

Man as Media: Man as interface, as extended by properties of media.

Chapter 5
Semantic Entities in Virtual Worlds:
Reasoning Through Virtual Content

Vadim Slavin
Lockheed Martin Space Systems, USA

Diane Love
Lockheed Martin Information Systems & Global Services, USA

ABSTRACT

This chapter explores how a parallel semantic knowledge base describing the virtual world can improve the utility of virtual world environment by enabling virtual agents to interact and behave like their peer human participants. To a computer, the virtual world is nothing but a set of triangles derived from tessellated shapes comprising the virtual environment. The burden of instrumenting physical interaction is placed on algorithms to check physical constraints and object properties for each time step. These 3D object properties can be defined in a domain knowledge base where semantic descriptions of these objects contain not only physical properties, but relationships, inheritance, polymorphic properties, past history, course of action, acceptable effects, desired objectives, and purpose, resulting in a much richer architecture for defining the behavior of and interactions between various virtual entities.

INTRODUCTION

In the context of virtual worlds the discussion of application of metadata has centered on the issues of interoperability for virtual worlds. Semantic tagging and classification of virtual agent actions, virtual objects, and other entities can certainly help define a common language or framework to enable interoperability between virtual environments and virtual world platforms (MMOX, 2009). The first step towards this has already been made as XML is becoming a tool of choice for describing 3D models and, therefore, virtual environments. (Soto, Allongue, Pierre, Curie, & Jussieu, 1997) Although virtual worlds seem to be a 21st century phenomenon, virtual world timelines (Ok, 2009) indicate that in fact virtual worlds have, over the past decades, enjoyed bursts of activity interspersed by "winters" where there was relatively little progress (Damer, 2008). Gartner (2009) indicates that present day public virtual worlds have passed the Peak of

DOI: 10.4018/978-1-60960-077-8.ch005

Inflated Expectations, plunged into a Trough of Disillusionment and are set to climb a Slope of Enlightenment towards a Plateau of Productivity. The strengths of virtual worlds platforms are now emerging in applications for government, defense, healthcare, education and business collaboration. Motivators for adoption of virtual worlds include cutting the various costs (monetary, time, quality of life, risk to life) of commuting and business travel, enabling meetings to be scheduled for the time they take rather than including days of travel time, and placing expensive people in realistic simulations of expensive, dangerous or impossible locations – from city center disasters, burning oil rigs and war zones to the interior of an atom, a brain or the sun.

Mass virtual world adoption has been enabled by the evolution of technology which has seen the user interface develop from the one dimensional command line, through two dimensional graphical user interfaces and the present day "flat web" internet to three dimensions. Once 2D user interfaces were established, there was no going back to the command line, because content was being created and consumed in 2D in ways which could not be supported by a command line user interface. By the same token, there will be no turning back from virtual worlds when users routinely create and consume valuable content in 3D.

For example, rather than loading a PowerPoint presentation into a virtual world slide viewer, sitting avatars on chairs and mimicking a real world presentation, virtual world presenters will create static, animated or interactive 3D presentations and lead audiences on tours around them (ThinkBalm, 2009). The following example from a ThinkBalm Data Garden tour illustrates the increased engagement offered by this approach.

The ThinkBalm Data Garden is a 256m x 256m island in Second Life®. Visitors land at a defined start point and are directed to follow a path which leads them on a circular tour of the island's exhibits. Circular blue decks are used to indicate resting points where the avatar may stop to view an exhibit. Figure 1 shows how a flat chart can be made more engaging by allowing the visitors to toggle data sets on and off using the red and green ball controls on the resting point.

Presented information can be animated to engage emotions. In the ThinkBalm Data Garden the possible alternatives to immersive technologies (e.g. web conferencing, in-person meetings, video conferencing and more) are represented by spheres sized according to merit (each sphere has

Figure 1. ThinkBalm Data Garden – flat data made engaging. © 2009 ThinkBalm. Used with permission.

Figure 2. ThinkBalm Data Garden – confusing alternatives. © 2009 ThinkBalm. Used with permission.

a name which is visible to the avatars but not shown in the figure). The spheres are scripted to fly randomly around the avatars, to indicate the confusion experienced by the person who is seeking the solution. (Figure 2)

In Figure 3, two avatars have reached a point where they are confronting the barriers to adoption of immersive technologies. It looks like they simply have to step off the cliff, fall down into a chasm and then climb up a steep slope.

However, once the avatars step off the cliff, a set of unexpected barriers rise up in front of them.

The barriers don't all appear at once – after the avatar succeeds in negotiating around the first set, a second set appears, then a third. Text on the physical barriers explains each problem – for example the first obstacles are redundant features with existing technology, corporate security restrictions and inadequate hardware. Once those are overcome, cost and budget approval unexpectedly arise. Figure 4 shows all of the barriers. ThinkBalm (2009)

These examples show how immersive technologies can use more engaging methods (interac-

Figure 3. ThinkBalm Data Garden – barriers to adoption. © 2009 ThinkBalm. Used with permission.

Figure 4. ThinkBalm Data Garden – unexpected barriers. © 2009 ThinkBalm. Used with permission.

tive exhibits, arousing emotions, metaphors, and creating physical challenges to the avatar) than the traditional "chalk and talk" flat presentations by a presenter in front of a seated audience. At the time this book is being published, the technology is in place and models for scalable mass user virtual worlds exist. What is arguably missing is the support for semantic relationships in virtual world content which can add a greater degree of interactivity to applications such as these to impact the user experience in a profound way. The underlying premise of semantic technologies is to apply semantic metadata to traditional relational knowledge bases to define rules and relationships between entities. The end result of a properly structured semantic knowledge base is that the contained data can be reasoned through to follow logic and temporal expressiveness towards building an incremental description(or definition) of a domain of interest.

The idea of converting data into information and subsequently to knowledge and wisdom (Ackoff, 1989) is being exploited in the intelligence community already where future systems will be able to browse through vast amounts of incoming data and information and fit it into an ever growing and changing repository of domain knowledge and wisdom, i.e. predicting future adversary actions based on available past and present intelligence. What will make this possible is the expressiveness of the knowledgebase where the meaning of data is encoded together with the data itself. This chapter will explore how enabling a parallel semantic knowledge base describing the virtual world can improve the utility of the virtual world environment and enable virtual agents to interact and behave like their peer human participants. At the present moment, to a computer, the virtual world is nothing but a set of triangles derived from tessellated shapes comprising the virtual environment. The virtual world 3D objects are created with defined properties to enable interactions between them. The burden of instrumenting physical interaction is therefore placed on algorithms to check volumetric boundary constraints and object properties to define the state of the virtual environment at a subsequent time step. These 3D object properties can be defined in a domain knowledge base where semantic descriptions of these objects would contain not only physical properties, but relationships, inheritance, polymorphic properties, past history, course of action, acceptable effects, desired objectives, and purpose. Note how these listed qualities, in the order above, progressively describe qualities attributed to a reasoning entity, i.e. an intelligent agent.

BACKGROUND

The quest for semantics in the "flat web" is often expressed from the perspective of the content consumer – e.g. search and recommendation. Taking the example of the NetFlix® prize (Netflix, 2009), Netflix offered $1,000,000 as the prize for improving movie recommendations by 10%. Competitors were able to use anonymized data from real customers, who had rated movies on a simple scale of 1 to 5. At the end of the competition, one of the leading teams published a 2D graph of relationships between 5,000 movies (The Ensemble, 2009; Monroe, 2009).

What value could be gained by rendering this graph in 3D instead of 2D? Perhaps the graph can be more meaningful if each axis has its own semantics. The reasons why a given individual likes or dislikes a given movie are complex, including their preference for the director, the stars, movie genres, and enjoyment of relationships between characters, plot twists and surprises. Distilling these factors into metadata and rendering them in 3D would allow the consumer to navigate a movie database in 3D. For example, movies could be mapped in 3D with genre on a horizontal plane and date of opening mapped to the vertical axis. The single sphere representing a particular movie might open out into a horizontal array of spheres for the cast, sized large to small according to their billing in the movie. Selection of any individual sphere could open up a vertical lifeline for that cast member with a sphere for each of their roles in previous movies as well as future movies known to be in production. Providing a persistent 3D rendering of movies would allow the consumer to navigate geographically and remember spatial locations where favorite movies could be found. Now switching to the perspective of the content creator, it would be theoretically possible to use the same knowledge base along with users' preferences, ratings, and interests to design movies (director, cast, genre, plot) which would appeal to specific demographic groups. Similar reasoning can apply to plots other than movie plots. Rendering a semantic knowledge base in 3D could improve the utility of the data to "consumers" searching for evidence of criminal or terrorist plots, but also provide the capability to "create plots" – i.e. identify potential threats and vulnerabilities, such as large amounts of resources (e.g. money, commodities or energy) lacking adequate physical protection.

A virtual world goes beyond 3D data visualization to provide not only an inworld point of view to the avatar but also a set of companions with whom to experience the virtual world. These can be real world friends or co-workers, or people who are brought together in the virtual world by shared interests. In the example of the 3D movie database, seeing a cluster of avatars in the distance may be a sign of an interesting new movie; an avatar friend may take you on a personal tour of their own movie favorites; and spending time at the entry for your own favorite movies will introduce you to other fans of the same movie. In summary, a multi-user virtual world environment can add social networking and the wisdom of the crowd to a dataset rendered in 3D. The example presented above serves as a junction point of the two technology areas: 3D, Virtual Environments and a set of technologies collectively known as Semantic Technologies.

Semantic technologies are a collection of software techniques that encode meaning of the data together with the data itself. If traditional relational databases systematically store data according to predetermined schema, the semantic knowledge bases present the functionality to redefine the structure of the knowledge base as appropriate.

The implications are already very powerful. No longer do you have to face the consequences of modifying schema and adjusting the table relationship constraints. In semantic knowledge bases, such adjustment would mean extending a part of an ontology – the building block of such knowledge bases representing a particular domain definition. Semantic knowledge bases are inherently extensible due to the possibility of object oriented inheritance hierarchy between domain concepts. Ontologies can import other ontologies and, furthermore, extend the definition of imported domain knowledge. Domain knowledge, in the context of semantic knowledge bases, is the description of known concepts surrounding entities. These could be tangible objects, intangible concepts and definitions, people, or other special topics. For example, capturing the definition of what it means to be a person would include the definition of a person and the types of attributes a person could have: name, height, age, etc. Further expanding the meaning of the knowledge about a domain is the description of relationship between entities. For people, this would define various relationship people may have: family, friends, romantic interests, business partners, etc. Each relationship is itself an entity fully extensible to be refined depending on the version of the domain knowledge. The good news is that the definition of entities can be built in terms of inheriting from higher order entities or even abstract entities adhering to the object oriented paradigm. A person could be defined as a mammal if such domain knowledge already exists elsewhere. This way, the domain knowledge of what it means to be a person extends beyond simple attributes about height and relationship about other persons and includes the mammalian traits as well.

What does this mean to artificial reasoning? Computer software can now be written to reason about entities described in domain knowledge once it is populated with instances of data. When Joe and David are described in the knowledge base as father and son and the definition of father-son relationship is such that father is always older than son, a software system can deterministically conclude that Joe (father) is older than David (son) even if their age attributes are not populated. Semantic technologies promise to bring structure and meaning to the data currently spread over a large number of data silos. Ontological knowledge bases can help formally express domain knowledge currently contained in unstructured data repositories (such as books, media, or even scholarly research).

APPLICATION OF SEMANTIC TECHNOLOGIES TO VIRTUAL WORLDS ISSUES

As an example of the potential of virtual world metadata, consider the metadata available for objects in a public virtual world platform (Figure 5).

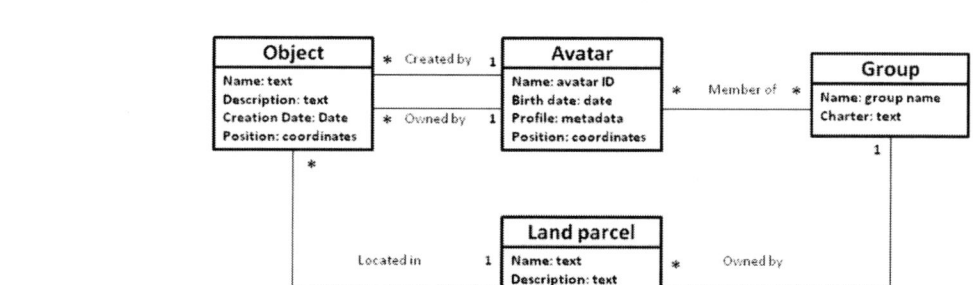

Figure 5. Example metadata for public virtual world objects.

Simply finding an object at a specific location in the virtual world enables another avatar to navigate this metadata in order to form conclusions about the object creator and object owner's relationship to the group owning the land. More can be deduced about the creator and owner if they share many groups in common. Further information about the owner may be gleaned from the metadata of other group members. In this case the metadata is solely related to the virtual world. It may be useful for investigating networks of people conducting unlawful activities within the virtual world, or using the virtual world as a communications medium for their real world unlawful activities. It can't easily be used to render a virtual world representation of real world semantic networks. A virtual world platform may not support user defined metadata for objects.

For example, the creator of a virtual fire engine may not be able to assign it a metadata value indicating that it's a vehicle, an emergency vehicle and specifically a fire engine. If the object name, a free text field, contains "fire engine", then a content consumer searching for a fire engine will also find irrelevant results such as a "fire engine red swimsuit" and "fire engine red manicure." Virtual world platforms may not support object relationships such as inheritance and assembly hierarchy. For example, if I create object A, and make copies A1 and A2, then change A's attributes (size, material, scripted behavior) to give A', the platform may not propagate these changes to A1 and A2. If objects B1 and B2 are both linked sets of objects and are then linked together to form B3, unlinking B3 may not return B1 and B2 as expected; instead the result may be unlinked set of objects. Lack of assembly hierarchy limits the capability to represent hierarchical structures composed of subunits which can move relative to each other and be disconnected and reconnected. Finally, although graphic limitations are not the focus here, they do have a bearing on the capability to render a semantic structure in 3D. Rendering a complex static semantic structure in a public virtual world may cause a temporary reduction in frame rates, experienced by users as "lag". Virtual world platforms may not efficiently support creation of a dynamic tree structure which can respond to inputs by growing, sprouting new links and nodes and transforming its geometry in response to environmental factors.

Example

Daden (2009) demonstrates Halo, an automated Second Life avatar or "chatbot" which references a semantic knowledge base, and various internet databases. She can answer questions about books and authors, make recommendations, and reply to simple questions such as "what is a comet?". She can also give human-friendly directions to named objects. The semantic knowledge base allows her to provide a correct answer to "is an ant bigger than Rigel?" despite having no precise knowledge of the size of either. She can explain that an ant is an insect, and therefore has a size within the range of insect sizes, while Rigel is a star and therefore has a size within the correct range for stars. She can react to nearby objects based on the free text in their object names. She likes rabbits, so applauds when rabbits appear and sags with disappointment when they disappear. She dislikes snakes, and so reacts with fear when a snake appears (Figure 6). And, she can learn from experience that an exploding bomb will injure her (reducing her health point count) and then react with fear when encountering a similar bomb in future.

Another application of avatars enhanced with semantic technology is the field of medical training and emergency first responder simulation. The Medbiquitous Virtual Patient standard (Medbiquitous 2009) defines the properties of virtual patients semantically to allow interoperability between different systems. Figure 7 shows Daden's emergency first responder training scenario with a virtual patient lying in the road.

Semantic Entities in Virtual Worlds

Figure 6. Daden – Halo, alarmed by a snake. © 2009 Daden. Used with permission.

Figure 7. Daden – virtual patient scenario. © 2009 Daden. Used with permission.

Recommendations

While many arguments can be presented for benefits of applications of semantic technologies in general, current state of the art is not without flaws. The underlying knowledge base construction consists of stringing together sets of domain knowledge definitions and their instantiations. These happen in ontologies which are essentially XML text files capable of importing each other by extending and redefining concepts within. This presents several scalability challenges. As the number and the scale of ontologies comprising a knowledge base grow, so does the complexity of querying proper metadata.

Ontologies were designed to be reusable and extensible by a group of authors. This presents a challenge also. Particularly, versioning of ontologies becomes important as many copies of the same domain description could potentially co-exist. Supposed two people define the domain knowledge of human relationships and make it available to the virtual world community. Which version should others extend? Implement? What if both are picked up by others for extension and implementation? The challenge then is to fuse the data to consolidate the discrepancies and allow the same reasoning algorithms to ingest data from both sources.

Furthermore, once additional ontologies become available and entities are described and implemented it is important to realize that not all data is to be trusted. Various degrees of trust ought to be applied to unconfirmed data according to the uncertainty associated with the data being factually correct. It is the community of users that would have to decide whether the information is acceptable. Beyond the notion of data reliability is the idea that not all reported facts are agreed upon and such differences need to be properly consolidated

A 3D semantic web would allow us to perceive semantic relationships via 3D position and orientation rather than text. We could recognize patterns by their shapes rather than by reading about them. We could find new meaning in the world, by creating virtual representations of semantic networks and simply viewing them from different angles and/or watching them evolve in response to environmental factors and over time. In distilling the quintessential patterns in semantic relationships, we could then recognize them in different contexts. We could create novel semantic patterns and then search for them in the real world.

This applies to content consumption (e.g. search and recommendation), but also to content creation, knowledge capture and representation. Reasoning about the environment is the objective of the Semantic Web built using Semantic Technologies. The "environment" is meant here as the transactional exchange of data streams as well as compounded knowledge base at the disposal of the reasoning agent, in this case an application running on the server or on the client's browser. However, the same could be said about a reasoning agent in a virtual world. Let's explore this analogy to guide our way of thinking of virtual worlds as applicable venue for applications of Semantic Technologies.

There are several ways that automated reasoning can be applied in the context of virtual worlds. One is the simplest analogy of an avatar controlled by an artificial intelligence software program tasked with performing certain tasks. A pet could be programmed to follow its owner. A swarm of bees could be programmed to follow a queen. A tour guide could be programmed to answer questions and give directions. All these agents need to reason about their environment by extracting data, context, and acting on the stimuli which execute specific behavior on the part of the avatar. The other example of reasoning is the interaction of entities within the virtual world. A wall is an object defined by avatar's inability to walk through it. There are other qualities and attributes a wall can have yet these attributes exist in a knowledge base and are not a part of the virtual world until these attributes influence the behavior of objects. When a user's avatar is in close proximity with the wall, a certain amount of reasoning needs to take place to instrument the observed behavior based on the relevant attribute of the wall. The reasoning agent here is the environment itself, the program that is responsible for enforcing boundary constraints. In modern computer graphics this is done by continuous checking for violation of certain rules.

Now suppose the wall should have many other properties: the sound it makes when an avatar hits it when unable to walk through it, the resulting deformation, etc. A complex collection of checks and scripts is required to program each behavior resulting from the interaction of the two entities: wall and user's avatar. All these must be run centrally to guarantee proper behavior in response to every action resulted from the avatar's movement.

Furthermore, the querying of the values of the attributes (wall's plasticity, density, size) requires definitions of these attributes for the wall object. This places a burden on the designer of this virtual wall as these attributes need to be explicitly defined or exceptions resulting from undefined values need to be handled.

Let's suppose that a wall could be defined in terms of the material it is made out of. Given that material properties are known for deformation and sound wave resonance all one would have to do is to define the dimensions of the wall and assign the material it is made out of. A reasoning entity, the virtual world environment could then reason that the material properties would be responsible for proper behavior of the wall given the user's avatar coming into contact with it. If another knowledge base was defined with the relationship between material properties and avatar's body then one would only need a swift definition of material property to define a wall, a door, or a box with the same properties. This is the premise behind the semantic reasoning that semantic technologies could add to the virtual worlds.

To generalize, the holy grail of the artificial intelligence research is the one where an artificial agent can reason through its surroundings and maintain a spatial dialog with its peer, a human agent. Semantic technologies bring virtual worlds one step closer to this grand vision by enabling a framework within which such advances artificial reasoning is possible. Several key enhancements to the virtual world architectures are needed to implement the vision outlined above. These include large-scale and open semantic infrastructures, capabilities to enable run-time semantic description and annotation as well as semantic query processing for reasoning (Hauswirth & Decker, 2007). These reflect the required capabilities to store, incrementally build, and query a semantically powered knowledge base.

FUTURE RESEARCH DIRECTIONS

Several research directions are apparent when keeping in mind the vision of a semantically powered virtual world. As powerful as semantic technologies already are, they have some way to go before semantic databases can rival their legacy relational counterparts when it comes down to scalability and complexity of querying data. (Lopez, Annamalai, Banerjee, Ihm, Sharma, & Steiner, 2009). The other challenge currently being addressed by the academic community is the probabilistic extensions to ontological knowledge bases to rectify the inherent uncertainty and varying degree of reliability in semantic knowledge base data. (Costa & Laskey, 2005)

Data fusion techniques also need to be applied to consolidate conflicts in data as multiple domain definition become possible due to the wiki nature of knowledge base construction.

Aside from semantic technology advances, virtual world platforms need considerable development to support the atomic reasoning kernels. If every object and every entity is to reason for itself, these would have to correspond to small programs running asynchronously in a distributed fashion. Supporting such a great number of distributed programs running simultaneously yet having the ability to interact with each other would signal a necessity for a different large scale simulation architecture.

CONCLUSION

The benefit of semantic description of the virtual world where every object is described in this knowledge base is that the interaction between entities is no longer limited to defined constraints. Instead, entities can reason for themselves and interact with other entities. A virtual apple fallen from a virtual tree would spoil if left alone on the ground according to the rules that define its origin and relationship to surrounding conditions – this information can be semantically inherited from high order classes that define fruits, carbon based entities, bacteria edible entities and so on. In other words, polymorphic properties of the instance of the object (apple) will define the behavior of the object to the most granular level limited not by the expressiveness of semantic domain knowledge but by the amount of data contained in the knowledge base. One can easily imagine defining the apple's properties at the atomic level where interaction of the molecules bound together will define the behavior of the entity.

REFERENCES

Ackoff, R. L. (1989). From Data to Wisdom. *Journal of Applies Systems Analysis*, 3-9.

Costa, P. C., & Laskey, K. B. (2005). PR-OWL: A Bayesian Ontology Language for the Semantic Web. *Proceedings of the ISWC Workshop on Uncertainty Reasoning for the Semantic Web*.

Daden (2009). Automated Avatars. http://www.youtube.com/watch?v=9hte2MJ54CA

Damer, B. (2008). A New Virtual World Winter? http://terranova.blogs.com/terra_nova/2008/06/possibility-of.html

Gartner (2009). Hype Cycle Special Report. http://www.gartner.com/it/page.jsp?id=1124212

Hauswirth, M., & Decker, S. (2007). Semantic Reality – Connecting the Real and the Virtual World. *Microsoft SemGrail workshop*. Redmond, WA.

Lopez, X., Annamalai, M., Banerjee, J., Ihm, J., Sharma, J., & Steiner, J. (2009). *Oracle Database 11g Semantic Technologies Semantic Data Integration for the Enterprise. Oracle Database 11g Semantic Technologies Semantic Data Integration for the Enterprise*. Redwood Shores, CA: Oracle Corporation.

MMOX. (2009), Massively Multi-participant Online Games and Applications http://wiki.secondlife.com/wiki/MMOX

Monroe, D. (2009, August). Just for You. *Communications of the ACM*, *52*(Issue 8). doi:10.1145/1536616.1536622

Netflix (2009). http://www.netflixprize.com/

NetFlix® is a trademark of NetFlix, Inc. The authors are not affiliated with or sponsored by NetFlix.

Ok, A. (2009). Virtual Worlds Timeline, http://www.dipity.com/xantherus/Virtual_Worlds

Second Life® is a trademark of Linden Research, Inc. The authors are not affiliated with or sponsored by Linden Research.

Soto, M., Allongue, S., Pierre, U., Curie, M., & Jussieu, P. (1997). A Semantic Approach of Virtual Worlds Interoperability. *Proc. IEEE IEEE International Workshops on Enabling Technologies: Infrastructure for Collaborative Enterprises WET-ICE '97* (pp. 173-178). IEEE Press.

The Ensemble. (2009). http://www.the-ensemble.com/content/netflix-prize-movie-similarity-visualization

ThinkBalm. (2009). ThinkBalm Data Garden. http://www.youtube.com/watch?v=-GBxafG-zZsY

ADDITIONAL READING

Burden, D. (2009, May 13). Toward Semantic Virtual Worlds - A Thinkpiece. Retrieved September 22, 2009, from Converj: http://www.converj.com/sites/converjed/2009/05/toward_semantic_virtual_worlds.html

Chen, C. (2000). Individual differences in a spatial-semantic virtual environment. *Journal of the American Society for Information Science American Society for Information Science*, 529–542. doi:10.1002/(SICI)1097-4571(2000)51:6<529::AID-ASI5>3.0.CO;2-F

Kleinermann, F., Troyer, O. D., Mansouri, H., Romero, R., Pellens, B., & Bille, W. (2005). *Designing Semantic Virtual Reality Applications*. In Proceedings of the 2nd INTUITION International Workshop, (pp. 5-10). Senlis.

Laskey, K. J., & Laskey, K. B. (2008). Uncertainty Reasoning for the World Wide Web: Report on the URW3-XG Incubator Group. URW3-XG Incubator Group.

Latoschik, M. E., & Frohlich, C. (2007). Semantic Reflection for Intelligent Virtual Environments. *IEEE Virtual Reality Conference*, (pp. 305-306).

Sokolov, D., & Plemenos, D. (2008). Virtual world explorations by using topological and semantic knowledge. *The Visual Computer: International Journal of Computer Graphics*, 173-185.

Stiles, R. (1995). Virtual Environments for Training Final Report. Prepared for Office of Naval Research, Contract N00014-95-C-0179. Retrieved Sepetember 22, 2009, from http://www.dtic.mil/srch/doc?collection=t3&id=ADA371968

Virtual World Engines. (n.d.). Retrieved September 22, 2009, from Foundation of American Scientists: http://vworld.fas.org/wiki/Virtual_World_Engines

KEY TERMS AND DEFINITIONS

Semantic Technologies: Set of technologies that help encode meaning of data together with the data and then enable extraction, manipulation, and reasoning through this 'semantic' content.

Ontology: Formal representation of domain concepts which includes semantic relationships between these concepts and entities.

Semantic Web: The set of features, content, and services of the World Wide Web that leverage Semantic Technologies providing applications to serve content requests from users.

Knowledge Base: A general term used to describe data storage for knowledge management which enables storage, organization, and retrieval of data.

Object Oriented Paradigm: Conceptual paradigm that uses objects as entities to model application processes, design specifications, etc.

Chapter 6
Humanness, Elevated Through its Disappearance

Jeffrey M. Morris
Texas A&M University, USA

ABSTRACT

Developments in electronic communications are drastically changing what it means to be human and to interact with humans. The value of recent technological developments to artists is more than doing more, faster and better; it is also the ability to highlight and elevate humanness in new ways through art, even by appearing to replace the real with the virtual. New tools don't simply replace humans, they allow human creators to shift into new realms of creation: creating dynamic systems and worlds instead of static products. This chapter will give consideration to the different types of presence manifest in various communications formats, stage presence in technology-mediated performance, and several artworks that bring new light to the artist's approach to virtual worlds as a kind of counterpoint with reality.

INTRODUCTION

Technology is changing so quickly that it is difficult enough to keep up with the changing landscape of digital communications for personal or business needs (McPhail, 2006). It is more difficult but crucial to take time to contemplate the impact besides new or improved abilities. Developments in electronic communications are drastically changing what it means to be human and to interact with humans, and it is critical to understand this impact, lest we all lose our humanness amidst so much progress.

It is important to remember that the idea of technology shaping art is not limited to the twenty-first or twentieth centuries. In my own discipline, music, steel frames and strings and standardized tuning systems did not just make music easier to perform or present to audiences, they significantly changed what composers wrote in past centuries. There has always been technology, and the same goes for virtuality. The idea of virtuality is essential to art: every work that is fictitious or demands attention for non-utilitarian purposes

DOI: 10.4018/978-1-60960-077-8.ch006

presents some virtual thing to be made whole by the viewer's "suspension of disbelief" (Coleridge, 1817/1985). Even as technology makes virtual experiences more realistic, the viewer is asked to bridge an *uncanny valley* (Mori, 1970) in order to fully enter the world of the artwork. Even though it seems reasonable to equate the virtual with the imaginary or synthetic in contrast with real or natural things, virtuality is in fact quite natural for human minds: Plato's forms, Immanuel Kant's a priori reasoning, and Erwin Schrödinger's cat are all virtual constructions.

So, basically all art is virtual. This is not the focus of most art, but many works from the twentieth century and later have explored this virtuality by challenging the basic assumptions of the artist and the audience, for example, John Cage's composition *4'33"* (1952/1967). It consists of three sections, all silent. The content of the work is void, and the only sounds heard are those sounds from the environment and audience that audiences have learned to ignore in live performance. In the virtual world of this composition, there is only a mirror that reflects things from outside its world. Such challenging works and overtly virtual technology-based artworks create tension between the viewer's sense of the virtual and the real. The value of recent technological developments to artists is more than doing more, faster and better in mundane activities; it is also the ability to highlight and elevate humanness in new ways through art, even by appearing to replace the real with the virtual. Rather than simply accepting technology-assisted abilities or lamenting technology replacing human, analog, and physical things, artists have the opportunity to explore a grand territory. New tools don't simply replace humans, they allow human creators to shift into new realms of creation: creating dynamic systems and worlds instead of static products. The artist can use algorithms to build a "painter" instead of directly placing each pixel in a single work; the composer can compose the *act* of composition itself instead of placing each note. This chapter will give consideration to the different types of presence manifest in various communications formats. A study of electronics and stage presence in live musical performance will illuminate a human sensitivity to authentic, live experience. This will bring new light to the artist's approach to virtual worlds as a kind of counterpoint with reality. A survey of approaches to virtuality will be discussed in terms of humanness lost and elevated and will be summarized in a basic classification.

BACKGROUND

Mediation refers to something coming between two other things, or for the purpose of this chapter, the intervention of communications technology in the path between sender (performer or artist) and recipient (viewer). The term *mediatize* was first used to describe the annexation of one nation by another, in which the leader of the annexed nation maintains his or her title and sometimes some authority. More recently, Jean Baudrillard (1981/1994) has adopted the term to discuss the transformation of events when they are recorded or transmitted by communications technology, originally to highlight more overt or intentional kinds of transformation of the symbols in play. It has come to be used by others in a more general way highlighting any result of the process of recording or transmitting once-live events. (Auslander, 1999) In this chapter, *mediation* refers to the intervening position of technology, and *mediatization* refers to the effects of that intervention (in the general way without Baudrillard's embedded implications).

It is interesting, however, to reconsider the original definition of mediatization. It already contains senses of both emasculation and preserved identity. In all senses of the word, the mediatized object becomes a reference to the original but loses its authenticity. In the mediatization of live events through technology, what is lost is the *aura* as Walter Benjamin (1936/1960) described it, the sense of authenticity that comes from witnessing the "real thing," whether it be the original *Mona Lisa* or an unmediated live performance. Through mediatization, real live events become *hyperreal* (Baudrillard, 1981/1994): they enter an artificial state in which once-live moments are frozen and are susceptible to endless manipulation by technology but still carry the reference to the authentic once-live event. Modern citizens have become accustomed to granting the same authenticity to mediatized objects as the realities they symbolize. Philip Auslander (1999) has carefully argued that contemporary society has become desensitized to the differences between live and recorded media, and so they are wholly equivalent. The sacrifice of quality and authentic experience for the sake of convenience is regrettable, but modern artists could view technological progress as shifting the playing field of expression into new territory instead of suffering a loss: as technology allows humans to communicate in new ways, structure and meaning can be manifest in new dimensions of a work. Auslander describes where audiences have placed value; it is not a prescription that inhibits artists from seeking new places where value could be found or created. In contrast with Auslander, Peggy Phelan insists on recognizing the difference between the live and the mediated:

Performance cannot be saved, recorded, documented, or otherwise participate in the circulation of representations of representations: once it does so, it becomes something other than performance ... To the degree that performance attempts to enter the economy of reproduction it betrays and lessens the promise of its own ontology. Performance's being ... becomes itself through disappearance. (Phelan, 1993, p. 146)

To "enter the economy of reproduction" is to do something that is *reproducible*, whether or not it is reproduced. The more that *would* be lost through reproduction, the more substantial it is, or the greater its aura is.

MEDIATION AND HUMANNESS

Let us consider familiar communications media, human situations, and works of art in these terms.

Presence

Weblogs (blogs) and e-mail can be thought of as one-actor plays, but they are experienced like reading a play, rather than watching it. Replies may be seen as separate scenes of the play, each with a different speaker. Presence is only felt by the reader when reading, and separately by the writer when writing, but these never happen at the same time. Micro-blogs and status updates on social media (e.g., Twitter [http://twitter.com], Facebook [http://www.facebook.com], and MySpace [http://www.myspace.com]) often function like miniature versions of weblog posts or e-mail messages, giving only a single-line summary of the state of the speaker that may invoke more vivid but varied pictures in the readers' mind. In this case, the brevity creates an incomplete image that could evoke a greater sense of presence in the mind of the reader if he or she knows the writer well enough to imagine what kind of situation might lead the writer to make such a statement. Instant messages and chat rooms are closer to being real-time, but the sense of presence slowly fades after each post and when a participant fails to respond quickly. Like e-mail, presence of a writer is not sensed until after the writer has completed a post unless the medium provides signals such as displaying, "[username] is typing...." Video chat provides a more authentic and consistent presence, but it can involve a passive togetherness as participants engage in other activities on their computers during the chat session, like being in the same room with the person without talking. In this case, presence (but not necessarily attention) can be felt when no one is speaking or looking directly at the chat window. Virtual world interactions can function like video chat, but with the pretense of user-selected avatars instead of real-time images of each participant. This aids in the theatrical suspension of disbelief in support of a story line (e.g., a team of warriors planning an attack in a game) or to allow participants to invent their own identities for purposes of creative expression or preserving privacy.

Humanness

In virtual interactions, certain aspects of reality and the human condition are set aside as part of the suspension of disbelief. Virtual systems, however, are only models of real (if sometimes hypothetical) phenomena, and at some point, every model will reveal its shortcomings. For example, a printed map is a model of a real portion of the planet. Its shortcomings include its limited ability to show changes in elevation and the necessity of using false colors to demonstrate political borders (i.e., the color of the ground does not really change when one crosses from one nation into another). Sometimes shortcomings are more often appreciated as features, for example the ability to simulate a bird's eye view over a large area without expensive aircraft: this is the main purpose for using this type of model in the first place. In mundane activities, electronic media are appreciated for their benefits, and viewers are conditioned to overlook their shortcomings as models of the real world. In general, the benefits of electronic media involve warping time or space, for example, enabling multiple people to share ideas as if meeting in person without requiring any of the participants to occupy the same physical space or even participate at the same time, as in an internet forum. This is a magnificent achievement, and it allows humans to get much more accomplished with fewer resources. However, as the convenience of electronic media makes them preferable modes of interaction, it is easy to lose sight about what human aspects of communication are being sacrificed. Humanness is more than corporeality. It is also temporality, morality, the sensitivity to aura, and the desire for connection.

Virtuality is what humans share with computers. Humanness, here, is the set of things that only humans can sense.

Stage Presence

Stage presence refers to the way a performer establishes a sense of presence during a performance. Even in non-theatrical music, this is a significant element of live performance. It is a special kind of presence: it is mostly one-sided, but its effects are amplified because of the intense attention directed toward the performer. A performer may choose to avoid showing affect or unnecessary motions in order to avoid distracting the audience from the sonic structure of the performance, or he or she may choose to embellish the sounds with exaggerated gestures and facial expressions. Much of this comes naturally for musical performers, however, and stage presence may not get much direct consideration. In acoustic musical performance, this may be passable, because every performer must move in order to play and acoustic instrument. Electronic instruments, however, enable performers to cause any number of events or changes to occur with minimal motions. In circles of computer-based musical performance, it is often difficult to tell just by looking whether the person on stage is giving a heartfelt virtuosic performance or doing office work. Julio d'Escriván (2006) agrees with Auslander (above): audiences will get used to this disparity between cause and effect. However, it is clear that humans still have a sensitivity to this tension between live actions and their technology-mediated results. I have seen discomfort result when a performer allows the technology-mediated sound to fall out of congruence with the live experience, even with contemporary art specialists and musicians in the audience, for example after watching a musical performance in which the artist used a ball clutched firmly in both hands to control all the sounds in the artwork. Discussions I heard afterward could not get past the question, "Which button triggered which sound?" The incongruity between sight and sound alienated the audience, and prevented them from appreciating the sonic structure of the work.

Another striking example occurred after a performance at the 2007 International Society for Improvised Music conference, in which Eric km Clark performed on electric violin and I performed with him using a computer to mediate his sound signal. The sound heard through the single instrument amplifier on stage did not always include the live violin signal. Sometimes the live violin playing only caused past violin passages to be heard, and at other times there was no obvious relationship at all between Clark's actions and the sounds and silences coming from the amplifier. One observer who was also performing at the conference expressed discomfort with this performance, but not with my stage presence: "If there's a violin there, I want to hear it," he said, and other performers separately made similar remarks. This is the opposite of the stone-faced laptop performer: the violinist was projecting a stage presence that was independent of and incongruent with the violin sounds heard. We had pared off the "liveness" from the performance and used it as an independent element. Members of our audience that came looking for meaningful relationships in only the usual places tended to see the overt mediation by the computer as a mistake or malfunction that only obstructed their experience of the performance.

A more Cagean mindset, ready to appreciate anything for what it is, would more readily notice and appreciate the tension and release in the counterpoint between the live events and the resulting sounds (whether or not they were intentional). The effects of mediation at play here are indeed salient; they are only overlooked as a result of social conditioning. The fact is that the relationship between the live and the mediated can establish tension. If this tension can be built up and resolved, liveness and mediation can be used as a new dimension for structure and expression in performance—one that can give new value to the authentic unmediated human experience.

Approaches to Virtuality in Terms of Humanness

The following is an examination of phenomena and artworks that each represent a loss of humanness in some way through mediatization but end up elevating humanness, either by manipulating the viewer's sense of presence, creating situations that manipulate the behavior of the viewer, transforming the self into the other, or making the self disappear.

Playing on Presence

Chatterbots are computer programs designed to imitate humans in chat rooms or instant messages. Chatterbots are always present and attentive (in that they are always ready to respond to input), but the question of presence becomes not whether one's partner in the chat is present but whether that presence is human (Auslander, 2002; Dixon, 2007). Further, if the thing that is present is not human, one may start to ponder the presence of the human that created the chatterbot. The chatterbot programmer may be watching a chat in order to monitor the chatterbot's performance, but it is more likely that the user will imagine the programmer working in the past, anticipating the various statements a user may enter and finding ways to answer them with the desired personality. One's contribution to the chat session may become guided by a curiosity about how the programmer achieved that, instead of thinking about the conversation with the chatterbot as such. A chatterbot isn't necessarily a work of art, but it highlights how people may tend to focus on the human element in an experience.

Ian Bogost and Ian McCarthy used text-based media for performance in *Twittering rocks* (2007), a representation of the chapter "Wandering rocks" from James Joyce's *Ulysses* (1922/2008) performed through Twitter micro-blogs for each character in the chapter, updated in real time according to the timeline of the story. The familiar human experience of reading the book is lost in this format, and these automated feeds are mixed in with the updates by each viewer's live human contacts. They don't convey the same sense of presence as the real Twitter contacts, but by juxtaposing the "tweets" from *Ulysses* with the tweets from ordinary life, the viewer gets to reflect on the fictitious events from a real-time human perspective as well as appreciating the artistic aspects of the tweets by real humans.

Ken Goldberg's web-based *Telegarden* (1995) removes humans from the picture and gives the user a chance to maintain a garden via remote control. The user can only experience the garden through this mediation, knowing he or she will never see the garden in person. A primary effect of this work is to cause the user to wonder, "What is the point?" Without a human to connect with or a chance of having a natural encounter with the garden, the user's mind may jump out of the premise of the artwork and begin to wonder if there even is a garden, or if his or her actions are really affecting it. Goldberg calls this form of inquiry *telepistemology*: "Media technology generally facilitates the suspension of disbelief. I'm trying to facilitate the resumption of disbelief" (Tribe and Jana, 2006, p. 46). The ability to "snap out" of the world that is originally presented and think about the system itself is important. It allows one to think critically about the effects of electronic media on humanness. Telepresence is explored in *Imaging Beijing* (2008) by John (Craig) Freeman, in which viewers in the virtual world Second Life [http://secondlife.com] can view panoramic photographs of real world Beijing. The human link is severed since humans in Second Life cannot interact with those in Beijing. However, multiple Second Life users can come together in Second Life to view Beijing at the same time. All Second Life users are already telepresent, coming together virtually from any place in the real world, but they create a more genuine kind of presence as their avatars stand together in Second Life and together experience telepresence in the real world site of the artwork. Telepresence in Second Life gains a measure of authenticity by spawning another telepresence in *Imaging Beijing*.

Telepresence is expanded into mixed realities in *The gate* by the Second Life performance art group Second Front (2007), in which a projection screen displays a site in Second Life, and a screen at the Second Life site displays images from the real world site like a portal between the two worlds. Here, both worlds are symmetrical and equal. Similarly, composer and performer Pauline Oliveros gave a mixed reality performance in which she improvised live music at the International Society for Improvised Music conference at Northwestern University in December 2007, accompanied by a team of dancers performing in Second Life. The real-world music and the virtual-world dance were equal components of the performance.

The mixed reality work *No Matter* (Figure 1) plays on the aura and authenticity of objects (Kildall and Scott, 2008). Artists were paid to render imaginary objects in Second Life, objects that never before existed in material form. Artist Victoria Scott explained, "These are not just everyday functional objects but special objects drawn from the well of shared cultural imagination (from myth, idioms, thought experiments, etc.)." (Scott, 2010) The digital objects were then translated into corporeal objects for the first time as three-dimensional paper models. As with *Telegarden*, the question, "What is the point?" easily comes to the mind of the viewer, giving him or her the chance to consider which is more valuable or authentic: the digital originals, the corporeal copies, or the ideas that inspired these objects. The digital versions came first, and artists were paid to make them, but the paper models are more real in a traditional sense. Additionally, Second Life does not allow exporting 3D models, so the paper models are "digital plunder" or "ontological treasures," in the words of the artists (ibid.). Would you rather see this exhibit in Second Life or in the real world?

Figure 1. No Matter (Kildall and Scott, 2008) in digital form (top) and corporeal form (bottom). © 2008 Scott Kildall and Victoria Scott. Used with permission.

Gazira Babeli's installation, *Save your skin* (2007) also plays on the preciousness of digital assets in Second Life by displaying avatar skins in a Second Life gallery. These are things of value in the virtual world, because Second Life charges money for new skins. However, this preciousness only results from an arbitrary rule, not a necessary property of the world. This makes it an arbitrary kind of preciousness. It is also striking to see the flat rectangular skin images, like the flayed hides of avatars. It is simultaneously a bit shocking to see the face of an avatar removed and "lifeless" but also a hollow feeling to know that this was not a violent act or a significant sacrifice. Either way, it draws on one's human sensibilities and lets one reflect on differences between worlds despite their constructed similarities.

Eva and Franco Mattes, also known as 0100101110101101.ORG presented *Synthetic performances* (Mattes and Mattes, 2008), Second Life versions of classic performance art works, including Valie Export's *Tapp- und Tastkino* (1968; Martin, 2006; see Figure 2) and *Imponderabilia* (1977) by Marina Abramović and Ulay. Both

Figure 2. Tapp- und Tastkino (Export, 1968) performed in Second Life by Eva and Franco Mattes as part of Synthetic performances (2007). Creative Commons 2007 Eva and Franco Mattes. Courtesy Postmasters Gallery, New York. Used with Permission.

works involve public physical contact between viewers and the nude performers. However, in Second Life, the human element is severed, because there is no sense of touch transmitted between worlds, and the performers are not exposing their real bodies. They are merely represented by avatars that resemble nude humans. The virtual performance of works like these force one to step out of the assumptions of the virtual world and wonder, "What is the point?" again. Since these performances are clearly representations of possible events in the real world, the viewer is also likely to imagine experiencing the work in the real world and reflect on the disparity in intensity between the virtual and real world versions of the works. Like Cage's *4'33"* (1952/1967), the void of affect where one would normally find it forces one's attention toward one's own assumptions.

Manipulating the Participant

More than just inviting viewers to reflect, some works can compel a viewer to change his or her behavior. Augusto Boal, in his *Theater of the oppressed* (1979), described a technique for live theater that gives the audience a significant role in the content of the performance. Any audience member may stop the improvised plays focusing on situations of oppression and give directions to the actors in an effort to escape the oppression while the actors try to steer the story toward its original goal. In this case, the story of the play is a virtual experience, but the audience has a real role in competing against the story, striving for a more pleasing resolution. The audience is not merely invited to contribute; the audience is compelled to struggle against the virtual world of the story. The real world version of Valie Export's, *Tapp- und Tastkino* (1968) creates a conflict for the viewer that may make him or her refrain from participating. In this work, the performer stands in a public place wearing a closed box around her otherwise nude torso and invites participants to insert their hands into the box so the "touch cinema" can begin. The premise of this work calls for a suspension of ordinary behavior, to do something that would normally be unacceptable in public. Whereas in Boal's theater of the oppressed the virtual situation is just a story that could easily be ignored or dismissed, *Tapp- und Tastkino* creates real, corporeal sensations that bypass the rational mind, creating a conflict between the participant's understanding of the premise of the work and his or her sense of acceptable public behavior. Going a step further, Marina Abramović's performance work, *Lips of Thomas* (1975) gives the audience no explicit role in the work, but the audience is compelled to stop the performance after the performer cuts herself and lies upon blocks of ice indefinitely, nude, bleeding, and shivering. The performance ends when the audience forces it to end. The viewers snap out of the virtual world of the performance and rescue the performer from her real-world peril. It doesn't take life-threatening situations to affect behavior, however. It only takes the knowledge that one's actions as a participant will affect the experience of another participant. The mixed reality installation *Remote* (Donaldson, et al., 2008) creates complex modes of interaction between Second Life and real world participants. Participants can create environments for their fellow participants in the other world or at least act with a sense of courtesy toward the other, with whom they share the space but whom they will never meet.

Virtual world actions can have lasting real world effects, as in *Velvet-strike* (Schleiner, et al., 2003; Tribe and Jana, 2006), which is an intervention performed within the virtual world of a video game. The game *Counter-strike* (Valve Software, 1999), a modified version of *Half-life* (Valve Software, 1998), gained popularity for its realism, i.e., fighting human terrorists instead of monsters. *Velvet-strike* consists of pre-made graffiti and intervention instructions to be used within *Counter-strike* to promote a pacifist theme, including anti-violence messages, standing with a group in the shape of a heart, or befriending inworld opponents. While the goal of the game is to defeat enemies, and the game has no direct effects on the real world, *Velvet-strike* participants have chosen to operate outside the premise of the virtual world and deliver a message promoting peace that can influence gamers in their real-world lives.

Cross-world influence goes in both directions in *GPS::Tron* (2004) by Thomas Winkler. Players use GPS-enabled mobile phones to operate virtual light cycles as in the film, *Tron* (Kushner and Lisberger, 1982). Players need not have any contact with each other; they may never meet or communicate in any way, so a human connection is blocked for the most part. However, the belief that other humans are playing imbues the animated onscreen opponents with some sense of presence. Worlds conflict when the optimal move within the virtual world (e.g., to avoid an opponent or obstacle) is blocked by an obstacle in the real world. Like *Velvet-strike* but in a more concrete way, *GPS::Tron* repeatedly reminds players that one realm is more real or important than another, and the virtual world must be sacrificed or compromised at times in order to meet more basic human needs. The game inevitably requires the player to refuse to play it (or play it well) at times. Corporeal limits also come into play in Paul Sermon's *Telematic dreaming* (1992), which causes the viewer physical discomfort after long periods, requiring him or her to leave the installation despite the attraction of the human connections made in the work (Dixon, 2007). It is discussed more in the next section.

On the contrary, the Contagious Media Project by Jonah Peretti (2001) creates virtual assets that are given life and perpetuated by the real world. It is a series of news-worthy electronic items such as e-mail messages or websites that rely on the viral spread of memes to initiate discussions at all levels of communication from word of mouth to broadcast and print media. Some items are fictitious, such as the home page of a white-skinned couple bragging, "Black people love us," and some items are real but created with the project in mind, such as an e-mail message asking Nike for a custom pair of shoes with the word *sweatshop* printed on them. Each experiment "starts small" and propagates with no promotion but makes its way across the Internet and into television and newspapers. Most experiments bring moral issues to light, but in their purest form, a fictitious piece of information acts like a dye test within real-world viral communications, highlighting the way any message can flow from one source to everyone's attention. In this way, a meaningless virtual asset can work like a mirror on the world's communication habits.

The real-world effects of virtual interactions are perhaps most intensely felt in Stelarc's *SPLIT BODY: VOLTAGE-IN / VOLTAGE-OUT* (1996; Figure 3) and the Centre for Metahuman Exploration's *Project paradise* (1998; Figure 4). In each work, viewers use a remote interface to control the bodies of the performers and watch on a video screen. The brain-body connection is severed and split across multiple people: the viewer makes decisions, and the performers (with no control over their own bodies) execute the actions dictated by the viewer, but the viewer does not receive any tactile feedback from the movement. Even though this basic human sensibility is severed, the viewer knows that his or her decisions result in real sensations for the performers in an extreme way, and this knowledge adds to the impact of the works.

Transforming the Self to Other

Works in the previous section are effective because of human connections between viewer and performer or between viewer and real-world society. However, Jean Baudrillard (1987) points out that in electronic communications, people do not interact directly with others. Instead, users interact with screens, and remote screens interact with others on their behalf. This mediation can be exploited to turn the self into an other, highlighting the effects of mediatization on humanness. Sam Taylor-Wood's installation *Pietà* (2001) is a "video statue" based

Humanness, Elevated Through its Disappearance

Figure 3. SPLIT BODY: VOLTAGE-IN / VOLTAGE-OUT (Stelarc, 1996). Gallery Kapelica, Ljubljana 1996. Photographer: Igor Andjelic. © 1996 STELARC. Used with permission.

Figure 4. A viewer controls the hand of a performer in Project paradise (Centre for Metahuman Exploration, 1998). © 1999 The Centre for Metahuman Exploration. Used with permission.

on Michelangelo's sculpture *Pietà* (1499) in St. Peter's Basilica in the Vatican, in which a woman sits, holding a dying man. Even though no interaction is possible with the video, the work conveys more presence than Michelangelo's version. The viewer vicariously feels the strain and awkwardness of the scene as the actors strive to hold their positions. Tension is created through the human sense of empathy. Joachim Koester's *Pit music* (1996; Riemschneider and Grosenick, 1999) is also a video installation, but at first it appears to be simply a replacement for what should have been a live string quartet performance. The video depicts a quartet performing a musical work in an art gallery, giving the user the chance to ask, "What's the point?" The viewer may notice that the video occasionally shifts focus to the audience within the video, standing as a group in front of the quartet, watching. He or she may then make the connection that he or she is part of the same audience, standing and watching the string quartet perform, and further, that other viewers see him or her watching the quartet along with the other audience members. In this way, the scope of the artwork can expand in the mind of the viewer, and the viewer may notice his or her role shifting from being an isolated viewer to being part of the installation—part of the audience inside and outside the video, being watched by the other viewers. The viewer becomes the viewed. In *Text rain* (1999) by Camille Utterback & Romy Achituv, the viewer is a part of the artwork in a more obvious way. Text falling from the top of the projection collects on the top surfaces of the viewers' shadows. The viewers can move and change the shape of their shadows to let the words fall down the screen. The viewers receive no tactile feedback as the words fall on their shadows, but the connection ends up seeming real enough for the viewer to interact naturally with the installation, because of its quick and consistent responsiveness. Viewers start moving with the purpose of controlling their shadows. For the moment, viewers leave their bodies and inhabit their shadows onscreen so they may enter and interact with the virtual world.

Paul Sermon's, *Telematic dreaming* (1992) creates an interaction between viewers at two sites. Each viewer lies on a bed and sees a viewer at the other site as a projection lying next to him or her on the bed. This is similar to the intimate-but-remote setting of *Project paradise* (Centre for Metahuman Exploration, 1998; Fig. 4) discussed above. In *Project paradise*, each viewer controls one of two performers, an "Adam" and an "Eve," who are standing nude, facing each other in a remote site. Each viewer controls the arm of Adam or Eve to make one performer caress the body of the other. The awkwardness of the interface, based on a touch-tone telephone, highlights the loss of the human connection between viewer and performers. The artists pose the question, "at what point can anyone ... determine when a pretend caress becomes a real one?" (Dixon, 2007, pp. 585–586) This question applies equally well to *Telematic dreaming*, especially since there is no interface, only a "screen" between viewers. Even though they cannot feel each other or cause each other to feel anything, Suzan Kozel, who performed on one side of the work describes a sensation of "little electric shocks" when a viewer is touched by the projected image of the other (ibid., p. 216). Not only are the viewers the only content in *Telematic dreaming*, but the viewer experiences the artwork vicariously by imagining the tactile sensations of his or her remotely projected image.

The self becomes its own voyeur in *The surveillance bed III* by Julia Scher (2000). Like *Telematic dreaming*, the viewer is invited to lie in bed. However, in this work, the viewer sees himself or herself on television screens posted around the bed. Multiple cameras and screens provide a variety of perspectives on the viewer, all from the perspective of an other. The viewer can simultaneously feel the sensations of watching someone and being watched in an intimate setting. He or she is simultaneously the subject and the outsider.

Electronic Shadow's installation, *3 minutes^2* (2004; Figure 5) takes this transformation from self to other one step further. The viewer's image is captured as in the artworks above, but the image is displayed in the environment as a silhouette after a delay, breaking the viewer's sense that he or she can control the shadow. Instead, the image of the viewer becomes an independent inhabitant of the space, and the viewer gets to live beside himself or herself within that space.

Disappearance of the Self

Once the self has been transformed into the other, the self is already halfway gone. Composers in the 1950s were interested in removing the self from composition by removing all external influences by past music. In search of Roland Barthes's *degree zero* of influence (Barthes, 1953/1977), John Cage delegated his compositional decisions to chance operations, and Pierre Boulez delegated his to rigorous mathematical structures (Emmerson, 2007). More recently, artists have turned to creating alternate identities that receive credit

Humanness, Elevated Through its Disappearance

Figure 5. A viewer inside 3 minutes² (ElectronicShadow, 2003). © 2003 Electronic Shadow (Naziha Mestaoui, Yacine Aït Kaci). Used with permission.

for artworks. Gazira Babeli (mentioned above) is such a creation, "born" in Second Life in 2006. In a way, every avatar in Second Life is such a creation. Even the viewers of artworks in Second Life are removed from their real selves. Composer Uwe Schmidt has taken this to an extreme, having created several dozen aliases with different narrowly specific genres such as Señor Coconut y Su Conjunto, who perform Latin American-style covers of popular music (originally just music by Kraftwerk). Schmidt, a German, has even spoken on a recording in the voice of the Latin American persona Señor Coconut (Hofer, 2006). Modern technology, the mediatization of the screen, has made it a trivial task to invent identities and live through them while neglecting the real-world self.

Technology also enables another method of eliminating the human in the artwork. Algorithms and randomized procedures developed since the mid-twentieth century allow artists to create computer programs that make artistic decisions on their behalf. As opposed to traditional painting, in which the artist directly places every drop of paint, computer-aided artworks are more like sculpture, in which the artist pares down an infinite set of possible outcomes like a block of granite until results are achieved that, while they can have infinite variety across iterations (like a sculpture has an infinite number of vantage points), each one is still true to the underlying character of the work. Here, the artist does not directly create works of art, he or she creates *systems* that generate works of art on their own. William Seaman describes the composer's new challenge: "The artist need no longer seek to define a singular artefact, but instead need develop systems that enable a series of sonic artefacts to become operational and polycombinational, thus engendering an emergent relational sonic artefact during exploration" (Seaman, 1999, p. 234). This type of approach is demonstrated in four technology-based compositions by the author.

Time is the substance of which I am made (Morris, 2007) is a structured improvisation involving live audio and video sampling. As pianist Andy McWain (University of Massachusetts, Dartmouth) and violinist Eric km Clark

113

(California EAR Unit) improvise in the moment, their sounds and images are captured by the author and video artist Gracie Arenas (Electronic Arts), transformed, and reintroduced into the performance to interact with the other performers (see Figure 6). In this performance, the sound and video performers have no voice, no material, other than material that can be copied from the other performers during the performance. The Now is the raw material, and it is twisted upon itself in counterpoint. By taking away the voice of the audio and video performers, they lose their primary source of identity, but as a result, the work highlights the human appreciation of the preciousness of Now. However, that preciousness is only appreciated after the Now has passed and reappeared in mediatized, distorted form.

This is not a guitar (Morris, 2007) deconstructs the performer and audience. It is performed by an electric guitarist (premiered by Chapman Welch, Rice University) with live electronic processing. The title is a reference to the semiotic play in René Magritte's famous phrase, "Ceci n'est pas une pipe" (Magritte, 1928–1929). The work embraces the electric guitar as an instrument that is necessarily mediated. Its sound is disembodied, isolated from the physical acts that cause the sound, it is transformed by a series of "black boxes," and it is only heard through one or many loudspeakers that may be any distance from the guitar and performer. Despite this disparity, audiences are accustomed to watching the performer and listening to the speaker(s), considering them to be one seamless whole. This work explores various notions of presence and causality embedded in this construction, including the hum of the amplifier as the audible-but-ignored sensation of presence. It plays with the distance between physical cause and sonic effect and stretches the bonds between what is seen and what is heard by letting the two fall out of synchrony. It presents physical events without their expected sonic results, and it presents recognizable sounds like pitch bends and familiar riffs in the absence of the physical actions an audience would expect to create them. This piece uses counterpoint in the dimension of liveness and mediation to create and release tension. Manipulating the physical location of the sound deconstructs the instrument and shows that the performer has no voice except

Figure 6. Jeff Morris uses Nintendo Wiimotes to capture and manipulate sounds played by Eric km Clark (violin) in an improvised performance with Andy McWain (piano; not shown) at Texas A&M University's Rudder Theatre, March 9, 2009. © Texas A&M University. Used with permission.

through the magic of the P.A. system. Manipulating the relationships between visible causes and audible effects deconstructs the audience: it highlights the artificial construction of the senses that humans accept as the real performance. Decrying the instrument and the audience allows the viewers to snap out of the usual assumptions of performance and consider the mediatization of the systems at work.

The voice of the composer is attacked in *Zur Elektrodynamik bewegter Musiker* (Morris, 2005), created for cellist Ulrich Maiß. It was composed for the centenary of Albert Einstein's paper introducing his theory of relativity (from which the title is adapted; Einstein, 1905), and it draws inspiration on Einstein's famous question about what it would be like to ride a beam of light and then hold a mirror before his face. It is an algorithmic composition, but it has no means to make sound on its own, nor does it dictate the sounds of others. Instead, it transforms and represents the sound captured from the human soloist who improvises in the moment. The composition exists as a form, void of content, like a hall of mirrors in the dark. The character of the voiceless form is only illuminated by the presence of the improvising human performer. More than just removing the voice of the soloist's accompaniment and replacing it through live sampling, the composer has written himself out of the piece by creating a computer program that makes musical decisions in each improvised performance. However, just like a sculpture has infinitely many vantage points in space, this artwork has a distinct character but it represents a different experience from each of an infinite number of vantage points across time. This opens a new dimension to the artwork: its identity exists across all possible performances. Every single performance is only a glimpse of the artwork's true nature. Umberto Eco, inspired by the avant garde composers of the mid-twentieth century, examines this property of artworks in *The open work* (1962/1989). Even though the experience of the artwork can vary widely from one performance to the next, Eco uses this fact to celebrate the human ability to interpret art:

In fact, the form of the work of art gains its aesthetic validity precisely in proportion to the number of different perspectives from which it can be viewed and understood. These give it a wealth of different resonances and echoes without impairing its original essence. (ibid., p. 3)

In artworks like *Elektrodynamik*, thwarting the sense of identicalness between one performance and the next challenges the human sense of recollection and recognition and gives the audience the opportunity to appreciate a new dimension of identity across time. Finally, both composer and performer are deposed when the computer program from *Elektrodynamik* is fed into its own input, creating a feedback loop. In these performances, called *Tappatappatappa* (Morris, 2005), the human only provides minute sounds (like taps and scrapes) to stimulate the environment, and the complex feedback system grows around its stimulus and consumes it like a pearl forming around a particle that agitates an oyster. The structural properties of *Elektrodynamik* are thrust into overdrive and create new forms like a video camera turned on its own display screen. The human stimulus erodes away, and the voice of the system emerges as if an independent entity. Even though the human element is almost completely annihilated, the experience of the artwork is arguably the most sublime.

An explanation for the affect of this work may be found in Stephen Wolfram's writings on cellular automata. Wolfram presents the principle of computational equivalence: "almost all processes that are not obviously simple can be viewed as computations of equivalent sophistication" (Wolfram, 2002, pp. 716–717). This suggests that beauty or intense feelings result from performances of *Tappatappatappa* because the complex feedback system at work approaches the level of complexity of the human mind. It may be a momentary glimpse of what it would be like to witness a new form of life, something apart from ourselves that overthrows the human monopoly on artistic interpretation. There is another phenomenon operating in these artworks. The concept of the uncanny appeared in the introduction to this chapter. The *uncanny* is the Freudian concept of that which is simultaneously familiar and foreign, self and other. Matthew Causey applies the uncanny to electronic media like the telephone answering machine and surveillance camera that make one ask, "Is that really me?" Causey explains why the uncanny moves people: "The ego does not believe in the possibility of its death. The unconscious thinks it is immortal. The uncanny experience of the double is Death made material. Unavoidable. Present. Screened." (Causey, 1999, p. 386) However, the experience does not end there. The effectiveness of these artworks lies in experiencing one's own annihilation *and surviving*: still being there to witness and ponder it. This kind of experience comes naturally for electronic media.

MODELS AND HUMANNESS

Throughout these works, there are three basic ways of developing tension between the viewer's sense of real and virtual, in terms of the model:

1. The model fails the human. This is common, because it is difficult to establish a robust model, but it is usually not as effective as the other ways. It can be found in many of the first works explored above and is perhaps most effective when it forces the viewer to snap out of the world and ponder, "What is the point?"
2. The human fails the model. That is, the demands of the model exceed the viewer's human capacity, as in *Telematic dreaming* (Sermon, 1992; in which the viewer's body becomes sore and needs to exit the model) or *Lips of Thomas* (Abramović, 1975; in which human sensibilities may force the viewer to stop the performance). Despite the strength of the model, the viewer still has some underlying self, some control, and therefore some responsibility regarding real world effects of the virtual experience, and the model forces the viewer to confront that fact. This is common in performance art, especially in challenging physical endurance or ability, because the connection with a real human is often obvious. The moral side of this is an effective field of play for many works using new electronic media, as with the Contagious Media Project (Peretti, 2001) and *Project paradise* (Centre for Metahuman Exploration, 1998).
3. The model erodes the human presence. The viewer is immersed in the model, but the viewer's self has eroded from the model constructed in the viewer's mind. This is less common, because it is the most difficult to achieve. However, it is perhaps the most effective in reflecting the situation of the modern human. Works that achieve this often play on typical assumptions about human roles and turn the model back on itself in place of human influence to highlight those assumptions while eliminating the human element at the same time.

CONCLUSION

The value in virtual worlds is not limited to increased realism and technology-assisted ability. Something is necessarily lost when technology mediates human experiences. These non-transparent effects are called mediatization. What is lost is akin to Walter Benjamin's concept of the aura, the difference between the experience of seeing a copy of famous painting and seeing the original. Benjamin celebrated the loss of the aura as if saying, "Good riddance." The artworks presented here allow the viewer to celebrate the *lost*, saying, "How precious is our humanness!" Besides the direct experiences programmed into it, a virtual world is a living memorial to the aspects of human experience that are lost through mediatization. Borges shed light on this property of humanness in his non-fiction writing:

Denying temporal succession, denying the self, denying the astronomical universe, are apparent desperations and secret consolations...Time is the substance I am made of. Time is a river which sweeps me along, but I am the river; it is a tiger which destroys me, but I am the tiger; it is a fire which consumes me, but I am the fire. (Borges, 1946/1975, p. 187)

Virtuality allows one to experience one's own disappearance for the sake of self-discovery.

REFERENCES

Abramović, M. (1975). *Lips of Thomas*. [Performance art].

Abramović, M. (1977). *Ulay*. Imponderabilia. [Performance art]

Auslander, P. (1999). *Liveness: Performance in a mediatized culture* (2nd ed.). New York: Routledge.

Auslander, P. (2002). Live from cyberspace: or, I was sitting at my computer this guy appeared he thought I was a bot. *PAJ a Journal of Performance and Art, 24*(1), 16–21. doi:10.1162/152028101753401767

Babeli, G. (2007). *Save your skin*. Retrieved on October 4, 2009 from http://gazirababeli.com/saveyourskin.php

Babeli, G. *Gazira Babeli: Code performer*. [Artist website]. Retreived on October 4, 2009 from http://gazirababeli.com

Barthes, R. (1953/1977). *Writing Degree Zero (Jonathan Cape Ltd.,Trans.)*. New York: Hill&Wang.

Baudrillard, J. (1981/1994). *Simulacra and simulation* (Glaser, S., Trans.). Ann Arbor, MI: University of Michigan.

Baudrillard, J. (1987). Le xerox et l'infinitiy. *Traverses, 44–45*, 18–22.

Benjamin, W. (1936/1969). The work of art in the age of mechanical reproduction. In Arendt, H. (Ed.), *Illuminations: Essays and reflections*. New York: Schocken.

Boal, A. (1979). *Theatre of the oppressed*. London: Pluto.

Bogost, I. (June 16, 2007). *Bloomsday on Twitter: A performance of "Wandering rocks" on Twitter, and a commentary on both*. Retrieved on October 4, 2009 from http://www.bogost.com/blog/bloomsday_on_twitter.shtml

Borges, J. L. (1946/1975). A new refutation of time. In R. Simms (Trans.), *Other inquisitions: 1937–1952*. Austin, TX: University of Texas.

Cage, J. (1952/1967). *4'33"*. New York: Edition Peters.

Causey, M. (1999). The screen test of the double: The uncanny performer in the space of technology. *Theatre journal 51(4)*, 383–394.

Centre for Metahuman Exploration. (1998). *Project paradise*. Retrieved on October 4, 2009 from http://cultronix.eserver.org/pparadise/happinessflows.html

Coleridge, S. T. (1985). *Biographia literaria: Biographical sketches of my literary life & opinions*. Princeton, NJ: Princeton University. (Original work published 1817)

d'Escriván, J. (2006). To sing the body electric: Instruments and effort in the performance of electronic music. *Contemporary Music Review, 25*(1–2), 183–191. doi:10.1080/07494460600647667

Dixon, S. (2007). *Digital performance: A history of new media in theater, dance, performance art, and installation.* Cambridge, MA: MIT.

Donaldson, N., Haque, U., Hasegawa, A., & Tremmel, G. (2008). *Remote.* Retrieved on October 4, 2009 from http://turbulence.org/Works/remote

Eco, U. (1962/1989). *The open work* (2nd ed.; A. Cancogni, Trans.). Cambridge, MA: Harvard University.

Einstein, A. (1905). Zur Elektrodynamik bewegter Körper [On the electrodynamics of moving bodies]. *Annalen der Physik und Chemie, 17,* 891–921.

Emmerson, S. (2007). *Living electronic music.* Burlingon, VT: Ashgate.

Export, V. (1968). *Tapp- und Tastkino.* [Performance art].

Foucault, M. (1973/2008). *This is not a pipe: With illustrations and letters by René Magritte* (2nd ed.). Los Angeles: University of California.

Freeman, J. C. (2008). *Imaging Beijing.* Retrieved on October 4, 2009 from http://turbulence.org/Works/ImagingBeijing

Front, S. (2007). *The gate.* Retrieved on October 4, 2009 from http://secondfront.org/Performances/The_Gate.html

Front, S. *Second Front: The pioneering performance art group in Second Life.* [Artist website]. Retrieved on October 4, 2009 from http://secondfront.org

Goldberg, K. (1995). *The Telegarden website.* Retrieved on October 4, 2009 from http://goldberg.berkeley.edu/garden/Ars/

Goldberg, R. (1998). *Performance: Live art since 1960.* New York: Abrams.

Hofer, S. (2006). I am they: Technological mediation, shifting conceptions of identity and techno music. In *Convergence: The international journal of research into new media technologies 12(3),* 307–324.

Hofstadter, D. (1979/1999). *Gödel, Escher, Bach: An eternal golden braid* (20th anniversary ed.). New York: Basic Books.

Joyce, J. (2008). *Ulysses.* New York: Oxford University. (Original work published 1922)

Kildall, S., & Scott, V. (2008). *No Matter.* Sponsored by Turbulence, a project of New Radio and Performing Arts, Inc. for the exhibition, Mixed Realities—An International Networked Art Exhibition and Symposium. Retrieved on October 4, 2009 from http://turbulence.org/Works/nomatter/

Koester, J. (1996). *Pit music.* [Installation].

Kushner, D. (Producer), & Lisberger, S. (Director). (1982). *Tron.* [Motion picture]. United States: Lisberger Studios.

Magritte, R. (1928–1929). *La trahison des images* [The treachery of images]. [Painting].

Martin, S. (2006). *Video art.* Los Angeles: Taschen.

Mattes, E., & Mattes, F. (2007). *Synthetic performances.* Retrieved on October 4, 2009 from http://www.0100101110101101.org/home/performances

McPhail, T. L. (2006). *Global communication: Theories, stakeholders, and trends* (2nd ed.). Malden, MA: Blackwell.

Michelangelo. (1499). *Pietà.* [Sculpture].

Mori, M. (1970). The uncanny valley (K.F. MacDorman & T. Minato, Trans.). *Energy, 7*(4), 33–35.

Morris, J. M. (2005). *Tappatappatappa.* [Musical composition].

Morris, J. M. (2005). *Zur Elektrodynamik bewegter Musiker*. [Musical composition].

Morris, J. M. (2007). *This is not a guitar*. [Musical composition].

Morris, J. M. (2007). *Time is the substance of which I am made*. [Musical composition].

Morris, J. M. (2008). Embracing a mediat[is]ed modernity: An approach to exploring humanity in posthuman music. In *Performance paradigm 4*.

Morris, J. M. (2008). Structure in the dimension of liveness and mediation. In *Leonardo music journal 18: Why live? Performance in the age of digital reproduction*, 59–61.

Morris, J. M. (2009). Ontological substance and meaning in live electroacoustic music. In *Computer Music Modeling and Retrieval, Genesis of meaning in sound and music*. New York: Springer. doi:10.1007/978-3-642-02518-1_15

Peretti, J. (2001). Contagious Media Project. Retrieved on October 4, 2009 from http://www.contagiousmedia.org/

Phelan, P. (1993). *Unmarked: The politics of performance*. New York: Routledge. doi:10.4324/9780203359433

Quaranta, D. (Ed.). (2008). *Gazira Babeli*. Brescia, Italy: FPEditions.

Riemschneider, B., & Grosenick, U. (Eds.). (1999). *Art at the turn of the millennium*. Los Angeles: Taschen.

Scher, J. (2000). *The surveillance bed III*. [Installation].

Schleiner, A.-M., Leandre, J., & Condon, B. (2003). *Velvet-strike*. Retrieved on October 4, 2009 from http://www.opensorcery.net/velvet-strike

Scott, V. (2010). Personal communication via electronic mail January 12, 2010.

Seaman, W. (1999). *Recombinant poetics: Emerging meaning as examined and explored within a specific generative virtual environment*. Unpublished doctoral dissertation, Centre for Advanced Inquiry in the Interactive Arts, University of Wales.

Sermon, P. (1992). *Telematic dreaming*. [Installation].

Electronic Shadow. (2003). *3 minutes²*. [Installation].

Stelarc. (1996). *SPLIT BODY: VOLTAGE-IN / VOLTAGE-OUT*. [Performance art].

Taylor-Wood, S. (2001). *Pietà*. [Installation].

Tribe, M., & Jana, R. (2006). *New Media Art*. Los Angeles: Taschen.

Utterback, C., & Achituv, R. (1999). *Text rain*. [Installation].

Valve Software. (1998). *Half-life*. [Computer game].

Valve Software. (1999). *Counter-strike*. [Computer game].

Winkler, T. (2004). *GPS:Tron*. Retrieved on October 4, 2009 from http://gps-tron.datenmafia.org/?page_id=2

Wolfram, S. (2002). *A new kind of science*. Champaign, IL: Wolfram Media.

ADDITIONAL READING

Auslander, P. (2005). At the listening post, or, Do machines perform? *International journal of performance arts and digital media 1(1)*. Avatar Orchestra Metaverse. [Artist website]. Retrieved on October 4, 2009 from http://www.avatarorchestra.org

KEY TERMS AND DEFINITIONS

Algorithm: A computerized procedure that, for artists, may be used along with random number generators to make artistic decisions or interact with participants on behalf of the artist, for the purpose of automation, variety, or to creating structures that are otherwise unfeasible.

Aura: A special sense of presence attached to authentic things or experiences, in inverse proportion to their reproducibility.

Humanness: Properties of humans outside of virtuality, including corporeality, morality, sensitivity to aura, and desire to connect with others.

Hyperreal: A technology-mediated representation of reality that is has lost its authenticity because of its perfect reproducibility and susceptibility to digital manipulation.

Mediatization: The effects of mediation by communications technology upon the content or aura.

Now: The present moment, especially appreciated for its ephemerality and authenticity in comparison with the remembered past and anticipated future.

Presence: A sense of connection with and attention from another human.

Telepresence: The use of technology to establish a sense of presence between remote participants.

Uncanny: Something that simultaneously evokes conflicting feelings of familiarity and foreignness.

Chapter 7
The Meditation Chamber:
Towards Self-Modulation

Chris Shaw
Simon Fraser University, Canada

Diane Gromala
Simon Fraser University, Canada

Meehae Song
Simon Fraser University, Canada

ABSTRACT

The Meditation Chamber is an immersive virtual environment (VE), initially created to enhance and augment the existing methods of training users how to meditate, and by extension, to realize the benefits from meditation practice, including the reduction of stress, anxiety and pain. Its innovative combination of immersive virtual reality (VR) and biofeedback technologies added interoceptive or dimensions of inner senses to the already sensorially rich affordances of VR. Because the Meditation Chamber enabled users to become aware of autonomic senses that they are not normally conscious of, and to manipulate them in real-time, we found that it did enhance users' abilities to learn how to meditate, particularly those who had never meditated. We describe the Meditation Chamber, scientific methods of evaluation and findings, and discuss first-person phenomenological aspects, its long-term applicability for users who have chronic pain, and future directions.

INTRODUCTION

The Meditation Chamber was a project in service of our long-term research interest in the biopsychosocial aspects of chronic pain (Gatchel, 2009). It is a training system and engineering artifact that can be evaluated on engineering terms – questions of effectiveness and efficiency can be answered using scientific methods, as we have outlined. However, it can also be viewed as a tool for exploring ideas of subjectivity as they relate to the physiological states that are inextricably intertwined with subjective experience. Meditation and pain are subjective experiences that take on manifold dimensions that are difficult to communicate or measure (Scarry, 1987). As a means of exploring the subjective aspects of chronic pain, we are embarking on a follow-up project, the Virtual Meditative Walk. It builds upon some of the techniques of the Meditation Chamber to

DOI: 10.4018/978-1-60960-077-8.ch007

enable people who endure chronic pain to be able to better manage it. These two systems form part of a larger agenda to help people express and potentially manage their bio-subjective experiences over time. The knowledge gained in developing and analyzing these two VEs may provide a baseline and framework for understanding VR experiences among diverse knowledge bases. Thus, our approach is a phenomenological one that builds upon the work of Maurice Merleau-Ponty and, more recently, Francisco Varela, Evan Thompson (Varela, Thompson & Rosch, 1992) and others[1]. Because this approach accounts for the interrelations among mind, body and world, it closely parallels the biopsychosocial approach that is at the core of current pain management (Gatchel, 2009). Because felt experience is the subject of this type of phenomenology, it offers a method that necessarily accounts for the subjective aspects of chronic pain, as well as the objective aspects that can be measured scientifically. The research described in the present chapter contributes a new approach to VR-based pain research, because it specifically focuses on the longitudinal aspect of persistent, chronic pain rather than on acute, short-term pain that is addressed by what is termed VR pain distraction. Thus, rather than characterizing the Meditation Chamber and the subsequent research it spawned as "pain distraction," we focus on the way it affords users the ability to manage their attention and awareness so that they may exert agency over their on-going experience of pain. We term this "self-modulation."

THE MEDITATION CHAMBER

The Meditation Chamber was an immersive virtual environment that was originally created by long-time VR researchers Larry Hodges, Diane Gromala, Chris Shaw, and Fleming Seay. It was subsequently refined and used at Virtually Better, a VR clinic that was founded by Hodges, and expanded upon by the Transforming Pain Research Group (Transforming Pain Research Group, 2010), directed by Gromala, a Canada Research Chair. Reported briefly at Enactive 2007 (Shaw, Gromala & Seay, 2007), the goal of the Meditation Chamber was to design, build and test an immersive VE that used biometrically-interactive visuals, audio and tactile cues to create, guide and maintain a user's meditation experience. It is not necessary to use technology to meditate of course. However, the widespread use of CDs, DVDs, and online resources suggests that technology may be a useful way to enhance and reinforce the practices of meditation. More importantly, we discovered that immersive VR, integrated with biofeedback technologies, offer something unique — it enables users to see their intentional efforts to affect their continuously changing autonomic states. While standard biofeedback techniques also offer visual and auditory feedback, the simplistic monotones or waveforms are not immersive or aesthetically engaging. Thus, VR and biofeedback technologies were combined to determine if the immersion and

Figure 1. The Meditation Chamber. Users wear a head-mounted display (HMD) that provides them with stereoscopic imagery and sound. Interaction primarily occurs as users strive to manipulate their physiological states via biofeedback. Biometric sensors are attached to two fingers with Velcro; these sensors tracked galvanic skin response and heart rate. A flexible chest band tracked respiration.

biometrically-driven real-time feedback could help users achieve a meditative state. Biofeedback was considered to be potentially useful for enabling users to get real-time feedback and to gain a sense of agency or control over three aspects of their autonomic functions: heart rate, respiration, and galvanic skin response. Although biofeedback cannot, of course, offer a confirmation of being in a meditative state, it can indicate relative changes in physiological arousal and, after decades of testing, is considered to be a reasonably reliable indicator of reaching a meditative state provided there is additional questioning of the participants.[2]

In the Meditation Chamber, users sat in a comfortable, semi-reclining chair and experienced a VE that took them through three phases of a virtual experience. Prior to the first phase, users were first fitted with a head-mounted display (HMD) and three biometric sensors that measured galvanic skin response (GSR), respiration and heart rate. Once seated comfortably, users entered the first phase of the meditation chamber: as they were presented with a visual display of a sun (Figure 2), the system's interactive "vocal coach" asked them to relax. The biofeedback device measured their GSR in real-time, which directly affected the imagery: as the user began to relax and their GSR declined, the rate at which the sun moved would increase until the sun went beneath the horizon giving way to a peaceful night scene, complete with chirping crickets. If the user was unable to become relaxed or their GSR increased, the sunset would slow down. The second part of this relaxation phase operated in the same way as the first, but depicted a moonrise instead of a sunset. As the user relaxed and lowered their GSR, the moon would rise higher and higher into the sky. The user's GSR measure determined the frame-rate at which the sunset / moonrise animation would play. In this phase, users reported that they became aware of their intentional efforts to relax because they understood that the visuals were responding to their continuously changing physiological state. In the second phase, users were taken through a set of muscle tension and relaxation exercises, again by the system's vocal coach. 3D graphics of a human body were rendered and displayed from a first-person perspective (Figure 3). Thus, the 3D body that users saw corresponded to their physical body. The user was coached to flex, hold, and release a set of eight different muscle groups including the legs, arms, abdominals, and shoulders. Each muscle group sequence was accompanied by gender appropriate visuals depicting the described motion, usually from a first person perspective. This phase was not interactive, but instead asked the user to listen to the narrator's instructions while mimicking the movement examples visually presented to them on the screen. The system's creators and users noted that this was a strong and compelling illusion. The authors intend to expand upon this by including sensors on the users' wrists, knees and feet to strengthen the illusion of a one-to-one correspondence, or an embodied "felt sense." In the third phase, users were taken through a guided meditation and breathing exercise, interacting with soothing visual imagery and ambient

Figure 2. The Meditation Chamber, phase one. Users' continually changing physiological states first affect a setting sun and then a rising moon.

Figure 3. The Meditation Chamber, phase two. A first-person display of the user, male or female, mirrors their actions during the progressive muscle relaxation.

sound. As users approached what is considered an acceptable biometric approximation of a meditative state, the volume of the sound decreased, while the interactive visuals dissolved to black; often, users simply closed their eyes. After a prescribed amount of time meditating, the vocal coach gently suggested that users end their meditative session. The system was installed at the Emerging Technologies exhibition at SIGGRAPH, a five-day conference. Since many users at this conference were familiar with technology, every effort was made to avoid situations in which users could "game" the system — from the video that explained the known benefits of meditation, viewed by users as they waited for their sessions, to the design of the system itself. 411 users filled in pre- and post-experience questionnaires that asked them to rate their level of relaxation before and after their session in the Meditation Chamber. Throughout their sessions, the system tracked users' biometric measures; this data was offered to users at the end of their session, in the form of a printout. Surprisingly, most users asked for their printouts and remained to study it, often asking questions and expressing recognition of when they "felt" or became aware of their abilities to lower their respiration, heart rate and GSR.

Findings indicated that the majority of the 411 users reported their levels of relaxation increased after experiencing the Meditation Chamber, especially users who had never meditated. Users rated their level of relaxation from 1 (very anxious) to 10 (very relaxed). The average pre-session relaxation self-rating was 5.63, with a Standard Deviation (SD) of 1.75. The average post-session relaxation self-rating was 8.00, with a Standard Deviation of 1.69. A t-test showed that post-session relaxation ratings (M=8.00, SD=1.69) were significantly higher than pre-session ratings (M=5.63, SD=1.75), t(410) = -24.45, p=.0001. This indicates that the *Meditation Chamber* is effective at promoting the kinds of relaxation that consistently parallels meditation.

The extensive amount of biometric data collected from the SIGGRAPH attendees (Shaw, Gromala & Seay, 2007) was subsequently subjected to analysis. The analysis revealed a distinction between novices (i.e., users who had never meditated or had attempted to do so only a few times) and experts (i.e., those who regularly meditate). Only 25% of users had biometric profiles that fell between the two. This was surprising, since we expected to find a smoother continuum. To use GSR data as an example, just over half of the participants exhibited what can be called a "novice" GSR pattern or profile. This means that their GSR level started relatively high, descended through the first phase of the experience, increased and showed peaks in the muscle relaxation phase, and then began to decline again in the final phase, ending up at or more frequently beneath the low established in the first phase. Breathing patterns in novice profiles tended to be steadier and deeper in the final phase than in the first phase. In the expert profile, precipitous drops in GSR occur during the first phase, entered a very low and often flat GSR state before the muscle relaxation phase began. This flat-line state was typically maintained throughout the remaining two phases, and was accompanied by a very steady but not necessarily deep breathing pattern. Individuals who exhibited the expert GSR profile also showed a very consistent respiration rate and amplitude throughout the experience.

Most user reports were positive: 30% had enthusiastic written comments. Twelve negative comments had to do with the heaviness of the HMD and the noise from the other exhibits. The positive comments were that users felt relaxed, even though many initially wrote that they did not expect to. Several VR experts expressed strong skepticism, including experts such as Randy Pausch. After their sessions in the Meditation Chamber, all but one of these vocal skeptics expressed strong enthusiasm; the remaining one was mildly positive. We also discovered that the standard profiles of expert meditators vs. novices were strongly correlated with the self-reported experience in the questionnaire. Finally, a skeptic might suggest that our findings were simply the result of giving conference-goers a place to sit and relax. To address this concern, we later tested a baseline condition. Here, users sat quietly in a room, wearing the biofeedback hardware for the same duration of time. In this condition, users did not wear a head-mounted display, nor did they receive the *Meditation Chamber* experience. We found that half of these baseline-condition subjects experienced increases in GSR. These tests confirmed that our findings were not simply the result of providing a place to sit and relax. These findings, along with our other work on chronic pain (e.g., Gromala, 2000) and VR (e.g., Gromala & Sharir, 1996), suggest that VR assisted meditation may enhance learning to relax and meditate, and by extension, may be particularly suited to the needs

Figure 4. The Meditation Chamber. Sample GSR graphs typical of the novice (left two) and expert (right) profiles. Vertical lines indicate transition points from one phase to the next.

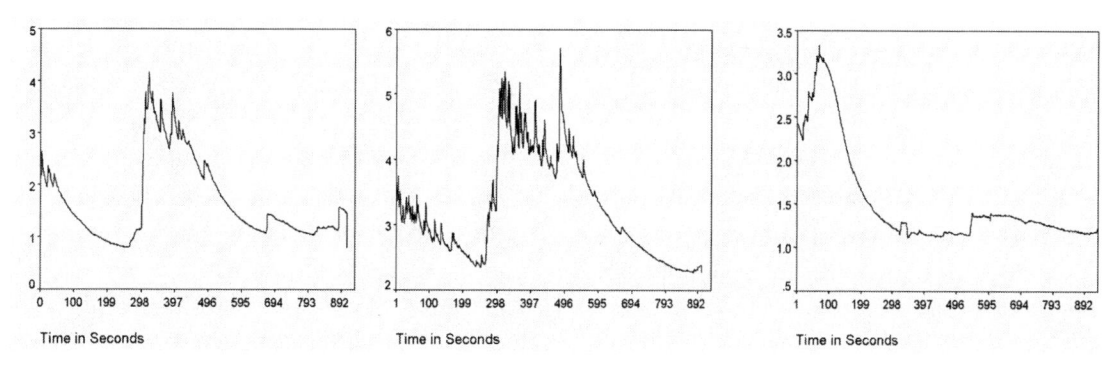

of those who have chronic pain. This is because methods of "self-modulation," such as learning to relax and to meditate, are the primary ways that chronic pain sufferers themselves can manage their own pain at will. All other methods of managing chronic pain — from pharmacology to physiotherapy, massage, acupuncture, and psychotherapy, for instance, — rely on the interventions and expertise of others. Furthermore, one of the most often reported aspects of chronic pain that lead to depression among this group of people is a sense of helplessness (Gatchel et al., 2007): that results from a lack of cure, and the difficulty of treating and managing this degenerative condition. Therefore, the ability to self-modulate or exert some form of self-directed control over chronic pain may afford those who live with chronic pain to gain a greater sense of agency. Indeed, subsequent informal reviews of these data by several experts in mindfulness meditation, some physicians who specialize in chronic pain, and some sufferers of chronic pain confirmed that the Meditation Chamber is a promising adjunctive tool for managing chronic pain. Thus, the Transforming Pain Research Group is in the process of refining the Meditation Chamber, and is developing a more rigorous approach to evaluation. Most importantly, we have now started working with a group of those who live with chronic pain, so that we might study them over longer periods of time.

APPLICABILITY TO CHRONIC PAIN

In medicine, immersive VR applications have proven useful for surgical training and planning, and as a therapeutic modality. Research that focused on VR as a therapeutic modality, showed that, if implemented correctly, it could successfully treat phobias (Brinkman, vander Mast, & de Vliegher, 2008), fears (Rothbaum, Hodges, Smith, Lee, & Price, 2000; Krijn et al., 2007), anxiety (Robillard, Bouchard, Fournier, & Renaud, 2003), post-traumatic stress disorder (PTSD) (Emmelkamp, 2006), and acute pain (Hoffman & Patterson, 2005), to name a few. Remarkably, an approach to reducing acute or short-term pain that utilizes VR has proven to be more effective than traditional pharmacological treatments using opioids (Hoffman & Patterson, 2005). Although the mechanism for the effectiveness of VR in addressing short-term pain is not well understood (Mahrer & Gold, 2009), it has been termed "pain distraction," and more recently, a "non-pharmacological form of analgesia" (Steele et al., 2003). While many forms of media such as video games may provide distraction, VR has been demonstrated to be more effective (Hoffman, 2009) at relieving acute pain. This might be because VR can isolate users from their everyday surroundings to a greater degree than a video game or headphones used during dental procedures. It may also be because VR affords a multi-sensory experience described as "presence" (Hoffman et al., 2004b). It is thought that the stronger the sense of presence, the stronger the effects of distraction. However, explaining why VR is more effective than opioids as "pain distraction" may be problematic according to studies of attention in neuroscience and psychology (Lutz, Slagter, Dunne, & Davidson, 2008). Whatever the mechanism of effectiveness may be, work in the domain termed VR pain distraction has almost exclusively addressed its effects on acute or short-term pain. The approach described as VR pain distraction bears limitations. Though it is demonstrably effective as a non-pharmacological analgesia during medical procedures, it is still in a nascent stage of development. Tested on small populations, the exact mechanisms for its effectiveness remain unclear, with methodological approaches that span computer science, psychology and medicine (Mahrer & Gold, 2009). Distraction is a short-term strategy for diverting attention that occurs and is measured during a VR experience itself (McCaul & Malott, 1984; Hoffman et al., 2004a). Current approaches do not account for what may persist beyond the VR experience, and do not track outcomes over time. Despite the growing number of hospitals and clinics that use VR, it remains relatively specialized and inaccessible when compared to desktop or laptop computers or mobile devices. In addition, according to studies of attention in neuroscience and psychology, the kind of distraction offered by most approaches to VR pain distraction do not meet the criteria known to be effective over time (Lutz, Slagter, Dunne & Davidson, 2008). In addition, differing approaches to pain modulated by VR have not been compared or evaluated. Finally, investigations concerning the design and effect of differing forms of "content" remain an under-examined area. For example, in the earliest work in pain distraction for those who suffered from burns, the "content", or what users saw and heard, was a VE involving snow and wind (Hoffman, Patterson, & Carrougher, 2000). While this appears to be obviously appropriate, the effects of this choice of "content" versus some other choice have not been studied. Similarly, in VEs designed to treat arachnophobia, the "visual rhetoric" of the spider has not been examined (Hoffman, Garcia-Palacios, Carlin, Furness, & Botella-Arbona, 2003). What roles might the kind of spider and its form of representation play? Is a photorealistic spider approach more effective than a cartoon-like rendering? Would the VE provide more therapeutic benefit if the spider morphed from more abstract forms to more photorealistic forms in parallel to the subject's increasing exposure times? VEs that progress from VR to AR suggest that this area of investigation is promising (Botella et al., 2005). Of course, one cannot expect all the implications of VR pain distraction to be addressed immediately or by one expert or field — VR pain distraction is still in its infancy. Therefore, because this area of research addresses issues of subjectivity (Scarry, 1987) and culture (Morris, 1993), scholars of Design and Media Studies (e.g., Bolter & Gromala, 2005), and other subdisciplines of the Humanities (e.g., Elkins, 1999) would appear to be well-suited to contribute to this research domain.

In contrast to pain distraction, we are investigating "pain self-modulation," defined as the ability of those who suffer from chronic pain to consciously and physiologically exert control over their experience of pain. Thus, we integrate VR and biofeedback in training users how to meditate, drawing upon the knowledge gained from decades of study of this learned ability in biofeedback and meditation practices. Long recognized in alternative and complementary medicine, this measurable ability to "self-modulate" pain and stress has been supported by standard medical research, particularly over the last decade. In the form of "mindful meditation," this ability has also been popularized by the work of Kabat-Zinn (2006), among other scholars and practitioners of biofeedback, pain medicine, prevention and wellness (Schatman & Campbell, 2007).

CHRONIC PAIN

Chronic pain is defined as pain that persists or recurs for six months or longer (Russo & Brose, 1998). According to conservative estimates, one in five North Americans is affected by chronic pain, and it ranks among the top five reasons for disability (Blyth et al., 2001). Similar rates are reported in Europe and developing nations. It is one of the most complex experiences that humans face, affecting bodies, minds and culture (Melzack, 1990). In recognition of this, as early as the 1950s, the practice of pain medicine made moves to a multidisciplinary approach attending to mind, body and social issues. (Schatman & Campbell, 2007). Chronic pain is notoriously difficult to diagnose and to treat, and becomes irreversible and often degenerative over time. The body itself changes, while psychological states are also affected (Melzack & Wall, 1996). Those who have chronic pain have difficulty in accessing treatment, and face social stigma and isolation as their abilities to work, play, socialize and maintain mobility diminish (Gatchel et al., 2007). High rates of depression are common, and when those who have chronic pain attempt suicide, they are "more successful [sic] than others at risk" (Gatchel et al., 2007). Thus, the approach to treatment in centres for pain medicine is generally not conceived of as a short-term cure, but as a long-term set of strategies for managing chronic pain, usually by multidisciplinary teams. These teams comprise physicians who represent various areas of expertise, nurses, psychologists and other healthcare practitioners. A review of centres for pain medicine in North America revealed that mindfulness meditation is a commonly recommended tool for self-managing pain (Pridmore, 2002). Moreover, because chronic pain is commonly considered a symptom rather than a long-term illness and because it does not immediately threaten life (Williams, Wilkinson, Stott, & Menkes, 2008), treatment and disability compensation remain elusive (Loeser et al., 2001); thus, it has been termed the "silent epidemic" (Canadian Pain Coalition, 2007). For these reasons, chronic pain remains an under-explored area in treatment and medical and wellness research.

FROM DISTRACTION TO SELF-MODULATION VIA MEDITATION

Extending discoveries that we made in the Meditation Chamber and in our other VR work, our Transforming Pain Research Group focuses on the long-term benefits of VR and pain "self-modulation," for those who suffer from persistent, chronic pain. It is more akin to exposure therapy, where the experiences in VR are considered to be part of a long-term process that is ultimately effective in day-to-day life. The goals for both approaches are long-term, with an emphasis on training mind and body, and developing a greater sense of self-agency.

Researchers in VR pain distraction describe the analgesic effect in terms of "modulation." The term modulation here refers to the effect VR has on the patient (Hoffman et al., 2004a) — the emphasis is on the technology's ability to produce a strong distraction, rather than on the users' abilities to alter their experience of pain through a process learned in VR. We posit a shift in approach and conceptualization, one that emphasizes the user's ability to learn how to self-modulate their experience of chronic pain, using VR as a training simulation, which enhances a user's ability to learn. In training, users may then take this new ability outside of VR and activate it at will. Thus, we define "pain self-modulation" as the ability of those who suffer from pain to consciously and physiologically exert control over their experience of pain, an ability achieved through VR, biofeedback and meditation. This is a measurable ability, increasingly supported and accepted by standard medical research. Though the long-term efficacy of training in VR simulations is well established, evaluation of the long-term effects of training users to self-modulate their pain is in a nascent stage. The goal of our work is to set a foundation for research in this under-examined area.

FUTURE RESEARCH DIRECTIONS

Through the Transforming Pain Research Group, we are continuing to explore the use of VR and biofeedback technology to address chronic pain. Dr. Pamela Squire, a physician whose specialty is in complex pain, and neuroscientist Dr. Steven J. Barnes have joined our group, adding significant expertise and enhancing our methodological approaches. Our current research initiatives include a VR application that extends the sitting forms of meditation used in the Meditation Chamber. In this work-in-progress, entitled the Virtual Meditative Walk (see Figure 5.), a self-regulated treadmill is added to our VR and biofeedback technology. Instead of sitting, users of the Virtual Meditative Walk learn how to meditate while they walk through virtual landscapes, which are displayed

The Meditation Chamber

stereoscopically and binaurally. Simultaneously, real-time feedback of users' physiological states alters these visuals and sound. This is arguably a more intense practice of meditation, as the meditators are in constant motion.

In one phase of this work, two major strategies for learning how to self-modulate one's experience of chronic pain are being compared and measured: the guided imagery and muscle tension and relaxation used in the Meditation Chamber with mindfulness meditation (Kabat-Zinn, 2006). In a second phase, the role of GSR is expanded in order to assess users' learned abilities to self-modulate their experience of chronic pain. This is in response to recent research in pain medicine (Gatchel, 2009). In addition, in order to assess whether users can increase their pain thresholds, and if so, to what degree, we are adding DNIC (diffuse noxious inhibitory controls), a measurement tool well-known in medical research (Villanueva, 2009). Both enhanced GSR and DNIC measures will be compared to users' self-reporting of their pain levels. In a third phase, the most suitable kinds of interaction techniques and comparisons of the roles of media forms will be investigated. These are important aspects, since focusing attention is key to learning how to mediate and self-modulate the experience of chronic pain. Many factors in designing the relations between technological performance and interactive media – such as lag time and optical distortions – affect proprioceptive and vestibular systems (De Boeck, Raymaekers, & Coninx, 2006). The result hinders the experience, leaving an immersant feeling displaced within the virtual environment (Song, 2009). In the context of meditation, such phenomena prove to be highly detrimental in self-directing and maintaining attention. Such displacement can also profoundly affect an immersant's consciousness of their body schema (Gromala, 2000), which is crucial for meditation. Finally, we are developing easily accessible ways to reinforce and track what users learn in VR through applications developed for computers and mobile devices.

The longitudinal aspects of this research are vitally important, since chronic pain is an ongoing and usually degenerative condition. Our research approach contributes to VR research in several ways. First, it uniquely focuses on long-term, chronic pain through an approach we term "self-modulation." Second, in the way we are integrating VR with biofeedback and meditation, we address the six ways that are recommended for coping with chronic pain: relaxation, biofeedback, cognitive restructuring, problem-solving, distraction and exercise (Turk & Nash, 1993).

Third, our phenomenological approach both provides an approach that can account for the multidisciplinary range of expertise in our group, and strongly parallels the biopsychosocial approach to chronic pain — the approach most widely used in pain management. Because of these alignments, our research may prove to be a useful tool in the long-term management of chronic pain.

Figure 5. Virtual meditative walk

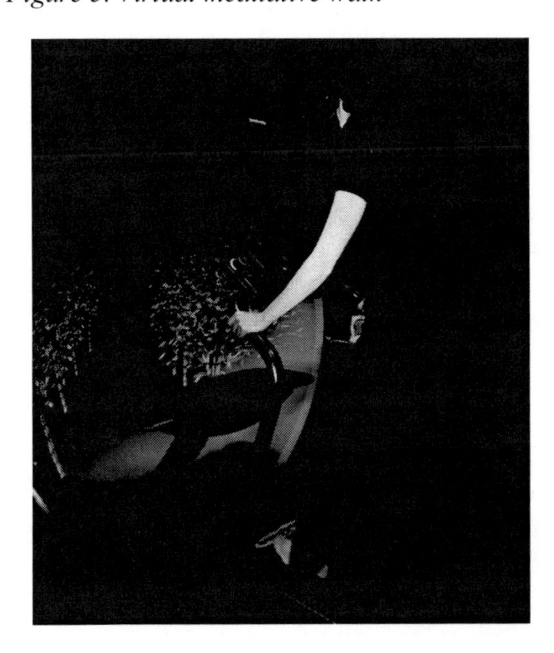

REFERENCES

Blyth, F. M., March, L. M., Brnabic, A. J., Jorm, L. R., Williamson, M., & Cousins, M. J. (2001). Chronic pain in Australia: A prevalence study. *Pain, 89*(2-3), 127–134. doi:10.1016/S0304-3959(00)00355-9

Bolter, J., & Gromala, D. (2005). *Windows and mirrors: Interaction design, digital art and the myth of transparency.* Cambridge, MA: MIT Press.

Botella, C. M., Juan, M. C., Baños, R. M., Alcañiz, M., Guillén, V., & Rey, B. (2005). Mixing realities? An application of augmented reality for the treatment of cockroach phobia. *Cyberpsychology & Behavior, 8*(2), 162–171. doi:10.1089/cpb.2005.8.162

Brinkman, W. van der Mast, C., & de Vliegher, D. (2008). Virtual reality exposure therapy for social phobia: A pilot study in evoking fear in a virtual world. *Proceedings of the 22nd HCI 2008 Workshop - HCI for Technology Enhanced Learning* (pp. 29-35), Liverpool, UK.

Canadian Pain Coalition (2007). Retrieved November 7, 2005, from http://www.rsdcanada.org/parc/english/resources/coalition.htm.

De Boeck, J., Raymaekers, C., & Coninx, K. (2006). Exploiting proprioception to improve haptic interaction in a virtual environment. *Presence (Cambridge, Mass.), 15*(6), 627–636. doi:10.1162/pres.15.6.627

Elkins, J. (1999). *Pictures of the body: Pain and metamorphosis.* Stanford: Stanford University Press.

Emmelkamp, P. (2006). Post-traumatic stress disorder: Assessment and follow-up. In Roy, M. J. (Ed.), *Novel approaches to the diagnosis and treatment of posttraumatic stress disorder* (pp. 309–320). Washington: IOS Press.

Gatchel, R. (2009). Biofeedback as an adjunctive treatment modality in pain management. *American Pain Society Bulletin, 14* (4). Retrieved March 10, 2009, from http://www.ampainsoc.org/pub/bulletin/jul04/clin1.htm

Gatchel, R., Peng, Y., & Peters, M. (2007). The biopsychosocial approach to chronic pain: Scientific advances and future directions. [Washington: American Psychological Association.]. *Psychological Bulletin, 133*(4), 581–624. doi:10.1037/0033-2909.133.4.581

Gromala, D. (2000). Pain and subjectivity in virtual reality. In Bell, D. (Ed.), *The cybercultures reader* (pp. 598–608). London: Routledge.

Gromala, D., & Sharir, Y. (1996). Dancing with the virtual dervish: Virtual bodies. In Moser, M., & MacLeod, D. (Eds.), *Immersed in technology: Art and virtual environments* (pp. 281–285). Cambridge, MA: MIT Press.

Hoffman, H. (2009). Virtual reality as an adjunctive pain control during burn wound care in adolescent patients. *Pain, 85*(1), 305–309. doi:10.1016/S0304-3959(99)00275-4

Hoffman, H. G., Garcia-Palacios, A., Carlin, C., Furness, T. A. III, & Botella-Arbona, C. (2003). Interfaces that heal: Coupling real and virtual objects to cure spider phobia. *International Journal of Human-Computer Interaction, 16*(2), 283–300. doi:10.1207/S15327590IJHC1602_08

Hoffman, H. G., & Patterson, D. (2005). Virtual reality pain distraction. *American Pain Society Bulletin 15* (2). Retrieved March 31, 2009, from http://www.ampainsoc.org/pub/bulletin/spr05/inno1.htm.

Hoffman, H. G., Patterson, D. R., & Carrougher, G. J. (2000). Use of virtual reality for adjunctive treatment of adult burn pain during physical therapy: A controlled study. *The Clinical Journal of Pain, 16*(3), 244–250. doi:10.1097/00002508-200009000-00010

Hoffman, H. G., Richards, T., Coda, B., Bills, A., Blough, D., Richards, A., & Sharar, S. (2004a). Modulation of thermal pain-related brain activity with virtual reality: Evidence from fMRI. *Neuroreport, 15*(8), 1245–1248.

Hoffman, H. G., Sharar, S., & Coda, B. (2004b). Manipulating presence influences the magnitude of virtual reality analgesia. *Pain, 111*(1–2), 162–168. doi:10.1016/j.pain.2004.06.013

Kabat-Zinn, J. (2006). *Coming to our senses: Healing ourselves and the world through mindfulness.* New York: Hyperion.

Krijn, M., Emmelkamp, P., Ólafsson, R., Bouwman, M., van Gerwen, M. L., & Spinhoven, P. (2007). Fear of flying treatment methods: Virtual reality exposure vs. cognitive behavioral therapy. *Aviation, Space, and Environmental Medicine, 78*(2), 121–128.

Loeser, J. D., Butler, S. H., Chapman, C. R., & Turk, D. C. (Eds.). (2001). *Bonica's Management of Pain* (3rd ed.). Philadelphia: Lippincott Williams & Wilkins.

Lutz, A., Slagter, H., Dunne, J. D., & Davidson, R. J. (2008). Attention regulation and monitoring in meditation. *Trends in Cognitive Sciences, 12*(4), 163–169. doi:10.1016/j.tics.2008.01.005

Mahrer, N., & Gold, J. (2009). The use of virtual reality for pain control: A review. *Current Pain and Headache Reports, 13*(2), 100–109. doi:10.1007/s11916-009-0019-8

McCaul, K. D., & Malott, J. M. (1984). Distraction and coping with pain. *Psychological Bulletin, 95*(3), 516–533. doi:10.1037/0033-2909.95.3.516

Melzack, R. (1990). The tragedy of needless pain. *Scientific American, 262*(2), 27–33. doi:10.1038/scientificamerican0290-27

Melzack, R., & Wall, P. D. (1996). *The challenge of pain* (2nd ed.). London: Penguin Books.

Morris, D. (1993). *The culture of pain.* Berkeley: University of California Press.

Pridmore, S. (2002). *Managing chronic pain A biopsychosocial approach.* London: Martin Dunitz Ltd.

Robillard, G., Bouchard, S., Fournier, T., & Renaud, P. (2003). Anxiety and presence during VR immersion: A comparative study of the reactions of phobic and non-phobic participants in therapeutic virtual environments derived from computer games. *Cyberpsychology & Behavior, 6*(5), 467–476. doi:10.1089/109493103769710497

Rothbaum, B., Hodges, L., Smith, S., Lee, J. H., & Price, L. (2000). A controlled study of virtual reality exposure therapy for the fear of flying. *Journal of Consulting and Clinical Psychology, 68*(6), 1020–1026. doi:10.1037/0022-006X.68.6.1020

Russo, C., & Brose, W. (1998). Chronic pain. *Annual Review of Medicine, 49*, 123–133. doi:10.1146/annurev.med.49.1.123

Scarry, E. (1987). *The body in pain.* Oxford: Oxford University Press.

Schatman, M., & Campbell, A. (2007). *Chronic pain management: Guidelines for multidisciplinary program development.* New York: Informa Healthcare.

Shaw, C., Gromala, D., & Fleming Seay, A. (2007). The Meditation Chamber: Enacting autonomic senses. *Proc. of ENACTIVE/07, 4th International Conference on Enactive Interfaces,* Grenoble, France, 19-22 November 2007, 405-408.

Song, M. (2009). *Virtual reality for cultural heritage applications.* Saarbrücken: VDM-Verlag Dr. Muller.

Steele, E., Grimmer, K., Thomas, B., Mulley, B., Fulton, I., & Hoffman, H. (2003). Virtual reality as a pediatric pain modulation technique: A case study. *Cyberpsychology & Behavior, 6*(6), 633–638. doi:10.1089/109493103322725405

Transforming Pain Research Group (2010). Retrieved April 01 2010 from http://www.transformingpain.org.

Turk, D. C., & Nash, J. M. (1993). Chronic pain: New ways to cope. In Goleman, D., & Gurin, J. (Eds.), *Mind body medicine*. New York, NY: Consumer Reports Books.

Varela, F., Thompson, E., & Rosch, E. (1992). *The embodied mind: Cognitive science and human experience*. Cambridge, MA: MIT Press.

Villanueva, L. (2009). Diffuse Noxious Inhibitory Control (DNIC) as a tool for exploring dysfunction of endogenous pain modulatory systems. *Pain, 143*(3), 161–162. doi:10.1016/j.pain.2009.03.003

Williams, N., Wilkinson, C., Stott, N., & Menkes, D. B. (2008). Functional illness in primary care: Dysfunction versus disease. *BMC Family Practice, 9*(30).

ADDITIONAL READING

Cahn, B. R., & Polich, J. (2006). Meditation states and traits: EEG, ERP, and neuroimaging studies. *Psychological Bulletin, 132*(2), 180–211. doi:10.1037/0033-2909.132.2.180

Gardner-Nix, J. (2009). *The mindfulness solution to pain*. Oakland, CA: New Harbinger Publications, Inc.

Gromala, D., Shaw, C., & Song, M. (2009). Chronic pain and the modulation of self in immersive virtual reality. *Biologically-Inspired Cognitive Architectures II: Papers from the AAAI Symposium.(FS-09-01)*, Association for the Advancement of Artificial Intelligence (www.aaai.org), Washington, D.C., U.S.A.

Ihde, D. (2002). *Bodies in technology*. Minneapolis: University of Minnesota Press.

Jackson, M. (2002). *Pain: The fifth vital sign*. New York, NY: Crown Publishers.

Kabat-Zinn, J., Lipworth, L., & Burney, R. (1985). The clinical use of mindfulness meditation for the self-regulation of chronic pain. *Journal of Behavioral Medicine, 8*(2), 163–190. doi:10.1007/BF00845519

Lynch, D. (2008). Do we care about people with chronic pain? *Pain Research and Management: The Journal of the Canadian Pain Society, 13*(6), 465–476.

Riva, G. (2002). Virtual reality for health care: The status of research. *Cyberpsychology & Behavior, 5*(3), 219–225. doi:10.1089/109493102760147213

Wall, P. (2000). *Pain: The science of suffering*. New York, NY: Columbia University Press.

KEY TERMS AND DEFINITIONS

Acute Pain: Pain that arises quickly, and that can usually be directly attributed to present injury or disease. Although it can be severe, it is generally defined to exist no longer than 30 days.

Biofeedback: A method of treatment that enables users to become aware of and influence various physiological functions of which they are normally unaware. Instruments monitor and provide visual or sonic information about the continuously changing functions, such as heart rate, respiration, galvanic skin response, blood pressure and brain wave activity, to name a few.

Chronic Pain: Pain that persists or recurs for six months or longer. In many cases, the cause of the pain is difficult to diagnose. Recent research suggests that chronic pain is not a symptom, but a systemic dysfunction or hypersensitivity. Over long periods of time, chronic pain is degenerative in complex physical, psychological and social ways.

Meditation: A practice in which the practitioner attempts to get beyond the reflexive, "thinking" mind into a deeper state of relaxation or awareness.

Pain Distraction: The act of directing conscious effort upon a task: thus directing attention away from a painful action or experience.

Pain Modulation: The reduction of the sensation of pain by the act of intervention by pharmacological or other methods.

Virtual Reality: An interactive, three-dimensional, computer simulated environment presented through a stereoscopic, wide field-of-view visual display such as a head-mounted display (HMD), or a multi-screen surrounding projection (CAVE). Spatialized sound and haptics are sometimes used to enhance the simulation.

ENDNOTES

[1] Contemporary phenomenologists who continue the work of Merleau-Ponty and are referred to in our research include: Patricia Benner, Robert Bosnak, Andy Clark, Thomas Csordas, Paul Dourish, Herbert Dreyfus, Natalie duPraz, Diana Fosha, Raymond Gibbs, Mark Hansen, Shaun Gallagher, Don Idhe, Mark Johnson, Drew Leder, Alva Noe, Gail Weiss, Iris Young, and Dan Zahavi, to cite a few.

[2] The most reliable measure is electroencephalography (EEG); however, this device requires 24 or more carefully measured points of direct contact on the scalp, and thus was not immediately viable for a large group of users. Further, EEG measures of a meditative state are very close to those of an incipient epileptic seizure.

Chapter 8
A Behavior Model Based on Personality and Emotional Intelligence for Virtual Humans

Héctor Orozco
Centro de Investigación y de Estudios Avanzados del I.P.N., México

Félix Ramos
Centro de Investigación y de Estudios Avanzados del I.P.N., México

Daniel Thalmann
École Polytechnique Fédérale de Lausanne, Switzerland

Victor Fernández
Centro de Investigación y de Estudios Avanzados del I.P.N., México

Octavio Gutiérrez
Centro de Investigación y de Estudios Avanzados del I.P.N., México & Grenoble Institute of Technology, France

ABSTRACT

In this chapter, we present a three-layered model based on the Triune Brain Model to simulate human brain functioning and human beings' behavior in realistic virtual humans. In order to implement this model, we use the ten personality scales defined by Minnesota Multiphasic Personality Inventory and the Emotional Competence Framework defined in the Emotional Intelligence Model to endow virtual humans with a real personality and emotional intelligence. In this model, we apply a set of fuzzy rules to change and regulate virtual humans' affective state according to their personality, emotional and mood history, and events they perceive from the environment. We also implement an EBDI-based intention selection using the Event Calculus formalism. This intention selection mechanism allows virtual humans performing actions based on their current affective state, beliefs, desires and intentions. Thus, these intentions define virtual humans' behavior for each situation they experience in the environment.

DOI: 10.4018/978-1-60960-077-8.ch008

INTRODUCTION

In Psychology, one of the most studied human factors that influence people's behavior is Personality. The term personality is refers to *"the set of psychological traits and mechanisms within the individual that are organized and relatively stable over time"* (Atkinson & Hilgard, 1996). These traits influence individuals' interactions and their adaptation to the environment. Psychological or personality traits represent forms of persistent patterns of perceiving, relating and thinking about the environment and oneself. These features distinguish one person from another and they are reflected in individuals' behavior in a wide range of contexts such as social and personal. In this manner, the characterization of individuals' behavioral patterns is used to explain in a more clear way how and why people react differently to the same events in their environment. In this chapter, we propose a three-layered model based on the Triune Brain Model (MacLean, 1973) to simulate human brain functioning and human beings' behavior in virtual humans. According to this model, human brain is formed by three-layered brains in one. Where, each of these brains emerged successively in the course of time in response to evolutionary needs. These brains are labeled as *reptilian brain* or *R-complex*, *mammalian brain* or *limbic system*, and *neocortex*, respectively. Each layer is geared toward separate functions of brain without operating independently of one another, but they interact by means of establishing numerous interconnections through which they influence one another. With the implementation of this model, we endow virtual humans with the ability of interacting in their environment according to their personality and emotional intelligence by controlling their affective state (emotional and mood states) taking into account their beliefs, desires and intentions. In our model, we use Minnesota Multiphasic Personality Inventory (MMPI) (Tellegen et al., 2003; Simms et al., 2005; Osberg, Haseley & Kamas, 2008). This inventory is one of the most widely used in psychometric practice and Psychology, because it is the best tool to assess individuals' major personality traits and their emotional disorders. The main objective of applying this test is to identify both individuals' personality profile and the detection of their behaviors and possible psychopathologies. Thus, the proposed model is valid from a psychological point of view, because it is supported by studies on Personality and resources provided by these studies. We take into account the most important behavior tendencies in each level of intensity (*very high*, *high*, *medium*, *low* and *very low*) of ten personality scales defined by MMPI. These scales are the following: *hypochondriasis (Hs), depression (D), hysteria (Hy), psychopathic deviate (Pd), masculinity-femininity for female and male (MF), paranoia (Pa), psychasthenia (Pt), schizophrenia (Sc), hypomania (Ma),* and *social introversion (Si)*. We use the level of intensity of these scales to form different sets of fuzzy rules. These fuzzy sets are used to obtain an emotional influence, a mood influence over emotions, an emotional regulation, a mood influence, an emotional influence over moods, and a mood regulation, which modify and regulate virtual humans' affective state according to their personality, emotional history and mood history, taking into account the level of intensity of perceived events from their environment.

We also implement an Emotional-Belief-Desire-Intention-based action selection using the Event Calculus formalism to endow virtual humans with the ability to behave and perform intelligent actions based on their current affective state, beliefs, desires and intentions. These actions define virtual humans' behavior for each situation they experience in the environment. In this way, virtual humans' personality represents a set of personal characteristics that influence their cognition, motivation and behavior in different situations. Therefore, this set of values indicates the way virtual humans should react according to the situation. Our approach applies the principles and original concepts of Emotional Intelligence Model (Salovey & Mayer, 1990; Goleman, 1995; Mayer & Geher, 1996; Salovey & Sluyter, 1997) and it provides the necessary tools to endow virtual humans with an autonomous intelligence and other human beings' characteristics, such as personality, moods and emotions, with the aim of making them more believable and conceivable. Thus, Emotional Intelligence enables virtual humans to distinguish and manage their emotions and behave in a similar way as human beings do. For us, virtual humans' emotional intelligence represents their ability to perceive and express emotions, assimilate emotion-related feelings, understand and reason with emotion, and regulate emotions in themselves and other virtual entities. That is to say, it refers to *the ability to recognize the meanings of emotions and their relationships*. Thereby, this ability includes: self-consciousness, goal understanding, intentions, reactions and behaviors, also consciousness of other virtual humans and their feelings.

We also implement the *Emotional Competence Framework* (Goleman, 1998; Mayer, Salovey and Caruso, 2000) defined in the Emotional Intelligence Model. We use this framework to apply virtual humans' personal and social competencies on the basis of the following characteristics: *Personal Competencies*: *self-consciousness* (awareness of emotions, accurate self-assessment and self-confidence), *self-regulation* (self-control, self-consciousness and adaptability) and *self-motivation* (achievement tendency and commitment to the task), and *Social Competencies*: *social awareness* (empathy, service orientation, awareness of other avatars and political awareness) and *social skills* (communication, conflict management and cooperation with other entities).

The present chapter is organized as follows: Next section, presents a study of the most important emotional architectures, personality theories and behavior models applied to intelligent agents. In the third section, we propose a new three layered model to simulate human brain functioning and human beings' behavior in virtual humans. As case study, in this section, we present a scenario where a male virtual human with a depressive personality and a female virtual human with a paranoid personality are interacting in a realistic way according to their personality, emotional intelligence, beliefs, desires, intentions, and the events they perceive from their environment. Finally, in the last section, we give our concluding remarks.

RELATED WORK

Several considerations must be taken into account when human characteristics such as personality (Rizzo, Veloso, Miceli & Cesta, 1997; Schmidt 2000), moods (Bless & Schwarz, 1999) and emotions (Vélazquez, 1997; Martinez-Miranda & Aldea, 2005), are added into agents with the goal of making them more conceivable and believable. Hence, good behavior models based on personality and emotions can help us to design and build better software agents, which approach human beings' behavior (Ören & Yilmaz, 2004, Murakami, Sugimoto & Ishida, 2005), in order to adapt and survive with more success in hostile and unpredictable environments that change over time.

Emotional Architectures

Many psychological models have been proposed to describe the emotional functioning of human brain and the different human being's mental processes (McCauley, Franklin & Bogner, 2000; Gratch & Marsella, 2001; Imbert & de Antonio, 2005; Liu & Pan, 2005; Bosse, Pointer & Treur, 2007). Traditionally, the OCC model (Ortony, Clore & Collins, 1988) has been considered as the standard model for emotion synthesis and the best categorization of emotions available. In this model, a set of 22 emotions are interpreted as positive or negative reactions to either consequence of events, or actions of agents, or aspects of objects. Thus, the OCC model explains some human emotions and tries to predict under certain situations, what emotions can be investigated. Though this model is rather good, it does not explain completely the origin of emotional processes and does not present how to filter mixed emotions to obtain a coherent emotional state.

Another approach was presented in Frijda's theory of emotion (Fridja, 1987). Frijda's theory points out that emotion does not refer to a natural class and that it is not able to refer to a defined class of phenomena, which are clearly distinguishable from other mental and behavior events. The central idea of this theory is the term *concern*. A concern represents the predisposition of a system to prefer certain states of the environment, that is to say, a concern can produce goals and preferences for the system. Thus, the intensity of emotions is determined essentially by influences generated by relevant concerns.

Roseman, Jose and Spindel (1990) designed a model of emotion, where they described emotions on basis of distinct event classifications. In their model, is evaluated the certainty of the occurrence of an event considering its causes. Thus, software agents may use emotions to facilitate social interactions, relationships and communications among them. This fact helps to coordinate tasks, such as cooperating among communities of agents.

LeDoux (1996) presented in his book *"The Emotional Brain"*, a neurological model of human brain. This model allowed exploring emotional processes in human brain, and emotions were described in terms of desires and expectations. By taking LeDoux's ideas, emotional agents have been used to simulate human beings' reasoning by means of the influence and effect of primary and secondary emotions (Jiang & Vidal, 2006). These emotions are used in the decision making process of agent architectures. Some works have increased the interest for the use of computational models of emotion to incorporate emotions into rational agents. Thus, Emotional-Belief-Desire-Intention (EBDI) agents (Pereira, Oliveira, Moreira & Sarmento, 2005) have better performance than rational agents, because they are more flexible and adaptable, allowing them to survive in dynamic and complex environments. Searching for a better model of emotion, the Fuzzy Logic Adaptive Model of Emotions (FLAME) is used to produce emotions and simulate emotional intelligence processes (El-Nasr, Yen & Ioerger, 2000). This model is based on fuzzy rules used to explore the capability of fuzzy logic for modeling emotional processes. These fuzzy rules are used for mapping from events to emotions, and from emotions to behaviors.

Khulood and Raed (2007) proposed other emotional agent architecture based on a symbolic approach. This symbolic approach was implemented using symbolic emotional rule-based systems (rules that generate emotions) with continuous interactions with the environment considering inferences, evaluation, adaptation, learning and emotions.

Personality Theories

Although there are several personality theories that seem to be right, we support the idea that human beings have a limited set of personality traits that influence their behavior, so individuals may exhibit a different behavior for any experienced situation. Personality traits, moods and emotions are very essential to create more believable synthetic agents. Some research works in Artificial Intelligence have used different sets of personality traits to simulate intelligent agents able to behave in a believable and consistent way. One of these works is the OZ project (Bates, Loyall & Reilly, 1998). This project was created to exploit behavioral features expressing personality, which may be modified over time at different situations. These features are used to endow synthetic characters with a behavior. Nevertheless, the set of behavioral features used in this work does not have a logical structure and it is restricted to the exposed examples by the authors. Most of the proposed personality models are based on trait theories, because the conversion from trait dimensions to an efficient computational model is very easy. These models consist of a set of dimensions, where each dimension represents a set of personality traits. One of the most widespread is the OCEAN model (McCrae & John, 1992). This model groups different personality traits in five dimensions: *Openness, Conscientiousness, Extraversion, Agreeableness* and *Neuroticism*. Where, each dimension represents a set of specific personality traits that correlate together. Although this model is widely accepted, it has many criticisms, because it does not exactly indicate how Personality affects human behavior based on the obtained stimuli, perceived events and different experienced situations.

A computational model of personality with personality change is explained in (Poznanski & Thagard, 2005). This model uses a neural network for simulating personality over time and intends to be used as an application in a Sim-type video game. But, this work requires establishing a set of psychologically inspired rules to determine which situations change the personality and in what ways. Other different mechanism to add a dynamic personality and an openness personality trait into agents is presented in (Ghasen-Aghaee & Oren, 2007). This work is based on the fact that openness has implications on cognitive complexity and the decision making ability of agents. In addition, this chapter implemented a fuzzy agent to show personality descriptors, personality factors, personality style and problem solving success consequently. Moreover, it included a prototype system to demonstrate how openness affects agents' problem solving ability.

Behavior Models for Intelligent Agents

A social-psychological model was proposed by Rousseau and Hayes-Roth (1997). In this model, mood is divided into two categories: agent-oriented moods, and self-oriented moods. The first one is directed toward other individuals, whereas the second one is not. The application of these two categories may make an agent rather happy for a long time or remain angry with other agent, for instance, due to a cause that induces the second agent to the first one.

In (Liu & Lu, 2008) is presented a computational model of motivation. This model integrates personality, motivation, emotion, behavior and stimuli together. In spite of the fact that this model shows how motivation and Personality drive virtual characters' emotions, it only gives a primary outline for a motivation model and it is restricted to a 3D facial animation system.

A new framework based on Artificial Intelligence for decision making is introduced by Iglesias and Luengo (2007). This framework is used to produce animations of virtual avatars evolving autonomously within a 3D environment. The exposed animations in this framework lacks of realism, because avatars follow a behavior pattern from the point of view of a human observer.

A model of individual spontaneous reactions for virtual humans is proposed in (Garcia-Rojas, Gutiérrez & Thalmann, 2008). This model was defined by analyzing real people reacting to unexpected events. This model presents a semantic-based methodology to compose reactive animation sequences using inverse kinematics and key frame interpolation animation techniques. Nevertheless, this model was created in a subjective way in according to authors' personal judgment. In addition, the reaction types and animation sequences virtual humans perform are not validated from the point of view of psychology. However, obtained results are satisfactory.

An interesting approach that describes emotions, moods, personality and their interdependencies using vector algebra is proposed in (Egges & Magnenat-Thalmann, 2004). However, many of the revised works, which address the use of Personality in virtual creatures' behavior, make mistakes when assigning random values to different basic personality traits. The allocation of these values cannot be supported, because the used theoretical framework does not make sense from a psychological point of view, indicating that so far none of the existing works provides an accurate and reliable mechanism for modeling human behavior.

Due to the direct correspondence between emotions and facial expressions (Hong, 2008), many researchers prefer to employ Ekman's six basic emotions (Ekman, 1994) (*anger*, *disgust*, *fear*, *happiness*, *sadness* and *surprise*) for facial expression classification and the OCEAN model, or else the OCC model in combination with the OCEAN model. Thus, mutual dependence between emotions and personality is often represented by Bayesian belief networks (Ball & Breese, 2000; Kshirsagar & Magnenat-Thalmann, 2002).

A BEHAVIOR MODEL BASED ON PERSONALITY AND EMOTIONAL INTELLIGENCE FOR VIRTUAL HUMANS

In this work, we propose a new model with three interrelated layers based on the Triune Brain Model (MacLean, 1973). Our model permits generating more believable and realistic behaviors for virtual humans. These behaviors are based on Personality and Emotional Intelligence. A brief view of the operating cycle of our model (see Figure 1) is the following:

- Virtual humans perceive events from their environment through sensors. Sensors transmit values (positives or negatives) of all captured events to them.
- Thus, once an event has been perceived, this is unconsciously processed by virtual humans' reptilian brain generating an instinctive reaction that may be of two types: a reflex reaction or an instinctive reaction of protection.
- Next, once the event has been received, this unconsciously generates an affective state (emotional and mood states) update based on virtual humans' personality and events' intensity (positive or negative) in virtual humans' mammalian brain to generate an emotional response.
- Immediately, in parallel to the generated affective state update process, the emotional response and the instinctive reaction, virtual humans become aware of the perceived events and interpret their affective state to decide whether they have received events that catches their attention.

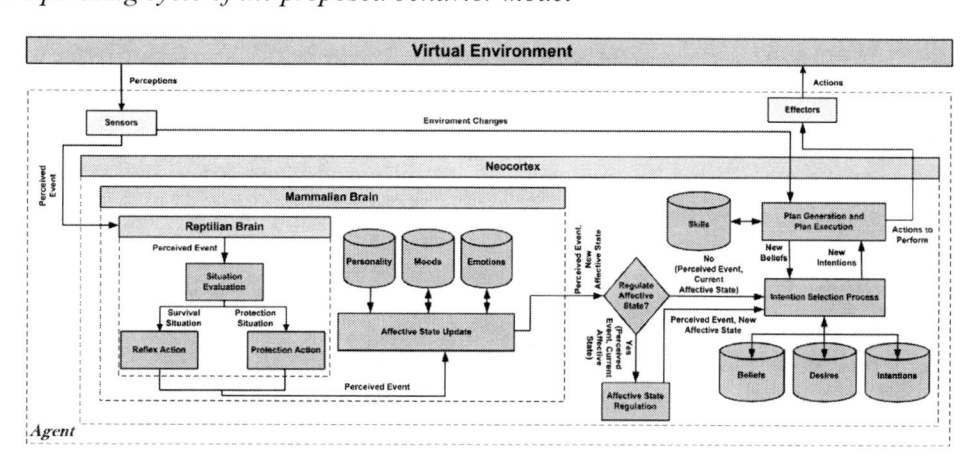

Figure 1. Operating cycle of the proposed behavior model

- Thus, virtual humans become aware of the perceived events and search for an explanation by means of looking for information from their beliefs and long and short term memories.
- Once virtual humans collect necessary information, they can regulate their affective state applying their emotional intelligence to control their emotional and mood states and then to show a better behavior that is consistent with their personality.
- Next, virtual humans update their desires and intentions.
- Based on new desires and intentions, virtual humans choose to execute a set of actions that considers the most appropriate to the situation. These actions are generated through a dynamic planning.
- Finally, virtual humans are capable of evaluating obtained results from their exhibited behavior and learn about them. For example, if certain beliefs result to be false in an experienced situation, then virtual humans update them.

Event Perception and Situation Evaluation Process

Virtual humans perceive through sensors, and the information provided by them is interpreted as the occurrence of a set of events. These events contain domain knowledge that provides the level of intensity (positive or negative) as well as environmental information. Then, in this way, perception represents the beginning of the primary cause of action and behavior in virtual humans. Thus, the way virtual humans perceive their environment affects their affective state, decisions and actions according to their current emotional state, beliefs, desires and intentions.

When a virtual human perceives an event from its environment, this event is firstly processed by the reptilian brain in a quickly and unconscious way, to generate an automatic reaction that could be either a reflex action or a protection action. Both reactions are predefined and their sensorial intensity is perceived from the captured event. The sensorial information associated to an event e, has two parameters: the level of intensity (int) and environmental information (inf), then a definition of a captured event is denoted as follows: $e(int, inf)$.

Virtual humans' perception is represented by a generator of events, whose formalization is given in Event Calculus (Kowalski & Sergot, 1986), as follows:

$$PERCEPTION = \sum Happens(e(int,inf), t_i)$$

That is to say, perception is the sum of all events occurred in the environment.

Further details of Event Calculus and how these events affect the virtual humans' affective state will be given in future sections.

Affective State Update Process

We propose a different way to update virtual humans' emotional and mood states using the following: MMPI's personality scales, Ekman's basic emotions, two basic moods (*good mood* and *bad mood*) and the level of intensity of perceived events (*positive* or *negative*). For this matter, we consider virtual humans as entities with a dynamic behavior, which is constantly changing over time t. Therefore, virtual humans' personality p is constant over time and it is initialized with the level of intensity of each personality scale defined by MMPI at time $t = 0$. Virtual humans' emotional and mood states, e_t and m_t respectively, are dynamic over time and these are initialized to 0 at time $t = 0$. We formalize these concepts as follows:

$$p = [Hs,D,Hy,Pd,MF,Pa,Pt,Sc,Ma,Si]$$

where: $Hs, D, Hy, Pd, MF, Pa, Pt, Sc, Ma$ and $Si \in [0, 1]$ are the ten personality scales defined by MMPI. The emotional state e_t represents the intensities of six Ekman's basic emotions at each time t. These emotions are grouped in a 6-dimensional vector in the following way:

e_t = [anger, disgust, fear, happiness, sadness, surprise] if $t > 0$ or $e_t = 0$ if $t = 0$

where: *anger, disgust, fear, happiness, sadness* and *surprise* $\in [-1, 1]$.

In a similar way, the mood state m_t represents the intensities of two basic moods (good mood and bad mood) at each time t. These moods are grouped in a 2-dimensional vector as follows:

m_t = [goodMood, badMood] if $t > 0$ or $m_t = 0$ if $t = 0$

where: *goodMood* and *badMood* $\in [-1, 1]$.

We also use an emotional history ω_t and a mood history σ_t, which contain the emotional states e_0 until e_t and the mood states m_0 until m_t, respectively. Once a virtual human has perceived an event, an affective state update process is carried out in order to change its emotional and mood states. These states are updated in two steps: the first step consists in updating the emotional state; and the second one consists of updating the mood state.

In order to obtain a new affective state, we use jFuzzyLogic (Cingolani, 2009), which is a java package that offers a complete fuzzy inference system (FIS). This package implements a fuzzy control language (FCL) specification according to (IEC 1131, 1997). We define in this language a set of input and output variables and a set of fuzzy rules used to update virtual humans' emotional and mood states. We use fuzzy logic, because events, personality and affective states can be modeled with fuzzy limits. This allows us to change quickly virtual humans' emotional and mood states in a more natural manner, generating more realistic behaviors in real time. Next, we describe how to update and regulate emotional and mood states.

Emotional State Update

A new emotional state (e_{t+1}) over time t is calculated taking into account virtual humans' personality, a perceived event and virtual humans' mood history as follows:

$$e_{t+1} = e_t + I_e(p, event) + M_e(\sigma_t, event)$$

We firstly compute the function $I_e(p, event)$ to obtain an emotional influence vector that contains a desired change of intensity for each of the six basic emotions based on virtual humans' personality and the perceived event. To obtain this vector we define a set of fuzzy rules. We use a set of ten input variables corresponding to each personality scale defined by MMPI, for these variables, we define five fuzzy sets: *very_low, low, medium, high*, and *very_high* (see Figures 2 and 3). An event is represented with the input variable *event*, which has seven fuzzy sets: *negative_high, negative_medium, negative_low, neutral, positive_low, positive_medium*, and *positive_high* (see Figure 4). Thus, the impact of a perceived event over each emotion in the emotional influence vector is described by six output variables associated to each one of the six basic emotions respectively. For these variables, are defined eleven fuzzy sets: *negative_very_high, negative_high, negative_medium, negative_low, negative_very_low, neutral, positive_very_low, positive_low, positive_medium, positive_high*, and *positive_very_high* (see Figure 4). In order to determine the emotional influence we have chosen the level of intensity of the most predominant virtual humans' personality scale and the level of intensity of the perceived event, to form fuzzy rules of the form given below:

IF *Predominant(p) = scale$_i$* IS A_j AND *Intensity(event)* IS B_k THEN

Influence(anger) IS C_1, *Influence(disgust)* IS C_2, *Influence(fear)* IS C_3,

Figure 2. Fuzzy sets used for the personality scales hypochondriasis, depression, hysteria, psychopathic deviate and masculinity/femininity (for female and male)

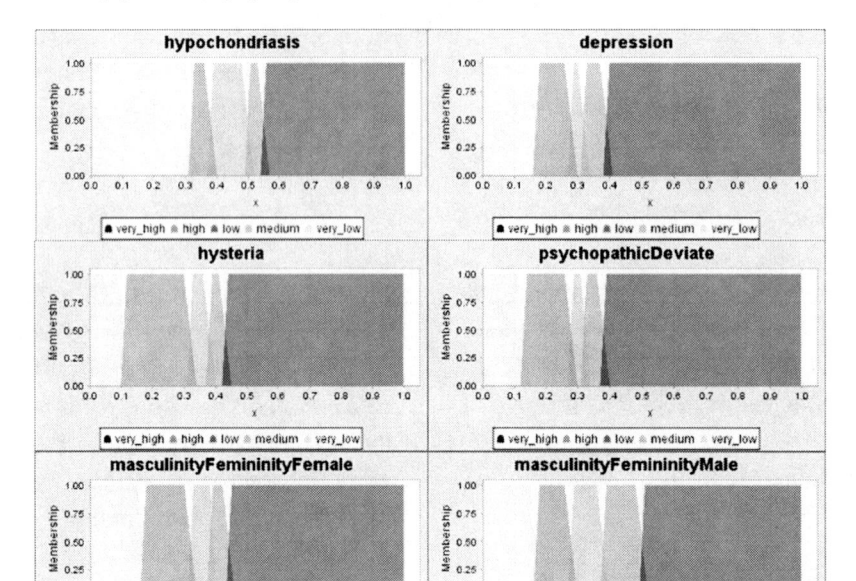

Figure 3. Fuzzy sets used for the personality scales paranoia, psychopathic deviate, schizophrenia, hypomania and social introversion

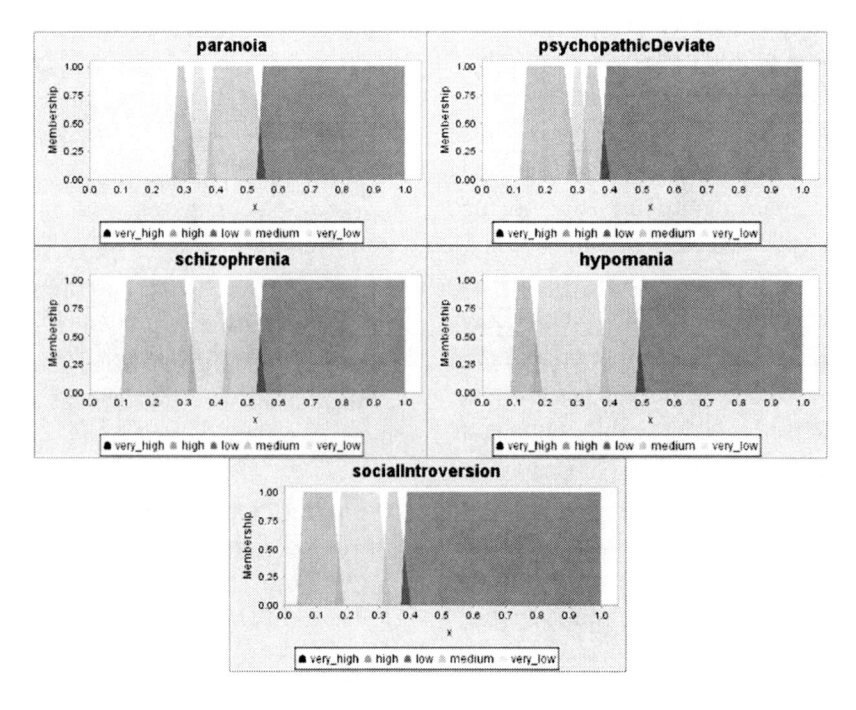

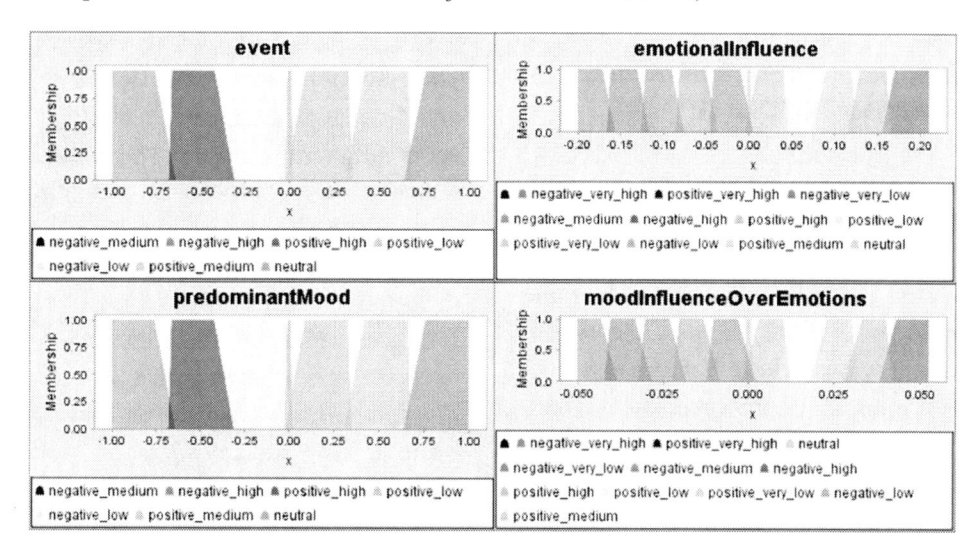

Figure 4. Fuzzy sets used to update the emotional state (perceived event, emotional influence over each basic emotion, predominant mood and mood influence over emotions)

Influence(happiness) IS C_4, *Influence(sadness)* IS C_5, *Influence(surprise)* IS C_6;

where A_j is the level of intensity of the most predominant personality scale $scale_i \in p$, B_k is the level of intensity of the perceived event, and C_1, C_2, ..., C_6 are the emotional influences to each of the six basic emotions. A_j, B_k, C_1, C_2, ..., C_6 represent the fuzzy sets previously described. An example of the above rules for a predominant medium psychopathic personality and an event perceived as high negative is the following:

IF *psychopathicDeviate* IS *medium* AND *event* IS *negative_high* THEN

anger IS *neutral*, *disgust* IS *neutral*, *fear* IS *positive_low*,

happiness IS *neutral*, *sadness* IS *neutral*, *surprise* IS *neutral*;

Next, we obtain other vector that represents the mood influence over the emotional state (how moods influence emotions) taking into account virtual humans' mood history (current mood state) and the perceived event. To do this, we define a function $M_e(\sigma_t, event)$, which applies a set of fuzzy rules. We use two input variables: *goodMood* and *badMood* that correspond to the two types of moods associated to virtual humans. For these moods we define seven fuzzy sets: *negative_high*, *negative_medium*, *negative_low*, *neutral*, *positive_low*, *positive_medium*, and *positive_high* (see Figure 4). The mood influence over the emotional state is described using six output variables for each one of the six basic emotions. For these variables, we define eleven fuzzy sets: *negative_very_high*, *negative_high*, *negative_medium*, *negative_low*, *negative_very_low*, *neutral*, *positive_very_low*, *positive_low*, *positive_medium*, *positive_high*, and *positive_very_high* (see Figure 4). In order to determine the mood influence over the emotional state, we have chosen the level of intensity of the most predominant virtual humans' mood and the level of intensity of the perceived event to form fuzzy rules as follows:

IF *Predominant(m)* = $mood_i$ IS D_j AND *Intensity(event)* IS B_k THEN

Influence(anger) IS E_1, *Influence(disgust)* IS E_2, *Influence(fear)* IS E_3,

Influence(happiness) IS E_4, *Influence(sadness)* IS E_5, *Influence(surprise)* IS E_6;

where D_j is the level of intensity of the most predominant mood $mood_i \in m$, B_k is the level of intensity of the perceived event (*event*), and E_1, E_2, ..., E_6 are the mood influences to each of the six basic emotions. D_j, B_k, E_1, E_2, ..., E_6 represent the fuzzy sets previously described. An example of the above rules for a predominant medium negative bad mood and an event perceived as high negative is the following:

IF *badMood* IS *negative_medium* AND *event* IS *negative_high* THEN

anger IS *positive_high*, *disgust* IS *positive_medium*, *fear* IS *positive_low*,

happiness IS *negative_high*, *sadness* IS *positive_low*, *surprise* IS *negative_medium*;

Once we have obtained the above two vectors, we add them to virtual humans' current emotional state e_t to obtain their new emotional state e_{t+1}. The above sets of rules are used to generate emotional and mood influences over the emotional state that change virtual humans' emotional state according to their personality and events perceived.

Mood State Update

A new mood state (m_{t+1}) is updated by a function that calculates the mood change based on a perceived event, virtual humans' personality and emotional history (new emotional state e_{t+1}) in the following way:

$m_{t+1} = m_t + I_m(p, event) + E_m(\omega_t, event)$

We first compute the function $I_m(p, event)$ to calculate a mood influence vector that contains a desired change of intensity for each of the two basic moods based on virtual humans' personality and the perceived event. To do this, we apply a set of fuzzy rules. In this way, based on virtual humans' personality the impact of a perceived event over each mood in the mood influence vector is described by two output variables associated to each one of the two basic moods respectively. For these variables, we define eleven fuzzy sets: *negative_very_high*, *negative_high*, *negative_medium*, *negative_low*, *negative_very_low*, *neutral*, *positive_very_low*, *positive_low*, *positive_medium*, *positive_high*, and *positive_very_high* (see Figure 5).

In order to obtain the mood influence vector we take into account the level of intensity of the most predominant virtual humans' personality scale and the level of intensity of the perceived event to define fuzzy rules in the following way:

IF *Predominant(p) = scale$_i$* IS A_j AND *Intensity(event)* IS B_k THEN

Influence(goodMood) IS F_1, *Influence(badMood)* IS F_2;

where A_j is the level of intensity of the most predominant personality scale $scale_i \in p$, B_k is the level of intensity of the perceived event, and F_1 and F_2 are the mood influence to each of the two basic moods. A_j, B_k, F_1 and F_2 represent the fuzzy sets previously mentioned. An example of the above rules for a predominant high paranoid personality and an event perceived as medium negative is the following:

IF *paranoia* IS *high* AND *event* IS *negative_medium* THEN

goodMood IS *neutral*, *badMood* IS *positive_medium*;

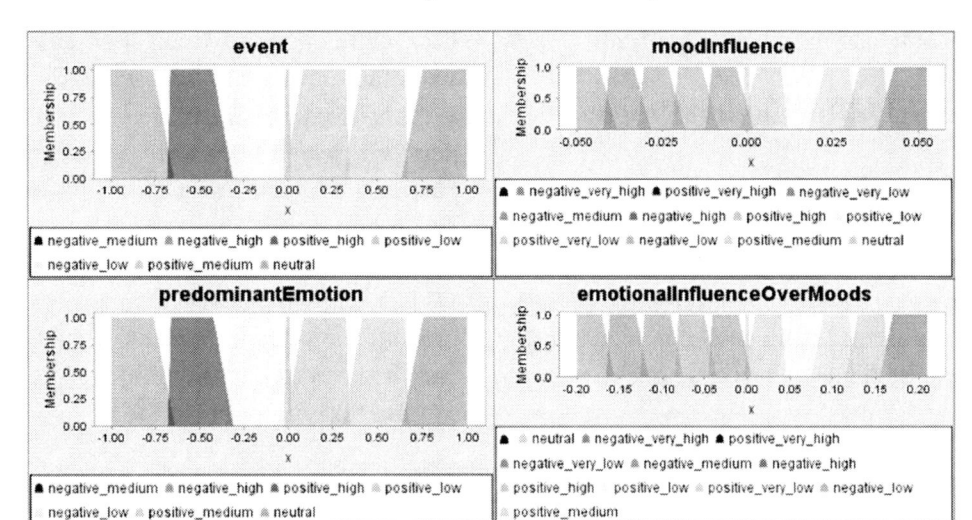

Figure 5. Fuzzy sets used to update the mood state (perceived event, mood influence over each basic mood, predominant emotion and emotional influence over moods)

Once we have obtained the mood influence vector, we calculate a second vector that indicates the emotional influence over the mood state (how emotions influence moods) considering the virtual humans' emotional history (current emotional state) and the perceived event. To calculate this vector, we compute the function $E_m(\omega_t, event)$ applying a set of fuzzy rules. We use six input variables that correspond to the six types of basic emotions associated to a virtual human. For these emotions, we define seven fuzzy sets: *negative_high, negative_medium, negative_low, neutral, positive_low, positive_medium,* and *positive_high* (see Figure 5). The emotional influence over the mood state is described by using two output variables for each one of the two basic moods. For these variables, we define eleven fuzzy sets: *negative_very_high, negative_high, negative_medium, negative_low, negative_very_low, neutral, positive_very_low, positive_low, positive_medium, positive_high,* and *positive_very_high* (see Figure 5).

Now, to obtain the emotional influence over the mood state, we have chosen the level of intensity of the most predominant virtual humans' emotion and the level of intensity of the perceived event to form fuzzy rules in the following way:

IF *Predominant(e) = emotion$_i$* IS G_j AND *Intensity(event)* IS B_k THEN

Influence(goodMood) IS H_1, *Influence(badMood)* IS H_2;

where G_j is the level of intensity of the most predominant emotion *emotion$_i$* $\in e$, B_k is the level of intensity of perceived event (*event*), and H_1 and H_2 are the emotional influence to each of the two basic moods. G_j, B_k, H_1, H_2 represent the fuzzy sets previously described. An example of the above rules for a predominant high paranoid personality and an event perceived as medium negative is the following:

IF *paranoia* IS *high* AND *event* IS *negative_medium* THEN

goodMood IS *neutral*, *badMood* IS *positive_medium*;

Next, once we have calculated the above two vectors, we add them to virtual humans' current mood state m_t to obtain their new mood state m_{t+1}. The above sets of rules are used to generate mood and emotional influences over the mood state that change virtual humans' mood state according to their personality and the events perceived.

Affective State Regulation Process

Based on their personality and emotional intelligence virtual humans can decide to regulate their affective state according to the level of intensity of a perceived event. This regulation process is done in two steps: the first step consists in regulating the emotional state; and the second one consists of regulating the mood state. Next, we show how to update emotional and mood states.

Emotional State Regulation

Current emotional state e_t is regulated taking into account virtual humans' personality and the intensity of the perceived event as follows:

$$e_t = e_t + R_e(p, event)$$

We compute the above function $R_e(p, event)$ to obtain an emotional regulation vector that contains an increment for some basic emotions and an decrement for the rest. This vector is used to regulate the level of intensity for each of the six basic emotions based on virtual humans' personality and the perceived event. To obtain this vector we define a set of fuzzy rules. In this way, the level of regulation for each emotion in the current emotional state is described by six output variables associated to each one of the six basic emotions respectively. For these variables, we define eleven fuzzy sets: *negative_very_high, negative_high, negative_medium, negative_low, negative_very_low, neutral, positive_very_low, positive_low, positive_medium, positive_high,* and *positive_very_high* (see Figure 6). In order to determinate the emotional regulation vector, we have chosen the level of intensity of the most predominant virtual humans' personality scale and the level of intensity of the perceived event to form fuzzy rules in the following way:

IF *Predominant(p) = scale$_i$* IS A_j AND *Intensity(event)* IS B_k THEN

Regulate(anger) IS I_1, *Regulate(disgust)* IS I_2, *Regulate(fear)* IS I_3,

Regulate(happiness) IS I_4, *Regulate(sadness)* IS I_5, *Regulate(surprise)* IS I_6;

where A_j is the level of intensity of the most predominant personality scale $scale_i \in p$, B_k is the level of intensity of perceived event, and $I_1, I_2, ..., I_6$ are the increment or decrement to each of the six basic emotions. $A_j, B_k, I_1, I_2, ..., I_6$ represent the fuzzy sets previously described. An example of these rules for a predominant very high hysteric personality and an event perceived as low negative is the following:

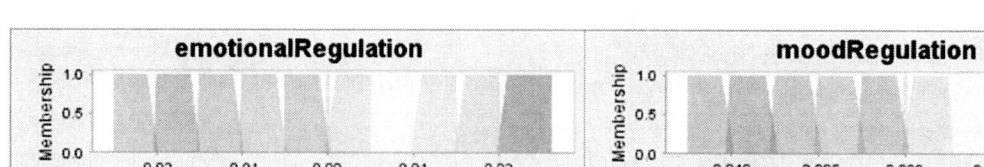

Figure 6. Fuzzy sets used to regulate the emotional and mood states

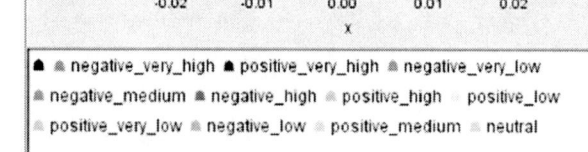

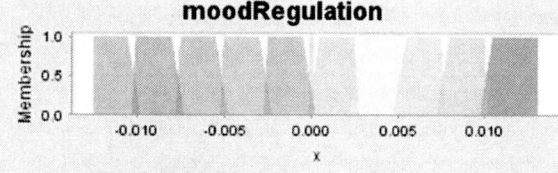

IF *hysteria* IS *very_high* AND *event* IS *negative_low* THEN

anger IS *negative_medium*, *disgust* IS *negative_low*, *fear* IS *negative_very_low*,

happiness IS *positive_very_low*, *sadness* IS *negative_low*, *surprise* IS *positive_very_low*;

Once we have obtained the above vector, we add this to virtual humans' current emotional state e_t to regulate the level of intensity of virtual humans' emotions, and thus to obtain their new emotional state $e_{t'}$.

Mood State Regulation

Current mood state m_t is regulated by a function that changes the level of intensity of virtual humans' moods taking into account virtual humans' personality and the intensity of the perceived event as follows:

$$m_{t'} = m_t + R_m(p, event)$$

We compute the above function $R_m(p, event)$ to calculate a mood regulation vector that contains an increment for one of the basic moods and an decrement for the other. This vector is used to regulate the level of intensity for each of the two basic moods based on virtual humans' personality and the perceived event. To obtain this vector, we have defined a set of fuzzy rules. In this manner, the level of regulation for each mood in the current mood state is described by two output variables associated to each one of the two basic moods respectively. For these variables, we define eleven fuzzy sets: *negative_very_high*, *negative_high*, *negative_medium*, *negative_low*, *negative_very_low*, *neutral*, *positive_very_low*, *positive_low*, *positive_medium*, *positive_high*, and *positive_very_high* (see Figure 6). In order to determinate the mood regulation vector we have chosen the level of intensity of the most predominant virtual humans' personality scale and the level of intensity of the perceived event to form fuzzy rules as follows:

IF *Predominant(p)* = $scale_i$ IS A_j AND *Intensity(event)* IS B_k THEN

Regulate(goodMood) IS J_1, *Regulate(badMood)* IS J_2;

where A_j is the level of intensity of the most predominant personality scale $scale_i \in p$, B_k is the level of intensity of perceived event, and J_1 and J_2 are the increment or decrement to each of the two basic moods. A_j, B_k, J_1, J_2 represent the fuzzy sets previously described. An example of these rules for a predominant very high depressive personality and an event perceived as medium positive is the following:

IF *depression* IS *very_high* AND *event* IS *positive_medium* THEN

badMood IS *positive_high*, *goodMood* IS *negative_very_low*;

Once we have obtained the above vector, we add this to virtual humans' current mood state m_t to regulate the level of intensity of virtual humans' moods, and thus to obtain their new mood state $m_{t'}$.

Intention Selection Process

Action selection process in agent-based systems is normally approached through an optimization perspective, selecting the action that maximizes a utility function. Nevertheless, realistic simulations of human avatars must consider an action selection process affected by emotions, in the same way emotions affect humans' decisions. The psychological foundations, provided herein, permit representing how emotions affect the selection of the next action to be performed. In order to represent this, we define an EBDI (Emotion-Belief-

Desire-Intention) architecture similar to the one presented in (Jiang, Vidal & Huhns, 2007), but we consider time constrains and internal and external events that affect virtual humans' emotional state.

Internal or motivational events are performed by virtual humans' free will in the pursuing of their goals. These events affect agents' desires and intentions. External or informational events represent agents' sensors for perceiving the virtual environment. These events only have influence over agents' beliefs. With the aim of formalizing the EBDI-based intention selection process, we have selected the Event Calculus (Kowalski & Sergot, 1986), which is an action formalism that provides a logical machinery to represent events and their effects over time. The event calculus is selected by its intuitive definition of internal and external events. The formalization of events endows agents with cognitive skills to assess events and properly react according to them. Next, the list of the event calculus predicates utilized to define our EBDI architecture:

Initiates(e(), f, t): A fluent f holds, after an event $e()$ is perceived at time t.

Terminates(e(), f, t): A fluent f does not hold, after an event $e()$ is perceived at time t.

HoldsAt(f, t): A fluent f holds at time t.

Happens(e(), t): An event $e()$ is perceived at time t.

InitiallyP(f): A fluent f holds from $t = 0$.

Fluents are variables that change over time. Herein, only boolean fluents are considered. For details of these predicates see (Shanahan, 1999). Three different sets of fluents are considered:

1. A set of emotions E, fluents corresponding to Ekman's emotions: $E = \{anger, disgust, fear, happiness, sadness, surprise\}$.

2. A set of beliefs B, fluents that represents agents' assumptions about the state of the world and agents' possible skills: $B = \{b_1, b_2, ..., b_i\}$.

3. A set of Desires D, fluents mapped to agents' goals: $D = \{d_1, d_2, ..., d_j\}$.

Finally, we obtain that:

Fluents = $E \cup B \cup D$.

Additionally, we define a predicate *Predominant(E)* that receives as input parameter the set of Ekman's emotions and returns the most predominant emotion $emotion_k \in E$. The result of this predicate is computed from the fuzzy evaluation of events presented above. Assumptions of agents' initial state are made, with respect to initial beliefs (*IB*) and initial desires (*ID*):

IB $\subseteq$ B, $\forall f \in$ IB: InitiallyP(f)

ID $\subseteq$ D, $\forall f \in$ ID: InitiallyP(f)

This means that some beliefs and desires are present from the beginning. Regarding intentions, we follow the definition given by (Georgeff et al., 1999) where an intention is seen as: "*a set of executing threads in a process that can be appropriately interrupted upon receiving feedback from the possibly changing world*". Agent's intentions establish the desires agents want to achieve in a given moment. The set of intentions is then a set of committed plans (Georgeff et al., 1999), this is formalized as follows:

I = $\{i_1$ = {Happens(action()$_g$, t$_1$) $\wedge$ Happens(action()$_{g+1}$, t$_2$) $\wedge$ ... $\wedge$ Happens(action()$_h$, t$_p$)},

i_2 = {Happens(action()$_i$, t$_1$) $\wedge$ Happens(action()$_{i+1}$, t$_2$) $\wedge$ ... $\wedge$ Happens(action()$_j$, t$_q$)},...,

A Behavior Model Based on Personality and Emotional Intelligence for Virtual Humans

$i_n = \{Happens(action()_k, t_1) \wedge Happens(action()_{k+1}, t_2) \wedge ... \wedge Happens(action()_l, t_r)\}\}$

It should be noticed that we do not attempt to provide a plan, but we provide a set of actions to be performed in order to attain a certain desire. An additional planning process should be carried out to check for time, actions, and ordering constraints. External events (*ee*) update agents' beliefs, without considering any emotional state; it just provides new beliefs about the virtual environment. We said that an external event initiates a set of beliefs and terminates another set of beliefs about the environment. This is formalized as follows:

$Initiates(ee(), b_1, t) \wedge Initiates(ee(), b_2, t) \wedge ... \wedge Initiates(ee(), b_i, t) \wedge$

$Terminates(ee(), b_{i+1}, t) \wedge Terminates(ee(), b_{i+2}, t) \wedge ... \wedge Terminates(ee(), b_j, t)$

$\leftarrow Happens(ee(), t)$

Now, in the case of internal events (*ie*), these are attached to agents' emotional state, that is to say, according to the most predominant emotion, agents update their beliefs, for example, supposing that the most predominant emotion of an agent is anger, and it could believe that all agents are against it. This update process is represented as follows:

$Initiates(ie(), b_1, t) \wedge Initiates(ie(), b_2, t) \wedge ... \wedge Initiates(ie(), b_i, t) \wedge$

$Terminates(ie(), b_{i+1}, t) \wedge Terminates(ie(), b_{i+2}, t) \wedge ... \wedge Terminates(ie(), b_j, t)$

$\leftarrow Happens(ie(), t) \wedge HoldsAt(Predominat(E) = emotion_k, t), \forall emotion_k \in E$

Both internal and external events were captured, now, based on current beliefs and predominant emotions agents update their desires, initiating new desires and terminating old desires, whenever an event, associated to an emotional change occurred. This is expressed as follows:

$Initiates(e(), d_1, t) \wedge Initiates(e(), d_2, t) \wedge ... \wedge Initiates(e(), d_j, t) \wedge$

$Terminates(e(), d_j+1, t) \wedge Terminates(e(), d_{j+2}, t) \wedge ... \wedge Terminates(e(), d_k, t)$

$\leftarrow Happens(e(), t) \wedge HoldsAt(b_1, t) \wedge HoldsAt(b_2, t) \wedge ... \wedge HoldsAt(b_j, t) \wedge$

$HoldsAt(Predominat(E) = emotion_k, t), \forall emotion_k \in E$

Plan Execution Process

Given current desires and after a deliberative process denoted by event *selectIntention()*, an agent commits to a plan to attain a goal. Next its event calculus formalization:

$Initiates(selectIntention(), decision, t) \wedge$

$Intention_k = \{Happens(action()_1, t_1) \wedge Happens(action()_2, t_2) \wedge ... \wedge Happens(action()_m, t_n)\}$

$\leftarrow \neg HoldsAt(decision, t) \wedge HoldsAt(d_j, t) \wedge Happens(selectIntention(), t), \forall d_j \in D$

As defined previously, plans can be interrupted. A plan interruption is managed by means of selecting a new intention according to the event occurred. Whenever events occur, agents reconsider their intentions. Finally, whenever a plan is satisfactory achieved, that is to say, it arrives to perform the last action *action$_m$()* of an intention, it releases the desire d_j that promoted the intention and enable the selection for a new plan to achieve a different goal.

Terminates(action()$_m$, decision, t$_n$)
∧ Terminates(action()$_m$, d$_j$, t$_n$) ←
Happens(action()$_m$, t$_n$)

The previous event calculus formulas provide a framework to support an EBDI-based action selection process in dynamic virtual environments.

Case Study

Our case study involves two virtual humans Brian and Helena. Brian has a depressive personality and Helena has a paranoid personality. The interaction takes place in a discotheque, Helena is dancing alone, meanwhile Brian is walking to her, with the intention of flirting with her, when she realizes Brian's intentions, she rejects him, gets angry and starts yelling at him, then Brian starts crying, and finally she tries to console him. The even calculus logic machinery defined in previous sections for intention selection reacts as follows:

Emotions$_H$ = {happiness}

Initial Beliefs$_H$ = {IamMarried, NiceParty}

Initial Desires$_H$ = {DanceAlone, BeAgressive, BePolite}

I={i$_1$={Happens(danceFavoriteSong(), t$_1$) ∧ Happens(waitForFriends(), t$_2$) },

i$_2$={Happens(yellsAtAggressors(), t$_1$)∧Happens(callMyFriends(), t$_2$) ∧Happens(goHome(), t$_3$)},

i$_3$={Happens(ConsoleSadPerson(), t$_1$)∧Happens(BeNice(), t$_2$) ∧ Happens(sayByeBye(), t$_3$)}}

Current Desire = DanceAlone

Emotions$_B$ = {happiness}

Initial Beliefs$_B$ = {IamHandsome, IdontHaveGirl}

Initial Desires$_B$ = {LookingForGirls, FlirtingWithGirl, KissGirl}

I={i$_1$={Happens(lookForGirls(), t$_1$) ∧ Happens(walkAtHer(), t$_2$)},

i$_2$={Happens(kneelDown(), t$_1$) ∧ Happens(declareMyLove(), t$_2$)}

i$_3$={Happens(cry(), t$_1$)∧Happens(waitUntilPity(), t$_2$) ∧ Happens(beginConversation(), t$_3$)}}

Current Desire = LookingForGirls

For this initial situation (see Figure 7.a and 7.b) both virtual humans have to commit to a plan to attain their current goal (intention), this is done through the execution of an internal event called *selectIntention()*:

Helena's case:

Initiates(selectIntention(), decision, t) ∧

i$_1$={Happens(danceFavoriteSong(), t$_1$) ∧ Happens(waitForFriends(), t$_2$) }

← ¬HoldsAt(decision, t) ∧
HoldsAt(DanceAlone, t) ∧
Happens(selectIntention(), t)

Brian's case:

Initiates(selectIntention(), decision, t) ∧

i$_1$={Happens(lookForGirls(), t$_1$) ∧ Happens(walkAtHer(), t$_2$) }

← ¬HoldsAt(decision, t) ∧
HoldsAt(LookingForGirls, t) ∧
Happens(selectIntention(), t)

Brian's current desire is looking for girls, and, Helena's current desire is dancing alone. When Brian gets close to Helena this is perceived as an external event for both virtual humans and it alters their beliefs:

Helena's case:

Initiates(walkAtHer(), AggressorInFrontOfMe, t) ∧ Terminates(walkAtHer(),NiceParty, t)

← Happens(walkAtHer(), t)

Brian's case:

Initiates(walkAtHer(), SheIsPretty, t) ∧ Terminates(walkAtHer(), IdontHaveGirl, t)

← Happens(walkAtHer(), t)

After the update of beliefs, an update of desires takes places (see Figure 7.c and 7.d):
Helena's case:

Initiates(walkAtHer(), BeAgressive, t) ∧ Terminates(walkAtHer(), DanceAlone, t)

← HoldsAt(Predominat(E) = happiness, t) ∧

HoldsAt(aggressorInFrontOfMe, t) ∧ HoldsAt(IamMarried, t) ∧ Happens(walkAtHer(), t)

Brian's case:

Initiates(walkAtHer(), FlirtingWithGirl, t) ∧ Terminates(walkAtHer(), LookingGirls, t)

← HoldsAt(Predominat(E) = happiness, t) ∧

HoldsAt(SheIsPretty, t) ∧ HoldsAt(IamHandsome, t) ∧ Happens(walkAtHer(), t)

With the update of desires comes the selection of intentions (see Figure 7.c and 7.d):

Helena's case:

Initiates(selectIntention(), decision, t) ∧

i_2={Happens(yellsAtAggressors(), t_1)∧Happens(callMyFriends(), t_2) ∧Happens(goHome(), t_3)}

← ¬HoldsAt(decision, t) ∧ HoldsAt(BeAgressive, t) ∧ Happens(selectIntention(), t)

Brian's case:

Initiates(selectIntention(), decision, t) ∧

i_2={Happens(kneelDown(), t_1) ∧ Happens(declareMyLove(), t_2)}

← ¬HoldsAt(decision, t) ∧ HoldsAt(FlirtingWithGirl, t) ∧ Happens(selectIntention(), t)

In this way, when new events are perceived by virtual humans these produce the acquisition of new intentions (see Figure 7.e and 7.f). Thus, the course of action is followed until the virtual humans release their current desires.

A possible ending for this scenario could be that finally:
Brian kisses Helena:

Terminates(BrianKissHer(), decision, t_n) ∧
Terminates(BrianKissHer(), KissGirl, t_n) ←
Happens(BrianKissHer(), t_n)

CONCLUSION

There exist many factors that influence individuals' behavior making impossible to predict with accuracy the behavior and actions they expose and perform in the face of certain events. These events trigger emotional influences that change individu-

Figure 7. a) Brian and Helena's initial states, b) Brian is walking to Helena, c) Brian is flirting with Helena and Helena is being aggressive, d) Brian declares his love to Helena and Helena yells at Brian, e) Helena rejects at Brian and Brian begins to cry, and f) Helena consoles Brian

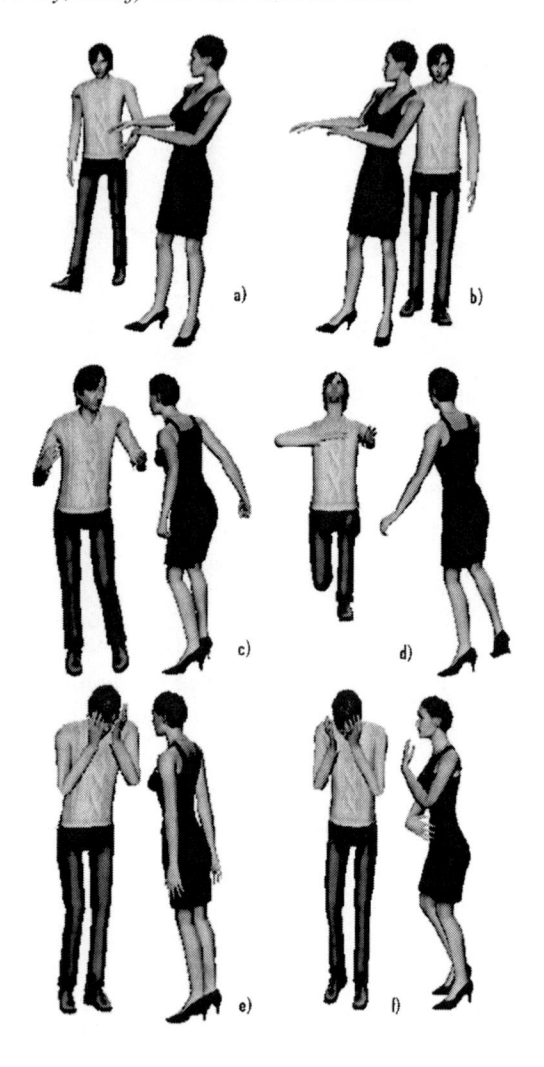

als' internal affective state, which endow them with the ability to generate emotional responses to different situations experienced in a real world. Emotional states can be viewed as internal states that promote or drive human decisions to take a specific action. These states have a tendency to interrupt the brain to call for an important need or action. Thus, these states have a major impact on the mind, including the emotional process and the decision making process, and hence on the behavior. In this way, virtual humans' dynamic behavior over a sequence of events is only apparent from the change in response to situations over time. Thus, expectation and resultant emotions are experienced due to a conditioned response. Some psychological models explicitly use expectations to determine the emotional state. However, classical conditioning is not the only type of learning that can induce or trigger expectations. Therefore, other types of learning must be incorporated to produce a believable adaptive behavior, including learning about sequences of events and other agents.

Intense emotions can affect significantly human beings' behavior, for instance, driving their behavior unstable as a consequence of a deep emotional influence. As a response to such influence individuals develop a skill to recognize and manage their emotions. This skill is called Emotional Intelligence and it permits them to have an emotional stability and self-control over time. In this work, we have endowed virtual humans with this skill.

In this chapter, our main objective is centered on simulating autonomous emotional virtual humans with a dynamic emotional behavior. Thus, we have presented a behavior model for realistic virtual humans. This model is valid from a psychological point of view, because it is supported by studies on Personality and Emotional Intelligence. In this model, we have considered all possible behaviors virtual humans must expose based on their personality, affective state, beliefs, desires, intentions and the events they perceive.

We have presented a three-layered model to simulate human beings' behavior in virtual humans. In this model, we have applied different sets of fuzzy rules to change virtual humans' affective state according to their personality, emotional and mood history, and the events they perceive. We have used fuzzy logic, because perceived events, personality and affective state can be modeled with fuzzy limits. This allows us to change quickly virtual humans' emotional and mood states in a more natural manner, generating more realistic behaviors in real time.

We also have implemented an EBDI-based intention selection using the Event Calculus formalism. This intention selection mechanism allows virtual humans performing actions based on their current emotional state, beliefs, desires and intentions. Thus, these intentions are used to define virtual humans' behavior for each situation they experience in the environment according to their personality, emotional intelligence, and the events they perceive.

REFERENCES

Atkinson, R. L., & Hilgard, E. R. (1996). *Hilgard's Introduction to Psychology*. Fort Worth: Harcourt Brace College.

Ball, G., & Breese, J. (2000). Emotion and Personality in a Conversational Agent. pp. 189–219.

Bates, J., Loyall, B. A., & Reilly, S. W. (1992). An architecture for Action, Emotion, and Social Behavior. *Modelling Autonomous Agents in a Multi-Agent World*, pp. 55-68.

Bless, H., & Schwarz, N. (1999). *Mood, its Impact on Cognition and Behavior. Blackwell Encyclopedia of Social Psychology*. Oxford, England: Blackwell.

Bosse, T., Pointer, M., & Treur, J. (2007). A Dynamical System Modelling Approach to Gross' Model of Emotion Regulation. *Proceedings of the 8th International Conference on Cognitive Modeling (ICCM'07)*, Oxford, UK: Taylor & Francis/Psychology Press, pp. 187-192.

Cingolani, P. (2009). jfuzzylogic: An Open Source Fuzzy Logic Library and FCL Language Implementation. http://jfuzzylogic.sourceforge.net/html/index.html.

Egges, A., Kshirsagar, S., & Magnenat-Thalmann, N. (2004). Generic Personality and Emotion Simulation for Conversational Agents. *Journal of Visualization and Computer Animation*, 15(1), 1–13.

Ekman, P. (1994). *Moods, Emotions, and Traits, the Nature of Emotion: Fundamental Questions*. New York, NY, USA: Oxford University Press.

El-Nasr, M. S., Yen, J., & Ioerger, T. R. (2000). FLAME: Fuzzy Logic Adaptive Model of Emotions. *Autonomous Agents and Multi-Agent Systems*, 3(3), 219–257. doi:10.1023/A:1010030809960

Fridja, N. (1987). *The Emotions: Studies in Emotion and Social Interaction*. New York: Cambridge Universty Press.

García-Rojas, A., Gutiérrez, M., & Thalmann, D. (2008). Simulation of Individual Spontaneous Reactive Behavior. *In AAMAS '08: Proceedings of the 7th International Joint Conference on Autonomous Agents and Multiagent Systems*, Richland, SC. International Foundation for Autonomous Agents and Multiagent Systems, pp. 143–150.

Georgeff, M. P., Pell, B., Pollack, M. E., Tambe, M., & Wooldridge, M. (1999). The Belief-Desire-Intention Model of Agency. *In Proceedings of the 5th international Workshop on intelligent Agents, Agent theories, Architectures, and Languages*. In J. P. Müller, M. P. Singh, and A. S. Rao, Eds. Lecture Notes In Computer Science, Springer-Verlag, London, 1555, pp. 1-10.

Ghasem-Aghaee, N., & Oren, T. (2007). Cognitive Complexity and Dynamic Personality in Agent Simulation. *Computers in Human Behavior*, 23(6), 2983–2997. doi:10.1016/j.chb.2006.08.012

Goleman, D. (1995). *Emotional Intelligence*. New York: Bantam Books.

Goleman, D. (1998). *Working with Emotional Intelligence*. New York: Bantam Books.

Gratch, J., & Marsella, S. (2001). Tears and Fears: Modeling Emotions and Emotional Behaviors in Synthetic Agents. *In Proceedings of the Fifth international Conference on Autonomous Agents* (Montreal, Quebec, Canada). AGENTS '01. ACM, New York, NY, pp. 278-285.

Hong, J. (2008). *From Rational to Emotional Agents: A Way to Design Emotional Agents*. Saarbrucken, Germany: VDM Verlag.

IEC 1131 (1997). Programmable Controllers, Part 7: Fuzzy Control Programming, Fuzzy Control Language. http://www.fuzzytech.com/binaries/ieccd1.pdf.

Iglesias, A., & Luengo, F. (2007). *AI Framework for Decision Modeling in Behavioral Animation of Virtual Avatars* (pp. 89–96). Berlin, Heidelberg: Springer-Verlag.

Imbert, R., & de Antonio, A. (2005). An Emotional Architecture for Virtual Characters. *ICVS'05, Third International Conference on Virtual Storytelling*. Lecture Notes in Computer Science, Springer. Strasbourg, France, 3805, 63-72.

Jiang, H., & Vidal, J. M. (2006). From Rational to Emotional Agents. *Proceedings of the American Association for Artificial Intelligence (AAAI). Workshop on Cognitive Modeling and Agent-based Social Simulation*.

Jiang, H., Vidal, J. M., & Huhns, M. N. (2007). EBDI: An Architecture for Emotional Agents. *In Proceedings of the 6th international Joint Conference on Autonomous Agents and Multiagent Systems*. AAMAS '07. ACM, New York, NY, pp. 1-3.

Khulood, A. M., & Raed, A. Z. (2007). Emotional Agents: A Modeling and an Application. *Information and Software Technology*, 49(7), 695–716. doi:10.1016/j.infsof.2006.08.002

Kowalski, R., & Sergot, M. (1986). A Logic-Based Calculus of Events. *New Generation Computing*, 4(1), 67–95. doi:10.1007/BF03037383

Kshirsagar, S., & Magnenat-Thalmann, N. (2002). A Multilayer Personality Model. *In SMARTGRAPH '02: Proceedings of the 2nd International Symposium on Smart Graphics*, New York, NY, USA. ACM, pp. 107–115.

LeDoux, J. E. (1996). *The Emotional Brain: The Mysterious Underpinnings of Emotional Life*. New York: Simon and Schuster.

Liu, Z., & Lu, Y.-S. (2008). A Motivation Model for Virtual Characters. *In Proceedings of the Seventh International Conference on Machine Learning and Cybernetics*, 5, 2712–2717.

Liu, Z., & Pan, Z.-G. (2005). An Emotion Model of 3D Virtual Characters. In Picard, R. W. (Ed.), *Intelligent Virtual Environments. In Affective Computing and Intelligent Interaction, Tao, JH* (pp. 629–636). Tan, TN: Springer-Verlag. doi:10.1007/11573548_81

MacLean, P. (1973). *A Triune Concept of the Brain and Behaviour*. Toronto: University of Toronto Press.

Martinez-Miranda, J., & Aldea, A. (2005). Emotions in Human and Artificial Intelligence. *Computers in Human Behavior*, 21(2), 323–341. doi:10.1016/j.chb.2004.02.010

Mayer, J. D., & Geher, G. (1996). Emotional Intelligence and the Identification of Emotion. *Intelligence*, 22, 89–113. doi:10.1016/S0160-2896(96)90011-2

Mayer, J. D., & Salovey, P. (1995). Emotional Intelligence and the Construction and Regulation of Feelings. *Applied & Preventive Psychology*, 4, 197–208. doi:10.1016/S0962-1849(05)80058-7

Mayer, J. D., Salovey, P., & Caruso, D. R. (2000). *Models of Emotional Intelligence. Handbook of Intelligence*. Cambridge, UK: Cambridge University Press.

McCauley, L., Franklin, S., & Bogner, M. (2000). An Emotion-Based "Conscious" Software Agent Architecture. *Affective Interactions*, LNAI, Springer-Verlag. Berlin Heidelberg, 1814, 107-120.

McCrae, R., & John, O. (1992). An Introduction to the Five-Factor Model and its Application. *Journal of Personality, 60*(2), 175–215. doi:10.1111/j.1467-6494.1992.tb00970.x

Murakami, Y., Sugimoto, Y., & Ishida, T. (2005). Modeling Human Behavior for Virtual Training Systems. *In Proceedings of the 20th National Conference on Artificial intelligence – Vol. 1. A. Cohn, Ed. Aaai Conference On Artificial Intelligence*. AAAI Press, pp. 127-132.

Ören, T., & Yilmaz, L. (2004). Behavioral Anticipation in Agent Simulation. *Winter Simulation Conference*, pp. 801-806.

Ortony, A., Clore, G. L., & Collins, A. (1988). *The Cognitive Structure of Emotions*. New York: Cambridge Universty Press.

Osberg, T. M., Haseley, E. N., & Kamas, M. M. (2008). The MMPI-2 Clinical Scales and Restructured Clinical (RC) Scales: Comparative Psychometric Properties and Relative Diagnostic Efficiency in Young Adults. *Journal of Personality Assessment, 90*, 81–92.

Pereira, D., Oliveira, E., Moreira, N., & Sarmento, L. (2005). Towards an Architecture for Emotional BDI Agents. *In EPIA'05: Proceedings of 12th Portuguese Conference on Artificial Intelligence*, pp. 40-47.

Poznanski, M., & Thagard, P. (2005). Changing Personalities: Towards Realistic Virtual Characters. *Journal of Experimental & Theoretical Artificial Intelligence, 17*(3), 221–241. doi:10.1080/09528130500112478

Rizzo, P., Veloso, M., Miceli, M., & Cesta, A. (1997). Personality-Driven Social Behaviors in Believable Agents. *AAAI, Fall Symposium on Socially Intelligent Agents*.

Roseman, I. J., Jose, P. E., & Spindel, M. S. (1990). Appraisals of Emotion-Eliciting Events: Testing a Theory of Discrete Emotions. *Journal of Personality and Social Psychology, 59*(2), 899–915. doi:10.1037/0022-3514.59.5.899

Rousseau, D., & Hayes-Roth, B. (1997). A Social-Psychological Model for Synthetic Actors. *Technical Report KSL-9707*, Knowledge Systems Laboratory, Stanford University.

Salovey, P., & Mayer, J. D. (1990). Emotional Intelligence. *Imagination, Cognition and Personality, 9*, 185–211.

Salovey, P., & Sluyter, D. J. (1997). *Emotional Development and Emotional Intelligence*. New York: Basic Books.

Schmidt, B. (2002). How to Give Agents a Personality. *In Proceeding of third Workshop on Agent-Based Simulation*, Passau, Germany.

Shanahan, M. (1999). The Event Calculus Explained. *Springer Verlag, LNAI, 1600*, 409–430.

Simms, L. J., Casillas, A., Clark, L. A., Watson, D., & Doebbeling, B. I. (2005). Psychometric Evaluation of the Restructured Clinical Scales of the MMPI-2. *Psychological Assessment, 17*, 345–358. doi:10.1037/1040-3590.17.3.345

Tellegen, A., Ben-Porath, Y., Arbisi, P., Graham, J., & Kaemmer, B. (2003). *The MMPI-2 Restructured Clinical Scales: Development, Validation, and Interpretation*. Minneapolis, MN, USA: University of Minnesota Press.

Vélasquez, J. D. (1997). Modeling Emotions and Other Motivations in Synthetic Agent. *In Proceedings of the AAAI Conference*, AAAI Press and the MIT Press.

ADDITIONAL READING

Bandura, A. (1977). *Social Learning Theory*. Prentice-Hall Inc.

Bar-On, R. (1997). *The Emotional Quotient Inventory (EQ-i): Technical Manual*. Toronto: Multi-Health Systems.

Batson, C. D., Shaw, L. L., & Oleson, K. C. (1992). Differentiating Affect, Mood, and Emotion. *Emotion (Washington, D.C.)*, 294–326.

Berkowitz, L. (2000). *Causes and Consequences of Feelings*. Cambridge, England: Cambridge University Press. doi:10.1017/CBO9780511606106

Brehm, J. W. (1999). The Intensity of Emotion. *Personality and Social Psychology Review*, *3*, 2–22. doi:10.1207/s15327957pspr0301_1

Butcher, J. N., Dahlstrom, W. G., Graham, J. R., Tellegen, A., & Kaemmer, B. (1989). *The Minnesota Multiphasic Personality Inventory-2 (MMPI-2): Manual for Administration and Scoring*. Minneapolis, MN: University of Minnesota Press.

Damasio, A. R. (1994). *Descartes' Error: Emotion, Reason and the Human Brain*. New York: Avon Books.

Damasio, A. R. (1995). *L'Error de Descartes: La Raison des Émotions*. Paris: Odile Jacob.

Downing, S. K., Denney, R. L., Spray, B. L., Houston, C. M., & Halfaker, D. A. (2008). Examining the relationship between the Reconstructed Scales and the Fake Bad Scale of the MMPI-2. *The Clinical Neuropsychologist*, *22*, 680–688. doi:10.1080/13854040701562825

Evans, D. (2001). *Emotion: The Science of Sentiment*. Oxford, England: Oxford University Press.

Eysenck, H. J. (1990). *Biological Dimensions of Personality. Handbook of personality: Theory and research*. New York: Guilford.

Eysenck, M. W., & Keane, M. T. (2000). *Cognitive Psychology: A Student's Handbook*. UK: Psychology Press.

Gardner, H. (1983). *Frames of the Mind*. New York: Basic Books.

Harkness, A. R., McNulty, J. L., Ben-Porath, Y. S., & Graham, J. R. (2002). *MMPI-2 Personality-Psychopathology Five (PSY-5) Scales: Gaining an Overview for Case Conceptualization and Treatment Planning*. Minneapolis, MN: University of Minnesota Press.

Hathaway, S. R., & McKinley, J. C. (1940). A Multiphasic Personality Schedule(Minnesota): I. Construction of the Schedule. *The Journal of Psychology*, *10*, 249–254. doi:10.1080/00223980.1940.9917000

Hathaway, S. R., & McKinley, J. C. (1942). A Multiphasic Personality Schedule (Minnesota): III. The Measurement of Symptomatic Depression. *The Journal of Psychology*, *14*, 73–84. doi:10.1080/00223980.1942.9917111

Kshirsagar, S. A. (2002). Multilayer Personality Model. In Proceedings of 2nd International Symposium on Smart Graphics, Hawthorne, New York.

Leary, T. (1957). *Interpersonal Diagnosis of Personality*. New York: Ronald Press.

McKinley, J. C., & Hathaway, S. R. (1940). A Multiphasic Personality Schedule (Minnesota): II. A Differential Study of Hypochondriasis. *The Journal of Psychology*, *10*, 255–268. doi:10.1080/00223980.1940.9917001

McKinley, J. C., & Hathaway, S. R. (1942). A Multiphasic Personality Schedule (Minnesota): IV. Psychasthenia. *The Journal of Applied Psychology*, *26*, 614–624. doi:10.1037/h0063530

Sellbom, M., Ben-Porath, Y., & Bagby, R. (2008). Personality and Psychopathology: Mapping the MMPI-2 Restructured Clinical (RC) Scales onto the Five Factor Model of Personality. *Journal of Personality Disorders*.

Sellbom, M., Ben-Porath, Y. S., Graham, J. R., Arbisi, P. A., & Bagby, R. M. (2005a). Susceptibility of the MMPI-2 Clinical, Restructured Clinical (RC), and Content Scales to Overreporting and Underreporting. *Assessment, 12*, 79–85. doi:10.1177/1073191104273515

Sellbom, M., Ben-Porath, Y. S., Lilienfeld, S. O., Patrick, C. J., & Graham, J. R. (2005b). Assessing Psychopathic Personality Traits with the MMPI-2. *Journal of Personality Assessment, 85*, 334–343. doi:10.1207/s15327752jpa8503_10

KEY TERMS AND DEFINITIONS

Emotion: It is a volatile feeling that affects our behavior and moods as a response to particular stimuli perceived from the environment.

Emotional Intelligence: It represents the ability to perceive, express, assimilate and understand emotion-related feelings with the aim of managing and regulating them in ourselves and others.

Intelligence: This is a global capacity that helps us to profit from experience, acting and thinking with the aim of adapting to changes in the environment.

Mood: It is a quasi-permanent feeling that affects our behavior and emotions.

Personality: It is a set of personal characteristics that influence our cognition, motivation and behavior in different situations.

Chapter 9
The Virtual World of Cerberus:
Virtual Singer using Spike-Timing-Dependent Plasticity Concept

Jocelyne Kiss
LISAA, University Paris East, France

Sidi Soueina
Strayer University, USA

Martin Laliberté
LISAA, University Paris East, France

Adel Elmaghraby
University of Louisville, USA

ABSTRACT

While exploring autonomous evolution concepts for virtual worlds, we will present a new design Cerberus an avatar singer who can accompany a singer, perform alone and make his song evolves using simple past events. This 3D interactive facial animated avatar was made thanks to virtools software. The main originality of Cerberus is to develop his own melody by using learning machines and constantly improve his musical style and emotions. Cerberus is implemented using competitive learning rules to trained artificial neural networks in order to perform these self-improvements. Self-improvements is a key of our learning capacity, The challenge of building a virtual singer that could promotes his own improvisation is an open research field (Minsky, 2000). We will expose the difficulties of the synchronization in real-time between the voice and animation to generate the right emotion, also the difficulties of establishing a classification which could be in contradiction with the ontology of the musical fact. Also we will expose the necessity of developing avatars that use the amazing potentialities of spike-timing-dependent plasticity concept (Abraham, & Bear, 1996), that hence metaplasticity. This powerful concept enhance the potentiality of avatar design and give the impression that the avatar has a memory and simulate "feelings" linked to a context.

DOI: 10.4018/978-1-60960-077-8.ch009

INTRODUCTION

Virtual worlds are looking for new socialization experiences, emotions sharing and avatars (Damer, 1997) are made to enhance virtually your skills and personality. Virtual world is a place where you can be someone else (Schroeder, 2008). Cerberus is based on this idea of building a game using a sort of karaoke, where you can sing even you are not able to in the real life. The user helps Cerberus to find its voice and to improve its creation of melodies. Singing is a way to express a feeling, to communicate a sensation, to feel something different. Also it is a way to extrapolate and to develop a sense of originality, a sense of self (Mingers, J, 1994). How to simulate a singer? How to translate this special feeling (Minsky, 1979) into a virtual world?

To minimize the tasks and the calculus in order to obtain an avatar working in real time, we will restraint our avatar to a limited musical context and we will provide a simple classification for the emotions, musical phrases and gestures, as suggest by the Oslen method (Olsen, & Belar, 1961). The most interesting part of this research is the potentialities that offer the STDP concept which could offer a sort of footprint, a selective memory of the event inside the neural network we used. STDP is inspired by biologic model, it is a form of associative synaptic modification which depends on the time of synaptic spikes (Gerstner & Kistler, 2002). This "selective memory" is built in the structure itself using this principle: If the connection decrease in efficiency then the connection may disappear. In a way the Neural network is reshaping itself depending of the information. This potentially offers great resources in term of memory and neural networks (Schacter, 1989).

After describing the Cerberus engine, the problem of classification of the events, melody analysis (Xu & al. 2009), synchronization and Neural network implementation, we will present the perspectives of metaplasticity concept to create an avatar. At the end of this chapter, we will propose in the future research section a discussion about the notion of avatar (Koda & al. 2009), feelings and expose our last problematic how could we claim that an virtual character could produce the impression to be self-conscious (Arrabales, & al. 2009), to generate creative gesture or self-design?

PROBLEM ANALYSIS

If we consider a simulation of a robot mimicking a singing gesture as mediating function between the actual act of singing and a more abstract to find inspiration using past events, is providing both a symbolic vocabulary and ideal objects, then it seems that it stands out as an effective way to know or even understand some of the music creation process (Barbaud, 1966). The main difficulty of such a system is the balance of shares that must be given to peculiarities inherent in this creation and the necessary generalization. As sketched out a first issue regarding the orientation of the program because it would, indeed utopian in the current research, as demonstrated in previous experiments Consider a system capable of generating imposed in a context, a complex musical phenomenon, regardless of any arbitrary choice beforehand. How then implement a decision unit capable of directing the genesis of an avatar engine?

Automated systems possess two qualities: the ability to deal quickly at the micro-compositional level, and that of acquiring a unit by the expression of certain 'musical intentions' (Miranda & Matthias, 2009). A singing avatar should have the abilities to "listen" to the proposition, to her public and to sing and improvise by itself. This virtual singer will be presented in the form of a visual interface to avatar 3D in order to articulate a phoneme while producing a sound (DiPaola, 1989). The main difficulty is to provide a real-time system which could evolve by itself, and simulate an accurate articulation and facial expression and a sound. Naturally, synthesis voice technical is used (it would be very difficult and time consuming

Figure 1. Cerberus performing

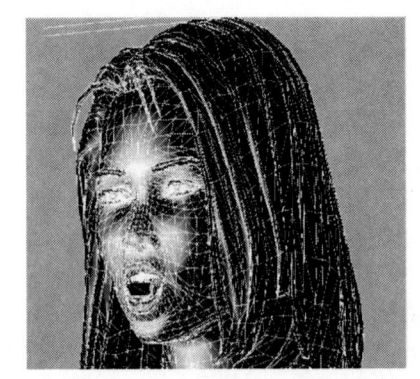

to use prerecorded sound using pattern matching method, (Efrat & al. 2004)). Also it seems that the first step will be to use an artificial neural network in order to manage the recognition of the events. It should be able to manage an "artistic choice".

How could we implement a decision system which could behave has a musician who can choose the right set of notes to play depending of the context ? How to preselect a set of features that could occur a accurate taxonomy. By accurate we mean that could be hear and be "correct" in the musical context. Should we enter every musical note? Which means generating interchangeability of musical concepts, such as treatment components of sounds, whose tone, length, height, which could in their combination of mathematical variables lose their intrinsic identity, placing itself in a continuum of entities equal or indistinct? This non-difference between elements, part of a principle of objectivity necessary for the development of a robot or a virtual building abstract language, which allows the adaptation of a compositional system to the computer, but appears in retrospect as a loss or dissolution of the meaning of the user's musical proposition (Hautop Lund & Ottesen 2008). Entering or consider the musical rules is also another option The ambiguities inherent from these rules are not sufficient to mask that they cannot describe all the existing musical processes. And for the facial articulation it would be as well difficult to make a list of all the possibilities.

Should we use a random and use a prerecorded answer when it seems to be the right moment. And how could we deal with the improvise articulation of the 3D model ? Here it seems that to consider not a specific time but an interval of time. In order to do that we build a database of ontologies which each of them will contain the animated articulation, link with a musical pattern (De Nicola, & al. 2009).So to define an ontology and to implement neural networks we use a group of curves that included time, musical pattern which could be transposed, group of translated points which could be interpolated. One of the main problem resides on this problematic, what could determine the ontology (Colace & al.2004) and especially what segment of time could we consider. The segment of time we use to define the ontology varies with the examples. There is no scale we could use and nothing we could define to produce a generalization. We cannot cut the time arbitrarily. Also we will see in the next section what kind of experiment we did to find a way to modelize the ontologies. Another problem was how to give the impression that the avatar sings continues a musical play articulated words that began by a user. In a way, we want to give to the public that Cerberus "has a memory" of what happened. The level of avatar's adaptivity is linked to the potentiality of self-improvements. We will see how metaplasticity which could be defines (by Fusi et al., 2005) as the plasticity of synaptic plasticity helps to create a automate that gives the impression to store past event.

RELATED WORKS

Since the birth of the computer we could find works about how to generate music automatically. Lejaren Arthur Hiller, Leonard Issacson collaborated on the first significant computer music composition, 1957's *Illiac Suite*. They proposed to implement the musical notes as variables and the musical rules as functions. It was amazing to see how the theory was far from what we could perceive. We cannot describe here all the computer music experiences. We just finish by a current example, *Melonet and Harmonet* (Feulner & Hörnel 1994) which use neural networks and simulate composers style; in this work the notion of past appears sometimes however it is not formalized (Laliberté, 1993). Avatars are also linked to a context, it is very difficult to "speak" to an avatar with no context and when it is the case you have reduce answer which could not be related to your question (Callaghan & al. 2008).

Our proposition will not solve this type of problem however it will help to consider that a certain way of storage of the information could help to give more elements of choice when you produce an answer. A cascade model, such as the metaplasticity concept could offer to solve this type problem (Fusi & Abbott. 2007). Naturally, this theoretical outline, which we conceive of many developments, can really lead to a precise definition of the architecture to implement the achievement of such a controller. There is indeed no predefined order or priority, or logical succession in the arrangement of the various levels of choices. However, the symbolic and random may help narrow the diversity of the features associated with learning phase and guide the nature of the test sets for connectionist applications. Bret and Tramus with their dancing avatar "Dance with me" used also a connectionist filter to perform (Bret, 2000). They placed a captor of movement on the user that will move the dancing avatar which has prerecorded movement. They also present the avatar that it could have a "memory" by bringing back the dancer to a centered position every now and then. That gives an impression of stability and confidence to the virtual dancer (Couchot & al. 1998). We will see how Cerberus is keeping the same idea and propose a treatment for managing past events.

PROPOSED SOLUTION

Cerberus is a 3D three faces avatar, the virtual camera inside a Daussault system technology, Virtools© is a platform for virtual reality world. We will focus on the faces only divided in three windows. User can interact with the avatar by choosing a mode "accompany me", play, chorus or improvisation, a face, an emotional state or you ask him to recognize yours using a web cam, a duration and setup the melody. Cerberus can sing with you by accompanying you using lower or upper sixth, which is a very simple way of doing it (Burkhart, 1979). We did not choose a complete harmonization and focus on this "secondary voice" self-developement. "The avatar could sing using a chorus form using the three head (Kiss, 2006), again the rules we set up are very simple. He can also improvise using the melodies you already taught him. The user enters a time frame, chooses an emotion or uses a webcam to show the emotion using is own face (this choice less accurate), and enters a melody using the microphone.

The system determines the emotion (Creed,. & Beale, 2008) and the melody using already made plug-in in C++ which we modified a bit to fit with our system, also it sorts the phonemes used if they are in the database otherwise the system uses the one who seems very close. When

Figure 2. Cerberus engine

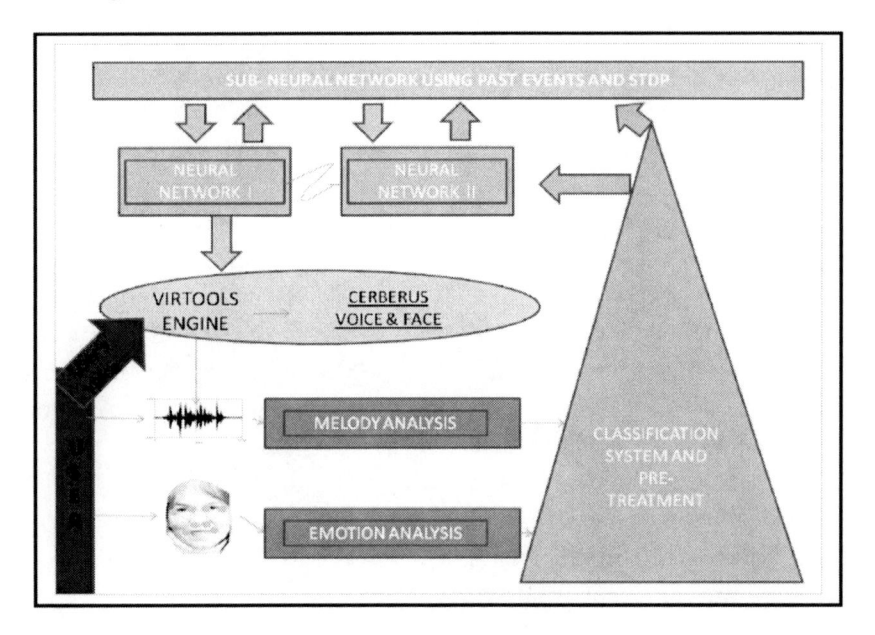

the data are analyzed, classified then be pre-treated during the learning phase. For this phase, the system takes the data one by one and propose a segmentation and a classification. This very delicate part is based on dynamics and simple neural networks which will elect (after the learning phase done) the segment by form recognition when the decision is made the segment will be stored in order to be use at the end in the patter matching selection system. To segment the sound we received we adapt to the sound the method developed for the image by (Bir & Lee 1994). The segment is sent in the in-use neural network who will elect and perform in real-time and also toward the sub-neural network.

The sub-neural network system is not linked directly to the real time system. It is made to "self-improve" the system after the first learning phase has been already setup the production quality of Cerberus. It is made to produce an non-ending learning phase. The sub-neural network is linked sometimes only when it finds an "interesting" way to evolve to some of the connections of the neural network II and does not propose real-time solution it is the reason why we cannot plug it directly to the engine. However, it is made to balance, improve the whole system and give new weights to the neural network connection depending on the time. The sub-neural network will refine the results of the learning phase by using the current user's production. Now, we will present further the details of this decision system using STDP, and present the nature of the ontologies which are elaborate to implement the decision system.

Building the architecture of the neural network which will receive the ontology (Pinto & Martins 2004) was the biggest part of the work done to implement the decision system. The purpose is to introduce inside our dynamic neural network the concept of metaplasticity to treat the ontology in order to later on elect the "good one" (Smirnov & al. 2007)". Ontology materialize by curves and represents all the features in a specific time. This time is not always the same and should be adapted to each example.

Our metaplasticity belongs to the learning concept, it will show during this phase of implementation. We follow the definition of Eriksson such that usually an internal variable $L_{pq}(t)$ used

Figure 3. Simplification of the schema of dynamic neural network. Time duration varies depending of the sample

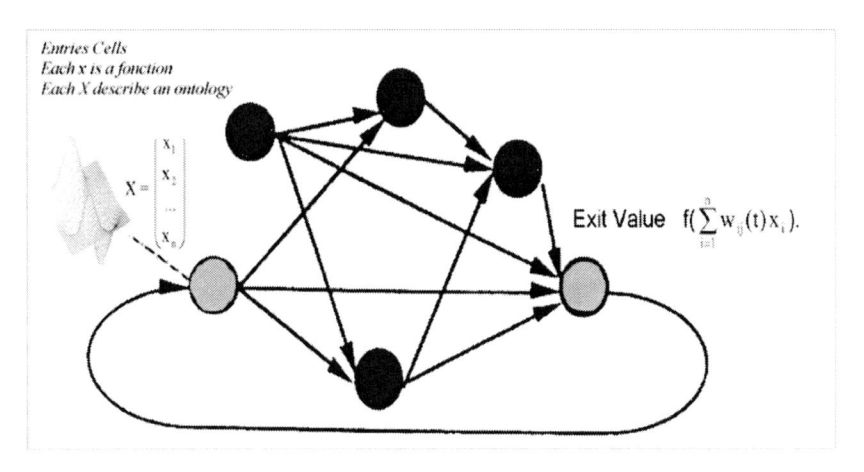

to control the discrete activation Apq(t) describing a level using spike-time-Dependent concept. User-defined boundaries of the moment, the interval of the singing action which means interval of pitch, time of events, articulation mouth position, at least 20 indicators. This action could be formalized such as L0 < .. < LN−1 < LN to represent the action and also a driving factor to force Lpq(t) towards the synaptic pruning. You can do it also continuously, with an interpolation between Lpq(t) and Apq(t) here decaying factor follow Lpq (t) to [(Lx−Lx−1) /2)] (Eriksson et al., 2003).Basically it will inhibit or promote one solution toward an another. Our system will also keeps all the good answer that was elect to not propose again the same if it was played more than three times. We had some difficulties with during the learning phase that we will describe in the last section.

OBTAINED RESULTS

Regarding the avatar's capability of building a melody by itself while using past events is working correctly. STDP concept is a great tool to store past events in a very creative way. Metaplasticity concept helps to erase, bold, keep and elect events which are relevant for the melody. It helps also to choose to articulation that will smooth softly the avatar's face. However this avatar was designed to work online and to be open by many users; such as one user begin a melody and another finish it. It seems that on that point we have an issue which was generate by the impossibilities of generalize the segmentation. In our implementation we had to consider only the way that one person build the melody and when we change the user we have to change the model. That means we train our neural network for one person only and we cannot use the same training phase for more than one. The segmentation is really a challenge for this type of engine. To adapt the segmentation system to one person to and other takes the time to do a unique training phase which could not included the examples of the previous user. so the memory of the past events has to be from one user. This is really limited for our model because at the very beginning we were hopeful to use the web to grow the memory of all the potential users (Mase, 2007) and like that the avatar would have enough experience to self-improve. The fact the system could not segment automatically is a real problem. What we produce is a system that is only linked to one personality, one identity and you could not layered two of them. We are quite far from the first will to design a avatar which could self design itself, nevertheless, despite the difficulties in our processes, we could see that thanks to the metaplasticity concept it becomes easier to propose a past depend solution model.

FUTURE RESEARCH DIRECTIONS

Cerberus needs to smooth some of its animations, to improve its real-time reaction and especially to generalize its model. This would be the main project like that we could try to develop Cerberus online and increase its samples resources for the general learning phase; It would be interesting to try to expand this automate toward musical polyphonic production while showing the 3D avatar playing violin, cello etc. The modelization and simulation of musical gesture is in the same field than the experience we did with Cerberus. To go a bit further, we would like to work on the modelization of the emotion, and simulation of expression of self and feelings for the avatar (Maturana, & Varela. 1980). We think that using this type of storage that the STDP provides it could give a strong impression of presence (Pollock, 1989). For developing this notion of presence, the choice of device is important; such as the spatialization of the sound, the screen. A last perspective would be working on the ontologies to describe musical phenomenon. Even if we are far from a generalization this technical way seems to be a very rich and productive fashion of modelization.

REFERENCES

Abraham, W. C., & Bear, M. F. (1996). Metaplasticity: the plasticity of synaptic plasticity. *Trends in Neurosciences, 19*(4), 126–130. doi:10.1016/S0166-2236(96)80018-X

Arrabales, R., Ledezma, A., & Sanchis, A. (2009), Establishing a Roadmap and Metrics for Conscious Machines Development, *The 8th IEEE International Conference on Cognitive Informatics*: 94-101

Barbaud, P. (1966). *Introduction à la composition musicale automatique*. Paris, France: Dunod.

Bir, B. Lee S. (1994) Genetic Learning for Adaptive Image Segmentation *The Springer International Series in Engineering and Computer Science,* 287

Bret, M. (2000) Virtual Living Beings, in Lecture Notes in Artificial Intelligence. *Virtual Worlds*, Ed. Jean-Claude Heudin, Springer. 1834: 119-134

Burkhart, C. (1979) *Anthology for Musical Analysis*, Holt, Rinehart and Winston, Toronto, Ontario, Canada: H BHolt.

Callaghan, M.-J., Harkin, J., Scibilia, G., Sanfilippo, F., McCusker, K., & Wilson, S. (2008) Experiential based learning in 3D Virtual Worlds: Visualization and data integration in Second Life, *Proceedings of Remote, Engineering and Virtual Instrumentation* (REV 2008) Conference, Dusseldorf, Germany.

Colace, F., Santo, M., Vento, M., & Foggia, P. (2004): A Semi-Automatic Bayesian Algorithm for Ontology Learning. In *International Conference on Entreprise Information* pp. 191-196.

Couchot E.Bret, M & Tramus M-H., (1998) Art virtuel, créations interactives et multisensorielles, *Beaux-arts Magazine*, hors série:11-18.

Creed, C. and Beale, R. (2008) Simulated Emotion in Affective Embodied Agents Affect and *Emotion in Human-Computer Interaction, From Theory to Applications. Lecture Notes in Computer Science Systems And Apllications*. Springer Berlin / Heidelberg. 4868: 163-174

Damer, B. (1997). *Avatars! Exploring and building virtual worlds on the Internet.* Berkeley, CA: Peach Pit Press.

De Nicola, A. Missikoff, M. Navigli R. (2009). "A Software Engineering Approach to Ontology Building". *Information Systems*, 34(2), Elsevier, 2009, pp. 258-275.

DiPaola, S. (1989) Implementation and use of a 3d parameterized facial modeling and animation system, *ACM Siggraph '89 Course Notes, State of the Art in Facial Animation. SIGGRAPH Conference 16th 1989 Boston, Mass 22:20-33.*

Efrat, A., Indyk, P., & Venkatasubramanian, S. (2004). Pattern Matching for Sets of Segments *Algorithmica. Springer New York*, *40*(3), 147–160.

Feulner, J., & Hörnel, D. (1994). *Neural Networks that Learn Harmony-Based Melodic Variations." Proceedings. ICMC, Århus* (pp. 121–124). MELONET.

Fusi, S., & Abbott, L. (2007). Limits on the memory storage capacity of bounded synapses. *Nature Neuroscience*, *10*, 485–493.

Gerstner, W., & Kistler, W. (2002). *Spiking neuron models: single neurons, populations, plasticity.* Cambridge: Cambridge University Press.

Hautop Lund, H., & Ottesen, M. (2008). RoboMusic: a behavior-based approach. *Artificial Life and Robotics Computer Science. Springer Japan.*, *12*(1-2), 18–23.

Khan, K. S., & Al-Khatib, W. G. (2006). Machine-learning based classification of speech and music. *Multimedia Systems. Computer Science*, *12*(1), 55–67.

Kiss J. & all. (2006) Voice interaction system for video games used within "virtual singer" computer interface. *Proceedings of CGAMES Conference. 8th International Conference on Computer Games: AI, Animation, Mobile, Educational & Serious Games. Dublin, Ireland.* Louisville. USA

Koda, T., Ishida, T., Rehm, M., & André, E. (2009). Avatar culture: cross-cultural evaluations of avatar facial expressions. *AI & Society. Computer Science. Springer London.*, *24*(3), 237–250.

Laliberté, M. (1993). Informatique musicale: utopies et réalités (1957-90) *Les Cahiers de l'Ircam no 4. Utopies*, *4*, 155–172.

Mase, K., Yasuyuki Sumi, Y., & Fels, S. (2007). Welcome to the special issue on memory and sharing of experience for the Journal of Personal and Ubiquitous Computing *Memory and Sharing of Experiences Personal and Ubiquitous Computing. Springer London.*, *11*, 213–328.

Maturana, H., & Varela, F. (1980). *Autopoiesis and Cognition: the Realization of the Living.* Robert S. Cohen and Marx W. Wartofsky (Eds.), Boston Studies in the Philosophy of Science 42. Dordecht: D. Reidel Publishing

Mingers, J. (1994). *Self-Producing Systems.* Kluwer Academic/Plenum Publishers.

Minsky, M. (1979). *Web of Thoughts and Feelings.* Sound Recording. Cornell University, Ithaca, NY. April 18th.

Minsky, M. (2000). Commonsense Based Interfaces: To Build a Machine that Truly Learns by Itself will Require a Commonsense Knowledge Representing the Kinds of Things Even A Small Child Already Knows. *Communications of the ACM*, *43*(8), 66–73. doi:10.1145/345124.345145

Miranda, E. R., & Matthias, J. (2009). Music Neurotechnology for Sound Synthesis: Sound Synthesis with Spiking Neuronal Networks. *Leonardo*, *42*(5), 439–442. doi:10.1162/leon.2009.42.5.439

Nierhaus, G. (2009). *Algorithmic Composition, Paradigms of Automated Music Generation.* Vienna: Springer.

Olsen, H. F., & Belar, H. (1961). Aid to Music Composition Employing a Random Probability System. *The Journal of the Acoustical Society of America, 33*, 1163–1170. doi:10.1121/1.1908937

Pinto, H, S, and Martins, J, P. (2004). Ontologies: How can They be Built? *Knowledge and Information Systems, 6*, 441–464. doi:10.1007/s10115-003-0138-1

Pollock, J. (1989). *How to Build a Person*. Cambridge, MA: MIT Press.

Schacter, D. L. (1989). On the Relation Between Memory and Consciousness: Dissociable Interactions and Conscious Experience. In Roediger, H., & Craik, F. (Eds.), *Varieties of Memory and Consciousness: Essays in Honour of Endel Tulving* (pp. 22–35). Hillsdale, NJ: Erlbaum.

Schroeder, R. (2008) Defining Virtual Worlds and Virtual Environments. In *Virtual Worlds Research: Past, Present & Future*. 1(1). 2-3. ISSN: 1941-8477. http://journals.tdl.org/jvwr/article/viewFile/294/248

Smirnov, A., Shilov, N., Levashova, T., Sheremetov, L., & Contreras, M. (2007). Ontology-driven intelligent service for configuration support in networked organizations. *Knowledge and Information Systems Springer London., 12*, 229–253. doi:10.1007/s10115-007-0067-5

Xu J, Yang Zhao Y., Chen, Z. and Liu, Z. (2009) Music snippet extraction via melody-based repeated pattern discovery. *Science in China Series F: Information Sciences*. Science in China Press, co-published with Springer-Verlag GmbH. 52(5): 804-812.

Chapter 10
Learning from Baroque

Carola Moujan
Université Paris 1 LETA/CREDE, France

ABSTRACT

The development of ubiquitous computing has brought up the emergence of a new type of space, sensitive and neither fully material nor totally virtual, within our environments. This chapter discusses, from a cultural perspective, aesthetic and philosophical issues related to what has been called "mixed reality spaces". It aims to show how early examples of interactive art can be found in Baroque architecture and, through analysis of the perceptual means used in some of those works, proposes a strategy for bringing aesthetic depth and relevance into mixed reality installations. Depicting philosophical implications between vision and touch and their consequences for aesthetics, this essay proposes that, while designing mixed reality installations, artist operate a radical shift from the vision to touch in order to create meaningful experiences and preserve freedom for the participant.

INTRODUCTION

Through pervasive technologies, environments have become sensitive. Body movements can be sensed and restored in real time, often through images projected onto the built space. As situated technologies continue to penetrate our environments, irreversibly changing the way we experience space, important questions arise: what kind of relationship do such images establish with their hosting space? How do they alter our perception of it? Do they enhance our experience of space, or disturb it? Are they works of art? Are they "contemporary"? Do we want them? The issues that lie behind these questions are whether or not the use of cutting-edge technology automatically makes a piece an expression of its time; and upon which criteria we might consider a specific technology to be relevant within the context of art. They question us in our relationship to space and change.

DOI: 10.4018/978-1-60960-077-8.ch010

In addition to what has just been said, the question of how these spaces affect our body needs to be asked as well, both from a physical and a from a philosophical standpoint. Indeed, the body is the main agent through which these virtual spaces emerge. Unlike what happens in other artistic languages, in an interactive installation we manipulate images directly and with the whole body -not just vision. This fact intensifies the impression of reality we get from these virtual spaces or, more precisely, creates a new perception of reality that includes them. Freed from mediation devices such as mouse and keyboard, we interiorize all the more so those spaces, as their transformation, sometimes even their generation, happens through our bodies. It is clear, though, that there are many possible ways to address the body, and that different ways will produce different aesthetic experiences. This chapter aims to expose different aspects of bodily implication in mixed reality installation, showing how they might produce radically different effects, among which some that might be desirable for us as participants, and others that might not. I will claim that if we want a mixed reality installation to be something more than a technological demonstration, its essence must be intimately tied to the technical apparatus, the images' content, and the physical space within which it exists. Making environments sensitive through computers is not enough, nor is the quality of the images alone. Mixed reality installations can be considered as works of art only from the moment they strike up a dialogue between the space, the projected images, and the viewer's body. It's in this way that this medium can do better than previous languages did. Yet, very often, such dialogue doesn't take place. If interactive installations often fail to embody the expressive capabilities of this new medium, it's mainly because they focus on technology rather than aesthetic experience. In the words of Anthony Dunne (2006, p.3), *"electronic art has become so technology-driven that it seems concerned only with the aesthetic expression of technology for its own sake"*. This is an important aspect of the problem, and one of the main obstacles to overcome, for when technological fascination takes over, the machine rules, producing superficially exciting pieces that mechanize the body and leave us unsatisfied. Instead, we need to consider what question mixed spaces are an answer to, and what examples in the history of art and design could help us unfold the opportunities that lie beyond the simple illustration of movement flow through space, which, as Usman Haque has pointed out, *"is just as representational, metaphor-encumbered and unchallenging as a polite watercolour landscape"*(2006). Instead, I will argue that Baroque, (with the term considered in a broad sense), provide us with a good paradigm onto which we can rely to build meaningful space events.

MIXED SPACES, REAL PLACES

Why Baroque? Much debate has been held around the term "Baroque". Even the best authors on the subject have had doubts about the notion's weight. Alternatively, either the concept was reduced to the sole realm of Architecture, or restricted to an increasingly limited period of time. Its elusive essence kept escaping categorization until in 1988 french philosopher Gilles Deleuze provided in his book *The Fold*, a definition that anchors the concept not in form or style, but in transformation. Rooting Baroque in Leibniz's thinking, Deleuze extracts the concept from its historical reference, and makes it a trans-temporal feature. He shows how artists with no apparent connection (such as Caravaggio and Paul Klee, for instance) are in fact linked by a secret tie: the fold. Deleuze understands it not as a result (the folded object), but, rather, as an operative mode.

Mathematically speaking, the fold lies in inflection, with the inflection point being the place of all potential places. Considered from a philosophical standpoint, this idea expresses a multi-dimensional conception of reality, where enabling co-existence of multiple and sometimes contradictory possibilities is preferred to solution and choice-making, expressing that, in Baroque aesthetics, essence lies in change, not in shape. The inflection point expresses an entity in between worlds; something that is neither only real, nor exclusively virtual, or, in the words of Paul Klee, *"a non dimensional point, between dimensions"*. An easy parallel can be established here between this "in-between dimensions" attribute and the nature of our mixed reality installation, making inflection particularly relevant when thinking about mixed spaces, and an appropriate metaphor to understand where, beyond outside appearance, might lie the artistic potential of this language.

What is the place for "virtual" spaces within our material world? In which sense can they add something to it, from an aesthetic point of view? Let's step back for a second, and consider the situation as a whole. We are at the heart of an mixed reality installation. Images are projected onto the space; they change according to our actions.

What are those images? What do they tell to the built space, and to our bodies? To answer these questions, we must first realize what psychologist Serge Tisseron calls *the essential spatial character of images*: *"images constitute a form of screen that invites us to explore and transcend it. In this sense, they only work as an opening to the hereafter if we first acknowledge them as a presence endowed with a form of body"* (2003, pp. 131-132).

This acknowledgement makes us aware of a major issue: the fact that images projected in our mixed reality installations are not really *images in space*, but should be considered as *spaces within the space* —spaces invested with a perceptual weight that is all but virtual. As we recognize the multiple spaces interlacing, the sensorial experience becomes more intense: what was originally only visual (an image) has become also tactile (a space). Ultimately, we can say that mixed reality installations fulfill an unfolded promise present in every image: the possibility of penetrat-

Figure 1. Francesco Borromini, dome of San Carlo Alle Quattro Fontane, Rome, 1667. (Photo by Marie-Lan Nguyen/ Wikimedia Commons)

ing and transforming it. Through these interactive images, a new dimension emerges, not just within the mind, neither inside a screen, but in physical space. This is a very powerful fact that artists can use, and arguably in itself a good-enough reason to create mixed reality installations– for, through it, hidden aspects of space may be revealed, enabling multiple readings of a single physical place.

Let's examine the meaning of the word "virtual". When used within this context, what is actually meant is "immaterial". Yet, rigorously speaking, the word it is not quite precise, for even though the matter of these artifacts is an extremely fluid one, nevertheless it still belongs to the material world. *"The CPU of an electronic object is, essentially, physically embodied symbolic logic or mathematics. [...] The algorithm is the logical idea behind the program, a strategy that allows symbolic logic to be translated into a programming language (such as C++) and run through the machine, controlling the flows of electrons through its circuitry. [...] Dematerialization, therefore, means different things depending on what it is defined in relation to: immaterial/material, invisible/visible, energy/matter, software/hardware, virtual/real. But the physical can never be completely dismissed [...]"* (Dunne, 2006, p. 11).

Words such as "real", "virtual" or even "hybrid" cannot accurately define the true nature of mixed reality, for those terms focus on the material aspects, while the essence of these propositions lies on the space-time relationship. In mixed reality installations, time and space are bound together in a very tight, intimate way, with the second being dependent on the first. What these artifacts really are, is a particular type of labyrinth: *"the labyrinth represents the ultimate figure of a space subject to time [...]. The maze's shape depends, however, on the type of relationship that knowledge will establish between time and space. Either the place pre-exists the path, then time adds on and unfolds space, [...] or time invents space, which doesn't exist by himself"*.(Ropars-Wuillemier, M.C., 2002, p. 119). In this new form of labyrinth, material aspects are not the essence, but means for existence.

The real challenge behind mixed reality pieces is neither technical performance nor message delivering, but a deeper, architectural problem: the creation of a place. And what is a place ? A moving entity, an essence rather than a specific form, that emerges in a given space and never stops changing: *"The real places no the earth, that is to say, are susceptible to continuous readings, which is almost certainly to say complex and ambiguous ones. It seems to be a characteristic of them, too, that they have extraordinary changeability, sometimes of use, almost certainly of size [...]"* (Bloomer, K. C., & Moore, C. W.,1977, p.107). In other words: though from a material perspective these spaces might be called "virtual", considered as places, they are real. It is interesting to observe that, as Deleuze shows, in this changeability lies also the core concept of baroque objects: *"The goal is no longer defined through an essential form, but reaches pure functionality, as if declining a family of curves framed by parameters, indivisible from a series of possible declinations[...]" (1988, p. 26)*. Following Bernard Cache, Deleuze points out that this statement also expresses a very contemporary vision of technological objects: *"the new status of objects doesn't refer to a spatial mould, that is to say, to a form-material relationship, but to a temporal modulation that implies a continuous variation of material as much as a continuous development of the form" (p.26)*.

DIALOGUES WITH SPACES

I stated earlier that mixed reality installations could only be considered works of art from the moment they were able to strike up a *dialogue* between the space and the participant, through the technical apparatus. What is a dialogue ? Can we consider any word exchange as "dialogue"? This question is obviously rhetorical. We all know from experience that many word exchanges in everyday life, such as commands and instructions, are not dialogues. However, while these extreme examples of non-dialogical structures stand out clearly, we might not be fully aware of all the issues implied, and of the fact that many of the word exchanges that we have each day might well not be what we think they are. In his book *Éloge de l'aspect*, Pierre-Damien Huyghe explains how while seemingly dialogical, many word exchanges that happen between people are merely communication. The distinction implied here is resumed by Huyghe as follows: while communication aims at a specific goal –the transmission of a precise message–, in a real dialogue, pre-determination is excluded. In the author's words, *""letting know", in fact, is not talking, it's implying in the human form of language a signal that doesn't open up to free response"*. (2006, p. 62). Huyghe's explanation makes us aware of two extremely important issues: first, that openness is essential aspect of dialogue, and second, the fact that, within dialogue, it is not a predefined result that matters, but the quality of the process in itself. Analogously, in baroque spaces –which, as Umberto Eco has shown, are good examples of *open works*–, determination of the effect is forsaken to the benefit of a perceptual experience where space reveals itself in ever transforming ways. This leads us to the question of interaction, where much of the relationship-generation capabilities lie.

It is important to elucidate the fact that behind a seemingly obvious meaning, the word *interaction* conceals contradictory views on the human-machine relationship. This is the main subject of a text by Usman Haque (2006), *Architecture, Interaction, Systems*. For Haque, "*at its fundamental, interaction concerns transactions of information between two systems (for example between two people, between two machines, or between a person and a machine). The key however is that these transactions should be in some sense circular otherwise it is merely "reaction".*" Directly speaking, this means that while an automatic door-opening system might be generally considered a form of interaction, because no feedback is involved in the process, it is, in fact, only reaction. Further ahead, Haque will show that how the quality of possible relationships within an interactive system depends on the number of feedback loops implied in the process –the more loops implied, the richer and subtler the quality of exchanges. "*Multiple-loop interaction does not depend upon complexity, it depends upon the openness and continuation of cycles of response. It also depends on the ability of each system, while interacting, to have access to and to modify each other's goals*". Haque's point of view is based on the work the Cybernetician Gordon Pask, whose 1976's *Conversation Theory* provides a precise framework for developing open and constructive relationships with machines. Brought to the context of this essay, and within the *Conversation Theory* framework exposed by Haque, this means that a dialogue between a participant and a mixed reality space is only possible through a multiple-loop interaction type of system, for, as Haque explains, *[such systems] rely on the creativity of the person and the machine as they negotiate across an interface, and it is this "conversational" creativity, I will argue, that makes these interactions the most desirable*. In addition, and importantly, Haque and Huyghe make us aware of the fact that within the two forms of interaction exposed above lie very different ideas of power exchanges; for while within *dialogue*, or *multiple-loop interaction*, the person and the machine are on a same hierarchical level, in *communication* —that is to say, *reaction* or *single-loop interaction*–, one participant dominates the other.

How and why are these artifacts pure products of our time? I would like to refer to the definition of contemporary art given by Op artist Carlos Cruz-Diez: "*I believe that for a work of art to be "contemporary" instead of taking into account any "traditional aesthetic" consideration, should contemplate the creation of an event, where the dialogue between real space and time is present*". (retrieved 2007). This interesting definition points out two things that are at the core of the formal discourse in mixed-reality installations: the space-time dimension, and the notion of event. But there is more to it: not only these works are space-time events; there are *unstable space-time events*. In these works, flow—not shape— is the key. Similarly, in Baroque architecture, illusions of movement in built masses are conveyed through the use of perspective. Baroque spaces are perceptual plays that recall body memories of haptic experiences, creating an event that is only possible with and through the participant's body. Deliberately ambiguous, they are an invitation to movement, making flow a central element and transformation, the only constant. Complexity, contradiction and paradox play a central role, while determination of the effects is abandoned to the benefit of event creation. By event, I mean a situation where the participant goes through a unique experience which is at once familiar and unexpected, as in, for instance, artistic *happenings* of the sixties, where the precipitated disruption was somehow already present at a latent state in the sociological situation. More generally speaking, we can say that with Baroque, the disruption takes place within a specific space-time situation that combines in a single experience contradictory feelings of expectation and surprise. In baroque architecture, a form of extension takes place: an extension of the space, from the built to the illusionary; and an extension of the illusion, which leaves the virtual to become actual. This extension quality is, as Gilles Deleuze shows, the first condition of the event. *"Extension exists when one element is stretched over the following ones, such that it is a whole and the following elements are its parts. Such connection of whole-parts forms an infinite series that contains neither a final term nor a limit (the limits of our senses being excepted). The event is a vibration with an infinity of harmonics or submultiples [...] "* (2006, p.87).

Deleuze's powerful image of an infinite vibration conveys two essential meanings: it makes us aware of the intrinsic openness of the event, which cannot, by nature, be limited to a pre-determined set of attributes (we will refer to this aspect later), but also, of the fact that an event is a moving entity in which equilibrium is never steadiness but, rather, an unstable state of counterbalancing tensions. Applied to mixed reality installation, this principle might reveal ways in which, following the example of Baroque, the complexity of existence could be expressed through contemporary means.

I would like to get deeper into the potential multi-dimensionality in mixed reality spaces. Earlier on, I mentioned that such installations carried within them the capacity to bring latent spaces into the material world. Yet, we find at work, in many contemporary pieces, what I will call the *visual* approach, in which the installation is considered either as an "interactive painting" or as an "interactive movie".

When mixed reality installations are designed from the *visual* standpoint, the surrounding environment is reduced to the sole role of "screen". This means that all the qualities of the built space that might disturb the main purpose —i.e., supporting the image– need to either be eliminated or ignored. Extreme examples of this can be found in virtual reality devices where, through special equipment (glasses, gloves, immersive rooms), the body landmarks are neutralized, thus allowing the program to take control. The participant does experience immersion, but to the prize of self-consciousness loss. Instead of fusion or dialogue, a form of alienation that takes place. Thus, beyond superficial excitement, it is important that, while designing computer-enabled illusions, we ask ourselves what kind of environments we want to evolve in, and which direction we want technology to take. As Rebecca Allen puts it, *"Living objects and sensitive spaces are provocative concepts, but ones that should be considered by artists, social scientists and technologists alike. Before ubiquitous computing irreparably weaves itself into the fabric of our everyday lives, we need to ask: How do we want pervasive computing to behave?"* (2005).

If we agree on the idea of the ultimate goal of mixed reality installations being their capacity to "augment" reality, to enrich life, we can see here matter for debate. Neither this type of vision-driven installation adds any new qualities to the built space, nor it enhances our experience of it. Compared to the initial situation (the space before anything was "installed"), the sensorial experience has been impoverished. The inadequacy of this type of proposal lies in the fact that, not only the opportunity of revealing latent spatial dimensions offered by the medium hasn't been seized, but something is lost in the process: what was originally three-dimensional (a space), has become two-dimensional (a screen). From then on, it becomes clear that before we make the built space "sensitive", it is important to identify the kind of transformation that would be meaningful within that given context.

SOLUTIONS AND RECOMMENDATIONS

I consider that, in order to avoid this pitfall, the nature of the built space hosting the installation is the first information to be contemplated - that

Learning from Baroque

the proposal ought to add something *within that specific framework*. Looking back into art history can help us understand what this might be about: *trompe l'œil* frescoes in baroque architecture for instance, such as Andrea Pozzo's dome painting for the Jesuit Church in Vienna *(1627)*, stand out as good paradigms. In this type of illusionary work, a visual impression of space is created through the association of the structural lines of the building with an image, and in reference to a given vantage point.

What makes those frescoes fascinating is the fact that we don't perceive them as added elements, but rather, as extensions of the place's essence. They explore what the space might have been, its potential for evolution or change, reinforcing the dynamics of it rather than focusing on its material aspects. They make multiple readings explicit, perhaps revealing original intentions or unconscious dreams that lie behind brick and mortar. Another important characteristic worth underlying, is the fact that, as participants, we are able to freely build our own experience, simply by changing positions within the space. We are not trapped in an illusionary world that looks real, but rather, uses his body as a cursor to walk in and out the illusion. Thus we can say that instead of an illusion of reality, it is the reality of an illusion that is presented to us. As Julian Oliver has pointed out, we can consider this type of work as early examples of interactive art: *"This work proposes a radical kind of interaction design precisely because it shifts the object of interaction from the corporeal into the perceptual: beyond augmentation of the environment it is an augmentation of perception itself, replaying belief against known doubt with the agent as both lens (eyes) and steering controller (body movement)"* (2008).

One might argue that applying this concept within the context of mixed reality installations is only possible in site-specific work, where the intervention takes place in a space which already has a strong character to rely upon; and it seems indeed easier to imagine this within a place with

Figure 2. Trompe L'Oeil on the Jesuit Church, Vienna, by Andrea Pozzo (1627). (Photo by Alberto Fernandez/ Wikimedia Commons)

clear structural lines, rhythm, and contrast, than in more neutral spaces such as art galleries' "white cubes". It has also been said that the site-specific approach is a more meaningful one: *Rather than relating the impact of technology to everyday life, art criticism in this area glamorizes technology as a source of aesthetic effect to be experienced in galleries*. (Dunne, 2006, p.3). While discussing these interesting issues goes beyond the scope of this essay, I will nevertheless emphasize the fact that even places as abstract as a white cube have material qualities, and that those qualities will play an important role in the final work. The white walls, for instance, will reflect light in a very different way than stone would, and this feature can deliberately be put into play by the artist. It is precisely what Japanese artist Mazaki Fujihata does in his installation *Morel's panorama*.

In this work, named after a masterpiece of fantastic literature, a series of cylindric images

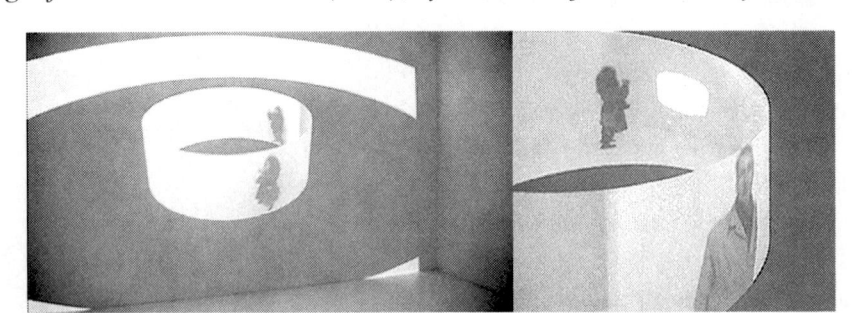

Figure 3. Images from Morel's Panorama (2003), by Mazaki Fujihata. Photo by Carola Moujan

in which we see ourselves as we are "from the exterior" are projected at the back of a white room. Inversion of the classical point of view of the portrait, with the subject facing the viewer at the center of the composition, creates an impression of strangeness –a feeling enhanced by the whiteness of the space which, as it reflects projected light, fills the room creating a direct impression of immersion.

This example shows well how if we design the object taking the environment into consideration, instead of focusing only on its superficial attributes, even a very simple space can become rich and meaningful. Closer in time, and without relying upon representation, Op artists such as Bridget Riley or Carlos Cruz-Diez have achieved similar goals through the use of colour and contrast. A virtual space is thus created *"a kind of painting that reaches for ambiance and only exists in the space between the painting and the viewer"* (Weinhart, 2007, p. 29). As in mixed reality installations, here the space-time couple plays the main role, and movement, a central theme. The material of these works is perception itself –as Bridget Riley's *"Perception is the medium"* quote admirably resumes. Like in mixed reality installations, often a big part of the effect is left to the participant's initiative:

Some artists prefer to use the viewer's own physical movement –walking past or moving the head–, to generate a large part of the effect of movement within the work. This can be seen in the work of Jesús Rafael Soto [...] Such images, like the installations, could be said to be completed physically, by the presence of the viewer –to exist only partially until they are engaged with by the spectator who creates that engagement physically, by moving in relation to the work. (Weinhart, 2007, p.50).

From a philosophical standpoint, this is of major importance: it shows us how, beyond the aesthetic purpose, mixed reality installations can become a tool for challenging the very notion of what we call "real". By shaking the conventional idea of an objective "reality" –upon which we continue to live in spite of scientific evidence– these objects can open up new ways of experiencing and understanding space. Boundaries between what can be considered "real" space and what would just be illusion start fading away, as do limits between the body and the space –between the subject and the object. A form of psychic fusion between the participant and the space takes place, and I believe it is precisely along those lines that mixed spaces can fulfill a specific role that hasn't been felt before: the transformation of the material world into a less static, more fluid environment. On a deeper level, we can consider that it is in this reality-destroying capacity that lies much of the artistic potential of these artifacts –for it is the role of contemporary art to set us free from the boundaries of social convention.

They are examples of what Michel Foucault has called *heterotopia*:

The heterotopia is capable of juxtaposing in a single real place several spaces, several sites that are in themselves incompatible [...] thus it is that the cinema is a very odd rectangular room, at the end of which, on a two-dimensional screen, one sees the projection of a three-dimensional space [...]..their role is to create a space of illusion that exposes every real space, all the sites inside of which human life is partitioned, as still more illusory. (1984).

It is interesting to observe the fact that this destruction of reality was also the role of cinema at its origins. Back in 1936, Walter Benjamin wrote:

Our taverns and our metropolitan streets, our offices and furnished rooms, our railroad stations and our factories appeared to have us locked up hopelessly. Then came the film and burst this prison-world asunder by the dynamite of the tenth of a second, so that now, in the midst of its far-flung ruins and debris, we calmly and adventurously go traveling. With the close-up, space expands; with slow motion, movement is extended. The enlargement of a snapshot does not simply render more precise what in any case was visible, though unclear: it reveals entirely new structural formations of the subject. (1936, p.13).

Like cinema, and with increased persuasion means, mixed reality installations have the power to set us free from conventional thought, from fixed space-time, from supposedly "objective" conditions of existence. And it is precisely through this feature that these works join in the general questioning of contemporary art. It is important to keep this in mind, for today we observe a tendency to confine mixed reality installations to specific digital and new media art events, dissociated from other disciplines of contemporary art. Yet, while from a material perspective these artifacts belong to the world of computers and technology, as aesthetic experiences they aren't, and shouldn't be, disconnected from the rest of the artistic production. It is important that, while experimenting with technology and innovation, artists working in this field should also look at art history, searching for ways in which mixed spaces can bring new answers to old questions. Only then will the medium win it's spurs as an art form, for while through technology new tools and languages emerge, fundamental questions remain.

Bloomer and Moore's description of real places quoted earlier clearly shows that their essence lies in changeability. If we agree that this is also the main attribute of mixed reality spaces, then we can say that, through ubiquitous computing, images can become places, built spaces which are not yet places, reveal as such, and real places express multiple, unfolded dimensions, enabling new and complex readings of space. And, beyond aesthetics, mixed reality installations hold the potential to develop a place's relationship-generation capabilities, whether by correcting or improving negative situations, by taking positive ones to the next level, or as research tools for solution testing.

Considered that mixed reality installations ought to create a dialogue between bodies, images and spaces, how does this articulation happen? As it was said before, the challenge is above all architectural; and it's no accident that, as Anthony Dunne (2006, p.16) points out through the example of Toyo Ito's *Dreams Room* (1991), *"the aesthetic possibilities of this form of dematerialization* [ubiquitous computing] *have been best exploited by architects"*. In France, a good example of architects successfully investing the realm of mixed reality can be found in the work of Electronic Shadow –a team composed by Naziha Mestaoui, architect, and Yacine Ait Kaci, new media and video artist, who say of their own work: *"Electronic Shadow proposes a new type of space that integrates from conception, its digital extension to create an entity that doesn't oppose real to virtual but combines them in the creation of a new perception of reality"*. (Ait Kaci and Mestaoui, 2006).

Figure 4. Electronic Shadow, SuperFluidity (2006). Photo ©Electronic Shadow

An important clue lies behind the seemingly obvious statement made above: what architects are more aware of than other kinds of designers, is the fact that the experience of space is mainly driven by the sense of touch, not vision. In the words of Kent Bloomer and Charles Moore (1977, p.34), *"no other sense deals as directly with the three-dimensional world or similarly carries with it the possibility of altering the environment in the process of perceiving it; that is to say, no other sense engages in feeling and doing simultaneously"*[7]. Images projected in mixed reality installations are generally created behind a computer screen, which filters tactile qualities. Yet, in material spaces, it is texture, movement, scale, temperature, and other sensations related to touch that play the main role. While the artist working on its computer pays much attention to the visual qualities of his work, once in the material world the images will not be perceived quite in the same way; and "collateral" aspects such as the temperature of the air or the distance from the room's entrance, might reveal to be more important for the shaping of the actual experience than the images alone.

Mark Weiser —considered the father of ubiquitous computing— and his colleague John Seely Brown once wrote that *"the most potentially interesting, challenging, and profound change implied by the ubiquitous computing era is the focus on calm"*(1996). By saying that, they were acknowledging the fact that while, in theory, technology is supposed to improve our quality of life, in practice, multiple solicitations in a world overloaded with high-tech devices often leads to stress and frustration. Foreseeing a near future when computers would massively embodied within the environment, the authors warn us about the potential dangers of such technological ubiquity, inviting us to take a step ahead and design them to be encalming rather than exciting: *"If computers are everywhere, they better stay out of the way, and this means designing them in a way that people [...] remain serene and in control"*. The authors argue that this desired calm can be achieved by *"engaging both the center and the periphery of our attention"*, pointing out the fact that very often, while designing electronic devices, the object's environment doesn't get the attention it requires and deserves. Already in *The Computer for the Twenty-First Century*, an essay written five years earlier, Mark Weiser stresses the importance of the periphery: *"too much design focuses on the object itself and its surface features without regard for context. We must learn to design for the periphery so that we can most fully command technology without being dominated by it."*(1991)

The periphery –which we could also call background– provides very important information to the participant. It subtly keeps him aware of events and things coming up or happening around, without occupying the center of his attention. This enables fluid transitions between perceptions, letting the participant adjust to change and therefore remain stable through it. Reaching beyond its specific technological context, Weiser's *The*

Computer for the Twenty-First Century heavily questions power relationships between humans and machines, implying that we must choose to address the periphery in our designs if we want to avoid being dominated by them.

How does this happen? If the center can be easily engaged through vision, how do we address the periphery? The answer to this question lies, as Walter Benjamin explains in his canonic 1939 text *The Work of Art in the Age of Mechanical Reproduction*, in the sense of touch:

Buildings are appropriated in a twofold manner: by use and by perception-- or rather, by touch and sight. Such appropriation cannot be understood in terms of the attentive concentration of a tourist before a famous building. On the tactile side there is no counterpart to contemplation on the optical side. Tactile appropriation is accomplished not so much by attention as by habit. As regards architecture, habit determines to a large extent even optical reception. (1939, p.14).

The last sentence of this quote is particularly interesting, for it states the implication between vision and touch, suggesting an overall presence of tactile sensations in images and spaces, and underlining the fact that touch precedes vision, as Bloomer and Moore's *Body, Memory and Architecture* clearly exposes:

The body image [...] is formed fundamentally from haptic and orienting experiences early in life. Our visual images are developed later on, and depend for their meaning on primal experiences that were acquired haptically. [...] Thus haptic experiences which include the entire body give fundamental meanings to visual experiences, while visual experiences serve to communicate those meanings back to the body." (1977, p. 44).

The hierarchy between the two senses having been established, we understand that, in order to create rich aesthetic experiences through mixed reality installations, we need to stop focusing on vision alone and start organizing the experience through the sense of touch. Significantly, this shift from vision to touch is precisely one of the defining features of Baroque: *"The openness and dynamism of the Baroque mark, in fact, the advent of a new scientific awareness; the tactile is replaced by the visual (meaning that the subjective element comes to prevail) [..]"* (Eco, 1989, p. 13).

An important aspect of touch, particularly relevant in our case, it is the fact that it is impossible to identify, within the act of touching, a touching object and a touched subject. This means that through touch boundaries between subject and object disappear; a psychic fusion takes place. Applied to mixed reality spaces, we understand that the more present and relevant tactile sensations are in an installation, the stronger the immersion. In other words: as tactile sensations become more present and relevant, feelings of fusion increase. An extreme example of this could be a steam bath, where the addition of heat and humidity make the quality of the air such that, when interacting with the skin, it becomes difficult to say where the participant's body ends and the steams begin. Similarly, in installations where images are projected, the special quality of the

Figure 5. Francesco Borromini, Sant'Ivo alla Sapienza (1642-1650). Photo © Joaquin Lorda/ Universidad de Navarra

air –hot and full of photons– plays an important role in the perception of the piece; yet, it is rare to find examples where this issue is addressed at all. A good exception can be found in the work of Toyo Ito, as Luigi Prestinenza Puglisi (2005, p.20) points out: *"Ito often works on an image drained of all meaning, almost reduced to an impressionist state, a stage that has reached the senses but has not yet made a formal impression on the intellect and in which space no longer appears to be a vacuum in which solid bodies live, but rather a medium through which information is diffused"*.

FUTURE RESEARCH DIRECTIONS

Transformation is an old, recurrent theme in architecture. From the baroque examples described in this essay, through Gerrit Rietveld' *Schröder House* (1924), to the work of contemporary architects such as Zaha Hadid, the will to overcome the limitations imposed by materiality into our places of dwelling has been a major utopia. From this perspective, mixed reality spaces appear as a possible way to start fulfilling this old quest, enabling a more fluid understanding of the concept of inhabitation, which would less rely onto habit and repetition to the benefit of transformation and experience. This could lead to increased freedom in very concrete ways, and to the enhancement of reality through openness. Instead of envisioning augmented reality as a superposition of information layers onto the environment, we could consider augmenting the experience of reality itself - that is, enabling within the built space, multiple and variable practices, preferably open ones defined through dialogues between the participant and the space. This, as Usman Haque has pointed out, would lead to a deep transformation of the concept of space design. Furthermore, instead of limiting the role of these artifacts to the sole realms of art and entertainment, such devices could also be envisioned as mediators between humans, buildings, and cities.

In many contemporary cities, for instance, poor articulation between the center and the suburban areas creates major social and architectural problems. Up until now, proposed solutions have often implied big structural works including massive demolition and re-building. It would be interesting to start thinking about mixed reality installations and their capacity to enable fluid transitions through perception instead of material, as alternatives to bold destruction. And, here again, there are interesting lessons that could be learned from baroque cities.

CONCLUSION

As situated technologies continue to expand, a new kind of space emerges within the material world —a space made of images, of computers, of illusion, of interaction and, importantly, of relationships. This new kind of space will have an increasingly dramatic impact in our everyday life; an impact that will radically change the way we perceive spaces and places. Belonging both to the world of technology and to the world of aesthetics, mixed reality spaces are fundamentally ambiguous and complex; and beneath their surface, there are forces at work that will have big consequences in the way we relate to the environment.

For artists and designers, this is of major importance, for it might lead either to the creation of pieces that contribute to a richer and deeper experience of reality or, inversely, lead to sensorial impoverishment and servitude. Being objects of a new kind, mixed reality spaces arrive with no previous cultural framework to refer to. This leads to situations in which technology for its own sake takes control, producing pieces that are both aesthetically superficial, and autocratic. Yet, the forms generated by that self-referred technological view slowly fixate within the vocabulary of this language under construction, and based on them, new generations of pieces are created, producing the cold, superficially exciting imagery that is so often associated with digital art. To overcome this state of things, it is important that we clearly identify what the core of mixed reality installations' aesthetics might be. Through these pages, I attempt to show the important analogies that exist between this new type of illusionary spaces and Baroque art, in order to suggest that Baroque –understood not as a style, but as an operation, *a fold*– can stand as a good paradigm when designing mixed reality spaces.

Starting from what could be called early examples of interactive art, this essay questions the widely-spread conception of mixed reality installations as "living paintings" (which implies contemplation and establishes static roles between the subject and the object) and reaches for a broader understanding of the relationships taking place. Between the participant and the work, by letting go of control, making place for experience. Between material and digital spaces, by making them talk instead of considering the first a mere container for the second. Throughout the paper, I point out the importance of touch, in an attempt to bring to the foreground a non-visual understandings of the forces that are at play.

Through opposition of vision and touch, I aim to underline two contradictory views on mixed reality installations: one, that we could call *visual*, which alienates the participant and impoverishes space, and another one, *tactile*, that preserves us as participants from alienation and enhances our relationships with the environment. I believe that only through full understanding of this notion we shall start foreseeing how these artifacts might, beyond simple novelty, open up new ways of inhabitation and bring poetry into everyday life.

By abandoning the fixed vantage point and bringing movement in front of the scene, baroque architects put the participant in the center, inviting him to build a personal experience of space. This might also well be the ultimate goal of mixed reality spaces.

REFERENCES

Ait Kaci, Y., & Mestaoui, N. (2006). *Electronic Shadow*. Retrieved September 15th, 2009 from http://www.electronicshadow.com/

Allen, R. (2005). Sensitive Spaces, in *Living Objects, Sensitive Spaces*, catalog of the *ArtFutura 2005* exhibiton, Barcelona: ArtFutura. Retrieved December 22, 2008 from http://www.artfutura.org/v2/artthought.php?idcontent=10&idcreation=37&mb=6&lang=En.

Benjamin, W. (1936), *Art at the Age of Mechanical Reproduction*. Retrieved on October 1st, 2009 from http://design.wishiewashie.com

Bloomer, K. C., & Moore, C. W. (1977). *Body, Memory and Architecture*. New Haven, CT: Yale University Press.

Cruz-Diez, C. (n.d). *My Reflexions on Color*. Retrieved January 9, 2007 from Carlos Cruz-Diez Official Site, www.cruz-diez.com

Deleuze, G. (2006). *The Fold. Translation by Tom Conley*. Minneapolis, MN: University of Minnesota Press.

Dunne, A. (2006). *Hertzian Tales. Electronic Products, Aesthetic Experience and Critical Design*. Cambridge, MA: MIT Press.

Eco, U. (1989). *The Open Work*. Cambridge, MA: Harvard University Press. Translation by Anna Cancogni.

Foucault, M. (1984). Dits et écrits 1984, Des espaces autres (conférence au Cercle d'études architecturales, 14 mars 1967), in *Architecture, Mouvement, Continuité*, n°5, october 1984, 46-49. Retrieved September 20th, 2009 from http://foucault.info/documents/heteroTopia/foucault.Hetero Topia.en.html. Translation by Jay Miskowiek.

Haque, U. (2006). Architecture, interaction, systems. First developed for an article in *AU: Arquitetura & Urbanismo*, N° 149 August 2006, Brazil. Retrieved Februray 25th, 2009 from Usman Haque Official Site, www.haque.co.uk.

Huyghe, P. D. (2006). *Éloge de l'aspect*. Paris: Éditions Mix. Translation by Carola Moujan.

Oliver, J. (2008). *Optical Illusion Art as Radical Interface*. Retrieved February 18th, 2009, from http://julianoliver.com.

Prestinenza Puglisi, L. (2005). *Hyperarchitecture: Spaces in the Electronic Age*. Basel: Birkhauser.

Ropars-Wuilleumier, M. C. (2002). *Écrire l'espace*. St Denis: Presses Universitaires de Vincennes. Translation by Carola Moujan.

Tisseron, S. (2003). *Le bonheur dans l'image*. Paris: Les Empêcheurs de penser en rond/Le Seuil. Translation by Carola Moujan.

Weinhart, M. (2007). In The Eye of The Beholder. In *Op Art* (pp. 18-38). Catalogue of the exhibition held at the Schrim Kunsthalle, Francfort, february 17th, 2007 - may 20th, 2007. Köln: Walter König.

Weiser, M. (2001). The Computer for the Twenty-First Century, *Scientific American*, September 1991, 94-100. Retrieved from http://www.ubiq.com/hypertext/weiser/SciAmDraft3.html.

Weiser, M., & Brown, J. S. (1996). The Coming Age of Calm Technology. Revised version of *Designing Calm Technology*. PowerGrid Journal, v 1.01, july 1996. Retrieved March 10, 2009. Website: http://www.ubiq.com/hypertext/weiser/acmfuture2endnote.htm.

ADDITIONAL READING

Argan, G. C. (1952). *Borromini*. Milan: Mondadori.

Bachelard, G. (1994). *The Poetics of Space*. Boston, MA: Beacon Press.

Baudrillard, J. (2001). *Impossible Exchange*. London: Verso. Translated by Chris Turner.

Bollnow, O. F. (1961). Lived Space, in *Philosophy Today 5*, Nr.1/4, 31-39. Translated by Dominic Gerlach.

Cache, B. (1995). *Earth Moves: The Furnishing of Territories*. Cambrige, MA: MIT Press.

Derrida, J. (2000). *Le toucher, Jean-Luc Nancy*, Paris: Gallilée, Deleuze, G. (1990). *The Logic of Sense*. New York, NY: Columbia University Press. Translated by Mark Lester with Charles Stivale.

Dubberly, H., Pangaro, P., & Haque, U. (2009). What is Interaction? Are There Different Types? in *Interactions*, Vol XVI.1, 69-75.

Ehrenzweig, A. (1967). *The Hidden Order of Art*, Berkeley, CA: Univ. of California Press.

Gibson, J. J. (1966). *The Senses Considered as Perceptual Systems*. Boston: Houghton Mifflin.

Greenfield, A. (2006). *Everyware: The dawning age of ubiquitous computing*. Berkeley, CA: New Riders.

Haque, U. (2007). The Architectural Relevance of Gordon Pask, in Lucy Bullivant (Ed.), *4dSocial: Interactive Design Environments* (pp. 54-61). London: John Wiley & Sons.

Klee, P. (1950). *On Modern Art*. London: Faber and Faber.

Manzini, E. (1989). *The Material of Invention*. Cambrige, MA: MIT Press.

Maturana, H., & Varela, F. (1980). *Autopoiesis and cognition: the realization of the living*. Berlin: Springer.

Merleau-Ponty, M. (1962). *Phenomenology of Perception*. London: Routledge.

Murray, T. (2008). *Digital Baroque. New Media Art and Cinematic Folds*. Minneapolis: University of Minnesota Press.

Pask, G. (1976). *Conversation Theory, Applications in Education and Epistemology*. New York, NY: Elsevier.

Riley, B. (2009). *The Eye's Mind, Collected Writings 1965-2009*, Robert Kudielka(Ed.), London: Thames & Hudson.

Simondon, G. (1958). *Du mode d'existence des objets techniques*. Paris: Aubier.

Vasseleu, C. (1996). Touch, Digital Technology and the Ticklish. In *Touch* (pp. 7–12). Artspace.

KEY TERMS AND DEFINITIONS

Augmented Reality: The expression generally refers to the localization of digital information layers within the built environment.

Mixed Reality Installation/Space: An interactive architectural proposal that combines built spatial elements with illusionary ones.

Periphery: The part of what we perceive that doesn't occupy the center of our attention.

Ubiquitous Computing: Technology that enables/enhances physical interaction with buildings, spaces, and objects. Also referred to as Pervasive technologies.

Chapter 11
Synthetic Worlds, Synthetic Strategies:
Attaining Creativity in the Metaverse

Elif Ayiter
Sabanci University, Turkey

ABSTRACT

This text will attempt to delineate the underlying theoretical premises and the definition of the output of an immersive learning approach pertaining to the visual arts to be implemented in online, three dimensional synthetic worlds. Deviating from the prevalent practice of the replication of physical art studio teaching strategies within a virtual environment, the author proposes instead to apply the fundamental tenets of Roy Ascott's "Groundcourse", in combination with recent educational approaches such as "Transformative Learning" and "Constructionism". In an amalgamation of these educational approaches with findings drawn from the fields of Metanomics, Ludology, Cyberpsychology and Presence Studies, as well as an examination of creative practices manifest in the metaverse today, the formulation of a learning strategy for creative enablement unique to online, three dimensional synthetic worlds; one which will focus upon "Play" as well as Role Play, virtual Assemblage and the visual identity of the avatar within the pursuits, is being proposed in this chapter.

INTRODUCTION

In his book "Exodus to the Virtual World" economist Edward Castranova predicts that a migration of considerable proportions from the physical realm to three dimensional, online synthetic worlds is to be expected within the next few decades. The anticipated outcome would be a demographic landslide of significant enough socio-economic impact to constitute a need for compelling changes in political, social, cultural and economic strategies not only in the virtual but also the physical realm (Castranova, 2007).

As opposed to a discrete, one way migration, as would be the case in population shifts in the physical world, the anticipated migration would be of a continuous nature, with migrants switching back and forth between the physical and the

DOI: 10.4018/978-1-60960-077-8.ch011

Figure 1. The Avatar "OrKa", by Alpho Fullstop, aka. Elif Ayiter, Second Life®, 2009

synthetic world. If, during this ebb and flow of time allocation more and more hours of activity become appropriated by the virtual world the physical world would suffer the consequences primarily through the loss of revenue generated by the consumption of (physical) goods. However, equally impactful would be the loss of interest and attention towards (physical) socio-cultural occurrences, events and policy. By looking at the current health indicators of virtual economies, the earnings of which can readily be translated into physical currencies such as the US Dollar, Castranova predicts that if a sufficiently large number of players migrate to virtual pastures the consequences upon physical economies, and by extension socio-political structures, will be powerful enough to instigate fundamental changes in (physical) public policies as well as a re-examination/re-definition of (physical) socio-cultural mechanisms globe-wide. Furthermore, Castranova sees this as a more than likely occurrence when viewed within the economic theory of human time use, the allocation of attention and the attractiveness of virtual worlds within its context, as well as the growth in the gaming industry coupled with the emergence of ubiquitous technologies.

Since creative practices are inextricably intertwined with the socio-cultural milieu within which they flourish, it would follow that vast change, not only in terms of the actual creative output itself, but especially and more importantly in terms of the contextual premises within which this creative output is generated should also be expected. Malcolm McCullough approaches the the process of virtual creativity through an examination of the virtual medium itself and the ensuing requirements of craftsmanship which this medium brings to bear upon the creative processes involved in the realization of virtual artifacts. According to McCullough, individuated human craftsmanship, a term largely overlooked by modernist art and design movements, is being, once again, brought very much to the fore by the practitioners of digital creativity: While the assembly line of mechanized industry predicates that the developer/conceiver of the design object is inevitably removed from the actual phase of its production; artists, by and large, seem to have embraced the distinction of labor between concept and realization as well, where pre-eminence would appear to have been allotted to the conceptual phase of the work. Today, the affordances of the digital

Figure 2. Virtual Assemblage: Skin created by Pandora Wrigglesworth and limb attachments created by Lucia Cyr are combined into a novel avatar by the author. Second Life®, 2008

medium, with its pronounced ease of instigating playful improvisations and the ability to produce many variations and iterations of a single artifact allow for a flexibility of output which is once again attracting virtual content creators to the very process of production itself (McCullough, 1996). When McCullough's observations are coupled with Castranova's description of the mechanisms of virtual economies however, we end up at a juncture where we may well be finding ourselves facing a mode of creativity which harkens back to the days before the Industrial Revolution where the creator of the artifact is not only the craftsperson thereof but also the merchant of his or her own output. As appears to be the case with higher educational activity in the metaverse in general, in the virtual extensions of most of the art educational institutions manifest in Second Life® today, teaching is usually seen as a mere extension of real life studio teaching, with assignments and teaching methodologies closely emulating what goes on in the physical campus. Again, in the case of a considerable number of art educational establishments in Second Life® the virtual campus at which the learning activity occurs is a very close adaptation if not exact replication of the real world campus.

However, the problem does not reside in how the learning process is approached within the boundaries of the virtual campus alone: Art Schools, be they physical or virtual, may need to adapt and change their learning content and strategies in the face of the oncoming landslide predicted by Castranova: While it may be premature to relinquish present day art educational methodologies in their entirety, the author nonetheless feels that provision for change, and on a noteworthy scale at that, needs to be incorporated into the present day art educational curriculum. It appears to be fairly evident from present day creative activity embarked upon by the Residents of a metaverse such as Second Life® that this change will not only involve the attributes and nature of the artistic output itself but will also need to take into account the changes in behavior, in usage, in utility and function on behalf of the creator/users of art objects and artifacts; as well as the locus of the usage itself; i.e., immersive, participative synthetic worlds, worlds which, according to Castranova, define their core premise of existence as embedded in the provision of "fun" (Castranova, 2007). Furthermore, an entirely novel student profile, comprised of mature player/learners who wish to further their creative abilities for enhanced participation in the ongoing metanomic game, may also manifest themselves at the doorsteps of art educational institutions within a foreseeable future; and again it would seem that provision for their specialized learning needs would need to be taken into account.

LEARNING AND LUDOS

The State of the Art of Art Education in Synthetic Worlds

Notwithstanding that much has been accomplished in online educational systems that address the needs of most mainstream educational needs (Hill, 2003), the open-ended nature of art education requires special solutions which involve opportunities of unpredictability, associative processes, perceptual transformation and ultimately behavioral change.

While web 2.0 domains have increased user interaction and participation, the metaverse has taken huge steps in the realization of a domain where awareness between participating agents is taken to an entirely new level providing not only the capability of social interaction and participation but also that of "presence", creating far deeper reaching implications than what a mere novel display system or tool would indicate: New forms of embodiment, of presentation as well as perception and indeed of autopoiesis are being materialized, as have also been the prior case in online games and simulations (Gredler, 2004).

Figure 3. The jellyfish avatar created by the author instigates novel artistic output in its turn: "Arrival, by Catarina Carneiro de Sousa, aka. CapCat Ragu. Second Life®, 2009

However, while computer games have the disadvantage of predefined structure and purpose, the metaverse poses the opportunity/challenge in that users define and create their own content and purpose. Thus, while having profound effects on every conceivable profession and walk of life, the effects of the metaverse on the creative arts does merit special consideration.

The highly socially interactive and emergent nature of the metaverse herself does indeed provide a satisfactory environment for the implementation of successful art educational strategies that position themselves in an open-ended discourse. It is thus surprising that very little usage seems to be made of these novel affordances by most art educational approaches at the present time. Instead, art institutions seem to be set upon following a far more pervasive and wide spread trend: Although Second Life® is used as a learning platform by hundreds of higher educational institutions worldwide (Lagorio, 2007), the general lack of concern over whether the unique properties of this thoroughly novel human condition can be exploited to develop entirely novel learning strategies is noteworthy. The overriding majority of Second Life® universities have appropriated dedicated sims upon which campuses in which learning activity that is entirely cut off from the rest of the metaverse have manifested. Most of these campuses have been built as exact replicas of their physical counterparts, thus metaverse learning activity is considered as a mere extension of education in the physical world. Indeed, in a considerable number of cases teaching is undertaken by faculty whose presence in the metaverse is limited to this activity alone:

What is it that we can do in a 3D virtual world that we cannot do out of it? I can attend a formalized teaching session but if all the time I want to build a castle or fly to the other side of the island I will not be too concerned with the contents of the lecture. However if the learning tasks we construct involve building and flying then the learning itself is embedded in the platform's unique attributes. This is the current challenge and our biggest questions surround how we might evolve these learning activities. Very few educationalists are currently involved in this. We need a way of assessing our impact in Second Life without influencing the process by the observation itself whilst yet allowing our assessments to be both valid and reliable. So many factors influence knowledge exchange and the learning of new skills when we start using immersive technologies to teach. (Kirriemuir, 008)

McPherson and Nunes (Mcpherson & Nunes, 2004) propose that the design of online learning environments should be based upon sound pedagogical models, appropriate to a specific educational scenario. For ground<c>, this pedagogical model is the Groundcourse (Ascott, 2003), a methodology, which through the emphasis it put upon behavioral change as an approach to the enablement of creativity, especially through the enactment of new personalities, i.e., role-play, is deemed to be particularly suited to the present quest of the author vis a vis the proposed realm of implementation, i.e. the metaverse.

The Precedent

Combining cybernetics and constructivist educational theory, The Groundcourse devised a flexible structure the aim of which was to create an environment which would "enable the student to become aware of himself and the world, while enabling him to give dimension and substance to his will to create and change"; achieved through a drastic breaking down of preconceptions related to self, art and creativity. Thus the operative tenet that was employed was one of providing an environment that fostered the rethinking of preconceptions, prejudices and fixations with regards to self, society, personal/social limitations, art and all the ensuing relationships through a carefully thought out, coordinated and orchestrated range of assignments and exercises that entailed behavioral modification and indeed change.

During one of the most important assignments of the Groundcourse the problem that students had to address was the task of acquiring and acting out a totally new personality, which was largely the converse of what they would consider to be their normal selves. These new personalities were monitored with calibrators that were designed to read off responses to situations, materials, tools, and people within a completely new set of operant conditions. These responses were then used in the creation of mind maps to be utilized as consultational charts enabling handy reference to behavior pattern dictated by change in the limitations of space, substance, and state. These "new" personalities were asked to form hexagonal groups which had the task of producing an ordered entity out of substances and space in their environment, with severe limitations on individual behavior and ideas, these forming the "irritants",i.e. the educational aids of limitation in the pursuit of creative enablement:

The Groundcourse places the student at the centre of a system of visual education designed to develop in him awareness of his personal responsibility towards idea, persons and the physical environment such that he may contribute to a social context within which his subsequent professional activity may become wholly creative and purposive. The intention of the Groundcourse is to create an organism which is constantly seeking for irritation. The term "organism" may be applied to both the individual student and to the Groundcourse as a whole. (Ascott, 2003)

Students were then invited to return to their former personalities, making a full visual documentation of the whole process in which they had been engaged, searching for relationships and ideas unfamiliar to the conventional definitions of art, reflecting and becoming aware "of the flexibility of their responses, their resourcefulness and ingenuity in the face of difficulties. What they assumed to be ingrained in their personalities they now tend to see as controllable. A sense of creative viability has been acquired". (Ascott, 2003) The Groundcourse, with its pivotal emphasis on behavioral change as a founding tenet for the enablement of creativity, utilized the creation and enactment of new personalities as an educational process. This corresponds to the present day phenomenon of role-play in MMORPG's and the metaverse. Research conducted in the emerging field of Cyberpsychology also substantiates the importance of role-play, the acquisition of alternative characters and indeed the acquisition of many alternative selves in the engenderment of behavioral change not only within the virtual environment itself but also, by extension, in real life. Thus, it is the position of the author that much insight and benefit can be attained from a critical examination and subsequent adaptation/re-interpretation of the Groundcourse's educational philosophy and premises as a pedagogical model aiding the enablement of creativity in a metaverse.

Transformative Learning and Constructionism

While Experiential Learning and Cybernetics were pivotal to the educational theory of the Groundcourse; relevant educational theory to the quest at hand that has been formulated between then and now. Two recent developments in adult education, namely Transformative Learning and Constructionism are also felt to be relevant subjects of inquiry.

As early as 1966 Ascott alerts readers to the emergence of "a new, leisured class" that will be involved in creative pursuits, furthermore a class which falls outside of the boundaries of traditional art educational practice. The current phenomenon of creative participation and sharing via www2 domains seems to amply validate Ascott's early claim who structured his learning system as a fluid, symbiotic construct within which diverse learner groups could be accommodated: Transformative learning (Sheared, 2001), specifically addresses adult education and lifelong learning, as a process of getting beyond gaining factual knowledge alone to instead become changed by what one learns in some meaningful way. It involves questioning assumptions, beliefs and values, and considering multiple points of view, coming out of Jack Mezirow's earlier theory of perspective transformation. In theorizing about such shifts, Mezirow proposes that there are several phases that one must go through in order for perspective transformation to occur. "Perspective transformation involves a sequence of learning activities that begins with a disorienting dilemma and concludes with a changed self-concept". While instrumental learning involves cause-effect relationships and learning through problem solving, communicative learning necessitates actively negotiating one's way 'through a series of specific encounters by using language and gesture and by anticipating the actions of others' (Mezirow, 1991). The former is about prescription whereas the latter is about 'insight and attaining common ground through symbolic interaction' with other persons.

Beyond role-play, the importance of playful activity itself as well as the building of concrete objects, i.e. toys, in the development of creative thinking is yet another concept that can be adapted with facility to the Groundcourse's key strategies: Seymour Papert's Constructionist learning is inspired by constructivist theories, as well some of the cognitive theories of Jean Piaget (Steffe, 1995). Constructionism holds that learning can happen spontaneously when people are engaged in actively making things (Kafai, 1996). To Papert, people best learn through constructing personally meaningful products that express something of importance to them. "To the adage 'you learn by doing' we add the rider 'and best of all by thinking and talking about what you do'" (Papert, 1990). It is thus, a grave mistake, in Papert's view, to forsake or cast off concrete thinking, in favor of purely abstract thought. Constructionism is a way of making formal, abstract ideas and relationships more concrete, and therefore more readily understandable.

Some of the research on which "Serious Play" is based has been charted into basic concepts such as play and identity; while the goals of the method are listed as social bonding, emotional expression, cognitive development, and constructive competition (Papert, 1980). Within this context play is defined as "a voluntary activity that involves the imaginary. That is, it is an activity limited in time and space, structured by rules, conventions, or agreements among the players, uncoerced by authority figures, and drawing on elements of fantasy and creative imagination" involving storytelling and metaphor. Emotions such as love, anger, or fear shape the different forms of play in which a player engages, as well as the symbolic expressions the player produces. Since play involves the capacity to pretend, and to shift attention and roles, it provides a natural setting in which a voluntary or unconscious therapeutic or cathartic experience may take place.

Play for Experience, Play for Learning, Play for Art

"Play is freedom. Play is extraordinary. Play is distinct from the ordinary both in locality and duration. Play is fun". (Huizinga, 1955)

Huizinga sets play and culture side by side, however insists that play is the primary force, since animal play pre-dates human culture; continuing that he does not propose to define the place of play among all other manifestations of culture, but rather to ascertain how far culture itself bears the character of play, thus setting an evolutionary framework for the concept, which is picked up again by Brian Sutton-Smith in the Ambiguity of Play (Sutton-Smith, 2001).

Sutton-Smith draws from the fields of animal play, psychology, folklore, literary criticism, biology and anthropology. The book considers seven major play rhetorics that cover play as progress, addressing the claims of research into animal and child development; play as power in sports and games; play in the construction of identity through cultural activities such as festivals; imaginary play in art and literature; the self in play from the perspective of individual psychology; and the frivolous as a deconstruction of play. Sutton-Smith invites us to look at a the separation of child play from adult play, juxtaposing play for adults, often altogether unrecognized as play, with play for children, usually recognized as play for progress; showing play to be a complex process, which is highly contributory to the cultural characteristics of a society. Much inspiration as well as clarity of purpose has been attained from reading John Dewey on the experiential qualities of aesthetics and art. In as synthetic a world as the metaverse of Second Life® where the bulk of art work presented is still housed in designated art spaces such as galleries or museums, Dewey's concern for the separation of art work from its experiential functions seem to be well founded, given the suitability of virtual worlds for an in-depth reexamination of the role of artistic output in (virtual) society. Drawing attention to the modernist practice of relegating art work to rarefied but sterile repositories where they pursue an existence essentially cut off from everyday usage and appreciation, as would indeed be the case with museums, Dewey draws attention to cultures, ancient as well as contemporary, where aesthetic appreciation is

Figure 4. Virtual "Objet Trouvé" as toy: A scripted pose stand initiates an elaborate yet entirely improvised game between avatars Alpha Auer and MosMax Hax. Second Life®, 2008

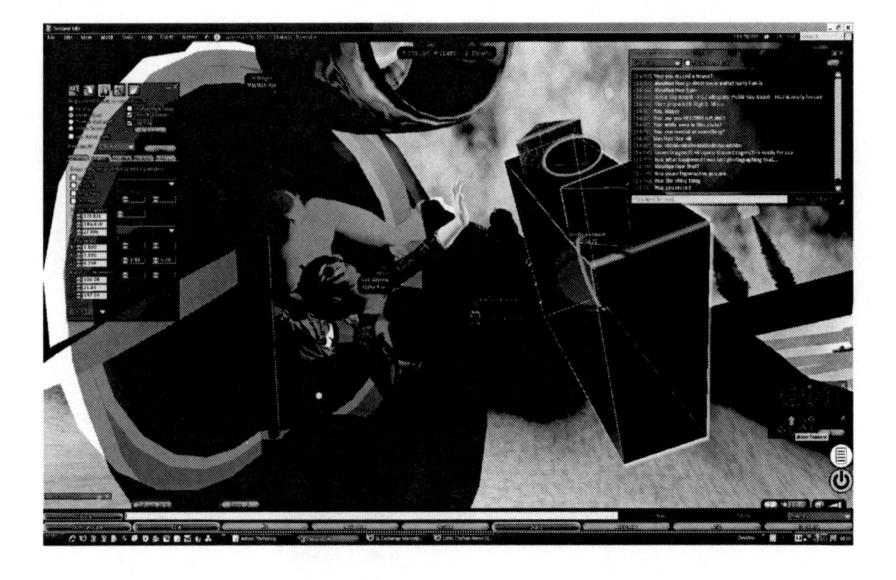

inextricably bound with day to day usage, saying that "we do not have to travel to the ends of the earth nor return many millennia to find peoples for whom everything that intensifies the sense of immediate living is an object of intense admiration", adding that the present task at hand "is to restore continuity between the refined and intensified forms of experience that are works of art and the everyday events, doings, and sufferings that are universally recognized to constitute experience", thus elevating art work from its current state of being the provider of mere "transient pleasurable excitations" into once again becoming the powerful carriers of experience.

Approaching play from the vantage point of the computational environment McCullough describes a circle:

A chain of developments should be clear: play shapes learning; learning shapes the mind; mental structures shape software; and software data structures afford play.

For McCullough improvisation is an intrinsic part play and where improvisation is concerned computational endeavor has a distinct advantage over its counterpart in the analog realm; one which is embedded into the very material difference

Figure 5. "The Bunny Girls" by Ravenelle, aka. Jennifer Olmstead, Second Life® and digital post-processing, 2009

between the two at that: The digital environment handles bits as opposed to material atoms. While atoms can only be manipulated to a certain degree before "material" starts breaking down, bits can be processed, reversed and manipulated infinitely with no loss. Computational creativity thus becomes an activity which can also be described as play with a perpetually evolving object which can endlessly be improvised upon. Although writing at a time predating the advent of fully operational metaverse such as Second Life®, which place their fundamental tenet solely in user created content, McCullough is nonetheless aware, through observing precursors of the genre such as SimCity, of the implications which these "builder worlds" have upon all digital creativty, when he notes that *"the popularity of these simulations without explicit winning conditions may reflect a constituency that also sustains the playful attitude in productive computing"*. (McCullough, 1996)

Returning once again to Castranova, we find that one of the major issues his hypothesis addresses is the entire notion of fun: While, Hedonics, as a field of psychological research investigating the grounds of human happiness and fun had already been proposed during the 1990's, it is the game industry of the 21st century which seems to have instigated resounding research in this area whereby it has been assessed that there exists a correlation between the production of endorphins and a sense of achievement, which in its turn, is a commodity easily attained in synthetic worlds through either game related success or in a builder world such as Second Life®, through the realization of creative activity. Indeed, so powerful seems to be the pull of "fun" provided by synthetic worlds that Castranova even foresees a considerable shift in future physical public policy and socio-economic strategies aiding the establishment of a more "fun" physical world as a means of competing for the attention of the migratory population whose exodus to synthetic worlds has been described above (Castranova, 2007).

SYNTHETIC EXISTENCE

The Body is the Message: Presence and the Avatar

The social and economic impact of the metaverse was at the core of Stephenson's fiction, and as such is still very much of an open question today. What differentiates the metaverse from online role playing games is that unlike games, the metaverse has no intrinsic rules that are game related: There are no scores to be gained, no levels to be attained. However metaverse activity can be thought of as a game on a very basic level: These are unstructured virtual environments where characters undertake activities for the sole purpose of personal enjoyment, i.e. play. What makes the metaverse particularly relevant when viewed in juxtaposition to creative activity however is that in order for this playful activity to commence the apparatus of play, i.e., the actors/avatars, the environments, apparel, vehicles and the like need to be created by the players themselves. And taking the act of creativity even beyond the rendering of all the accoutrements of the game, metaverse Residents are also called upon to create the very game itself.

A salient aspect of Ascott's methodology in the propagation of behavioral change and stimulating creative processes is Role Play, as well as "pure play", or "frivolous play" (Sutton-Smith, 2001). Since the author proposes to utilize the entire social and economic structure of the metaverse for creative learning activity, specifically through the enactment of different selves and persona through the agency of diverse play activity, this inevitably brings into prominence the role of the avatar.

Avatars are the all important, if not indeed sole agents of virtual social interaction since their inhabitants both consciously and unconsciously use them in ways very similar to their material body (Damer et al, 1997). They can be endowed with a wide range of physical attributes, and may be customized to produce a wide variety of humanoid and other forms. Furthermore a single person may have multiple accounts, i.e. "alts" and thus be represented through multiple identities in a synthetic world. Given that they visually portray an inhabitant and allow visual communication, Suler contends that avatar appearance is crucial for identity formation as well as attaining Presence in virtual worlds (Suler, 2007).

Presence is defined as a sense of "being there" in a mediated environment (IJsselsteijn, deRidder, Freeman and Avons, 2000). Lombard and Ditton (1997) define it as an illusion of non-mediation in which a user no longer perceives the display medium as a separate entity. A high level of presence will help users remember a virtual environment as more of "a place visited", rather than "a place seen" (Slater et al., 1999). A success indicator of the attainment of presence is also considered to be the realization of similar behavior patterns in virtual environments to those in the physical realm (Slater and Wilbur, 1997), and even the manifestation of similar physiological responses towards a given event to its approximation in the physical realm (Meehan, 2000). Various definitions of the term "Presence" and their relevance to the immersive virtual experience are discussed by Mantovani and Riva who challenge the notion that experiencing a simulated environment deals with the mere perception of its objective features;

Figure 6. Avatars Alpha Auer and Grapho Fullstop play a Norsk mythological game evoked by costumes created by June Dion. Second Life®, 2008

Synthetic Worlds, Synthetic Strategies

instead proclaiming that presence in an environment (real or simulated) means that individuals can perceive themselves, objects, as well as others not only as situated in that external space but as immersed in a socio-cultural web connected through interactions between objects and people (Mantovani, Riva, 1999).

Positioning the concept of Presence primarily within a socially interactive context finds further resonance in Castranova who places the social nature of online synthetic worlds, i.e., the agency of "social presence" as possibly the single most important contributing factor to the sense of joy and satisfaction that being immersed in these worlds brings to their inhabitants: Social interactions such as emotions and successes shared, the formation of close bonds and affections as well as wider peer groups and alliances, the attractions of social discourse, joint memories and experiences would seem to be vast enhancements which the social world of online gaming provides over the video game performed in isolation. Indeed such is the pull of the social factor in online "Presence" that when the future market strategies of video game producers are held under scrutiny it becomes quickly apparent that all major developers are planning upon bringing their consoles online within an immediately foreseeable future (Castranova, 2007).

Figure 7. The avatar Alpha Auer playing in the synthetic ocean of Second Life®. 2008

A rewarding overview of the literature on the avatar within the context of embodiment comes from Benjamin Joerissen (Joerissen, 2008) who directs us to the co-relation between the ideological affinity of the avatar and the human body: Drawing upon Plato, as well as the Sanskrit meaning of the word avatar itself, he points out that within these doctrines the human body itself can be identified as the disparaged, earthbound hybrid carrier/avatar of higher, divine, i.e., non-physical attributes. In a play upon McLuhan's famous statement Joerissen continues to say that if soul is indeed "form", the body is then the medium within which form becomes corporeal and as such the body becomes the very message which it carries. However, according to Joerissen, a recent, post-Cartesian shift in the attributes with which an avatar is endowed is also noteworthy: In the post-humanistic world of artist Stelarc the avatar is no longer the belittled, lesser manifestation of the higher "form" but rather the "upload" of a perishable, mortal physicality into the mundus possibilis of a virtual, non-corporeal space; an agent in the realization of a "cybernetic platonic" state (List 2001) wherein technology may overcome the shackles of mortality. According to Joerissen viewing avatars as mere representational agents in virtual realms has become increasingly problematic over the past decade. Instead a holistic approach which weaves together the human handler, the representation thereof and the medium within which this representation materializes seems to be called for: In describing this hybrid actor whose virtual sojourn is a two way experience which can have profound influences on the human behind the keyboard, Joerissen quotes Yee: "Just as we choose our self-representations in virtual environments, our virtual self representations shape our Real Life behaviors in turn. These changes happen not over hours or weeks, but within minutes". (Yee, Bailenson, 2007)

Drawing his conclusion Joerissen quotes Mark Hansen (Hansen, 2006) who points at a deep reaching biological/corporeal moment embedded within the virtual experience: Whilst placing the digital experience itself within the sensory organs of the biological body, Hansen ascribes a third element to digital embodiment, speaking of a "body submitted to and constituted by an unavoidable and empowering technical "deterritorialization", a "body–in–code", which can only be realized in association with technology, and which, in its turn, can lead to unexpected self-perceptions in the human handler. Indeed Hansen endows this novel constellation with the capability of increasing the field of influence of the human operator "as an embodied being". Thus, Hansen predicts a re-definition of the potential of the biological body through virtual embodiment.

Metanomics, Synthetic Objects and the (Virtual) 3D Collage

What makes a search for unorthodox strategies for art educational content particularly relevant at this juncture is the emergence and continued success of Second Life®, which made its debut in 2003 as the first 3D synthetic world that allows its users to retain property rights to the virtual objects they create in the online economy. Furthermore, everything created in Second Life, from the formation of its very terrain to the architectural constructs placed thereupon, and down to the vast array of objects and wearables on sale and in usage is user generated. Indeed it can be said that outside of taking advantage of the economic activity or the ensuing milieu of creativity provided by Second Life® there is really not much point to residing there at all. The synthetic world is not an easy environment to adapt to: The very absence of system defined content and purpose sees to it that a very steep learning and adaptation curve awaits the newly fledged Resident. However, although the growth rate of the population of Second Life is nowhere near those of multi-user online gaming environments whose content and purpose is largely defined by their game developers, there is still enough growth and success, manifest by the fiscal results of in-world trading activity (Linden, 2009), to assert that Second Life® is a noteworthy experiment in creative activity even when viewed from the standpoint of economic health indicators alone.

One of the founding strategies behind Second Life® was the notion that the world would draw a cadre of elite content creators whose endeavor would be noteworthy enough that it would attract sizable numbers of players into joining the world to make usage of their output (Castranova, 2007). While this early vision does indeed seem to have materialized, an unexpected development in terms of creative activity also seems to be in the offering: What makes the world particularly compelling as a platform of artistic expression to the author, herself a full time Resident of Second Life®, is the largely unstructured, indeed sometimes emergent, nature of the creative activity which the first order user generated content seems to breed quite spontaneously in its turn: Residents will combine output generated by others, sometimes with their own as well, to create extraordinary wearable collages and environments which have been assembled entirely or partially out of "objet trouvé". The conglomerated apparel and architecture, landscape as well as a diverse range of objects of usage will then be utilized as a point of trajectory in the creation of involved play/rituals, storytelling sessions, fantasy role play which then become the incubators for the generation of personal artwork by their participants. At the time of the writing of this text Second Life® video and photography output in www2 repositories is of sufficient creative and social impact to merit close scrutiny as a subject in its own right. Thus, far from being an activity held solely in the hands of an elite cadre, creative activity in the metaverse seems to be materializing as a mass pursuit, forging its own techniques, ways and means towards personal expression. Creative output in the metaverse becomes interactive in the truest meaning of the word: Far from being work meant to be viewed and admired but not to be interfered with in any truly fundamental way, design output as well as art objects are manipulated, re-structured and combined with others as fits the needs of the present owner; to suit specific purposes, such as props in playful activity, photography and video sessions, environmental decor and, of course, avatar appearance. What is however, most noteworthy in this second order creative content is that the roles and stories enacted very often find their origins in the combined visual elements of the initial assemblage/collage which seems to have instigated the very process of story-telling and of make-belief which these combined objects and imagery seems to evoke: Thus, akin to their Dadaist predecessors, the assembly of unrelated objects seems to go as noteworthy a distance virtually as they do physically in the instigation of associative creativity. Indeed quite possibly acceleratedly so in virtuality, when the contributions which a coded domain can bring to bear upon the enablement of the fantastical is also taken into account: Unconscious associative processes, which provoke further creative manifestations in the form of narrative/performative as well as visual output are evoked as par for the course in the synthetic lives of users and extended user groups, to the extent where it would be not much of an exaggeration to claim that the pursuit of these is at least one of the key joys and components of many a synthetic existence.

In "Art and Agency" (1998) Alfred Gell expounds upon the found object as part of the process of artistic activity stating that in the idea of the "found object" or the "ready-made", the artist does not so much "make" as "recognize" the particular cognitive index of the object. According to Gell, Western cultures seem to have a far more activist notion of artistic activity, whilst the Oriental approach esteems far more the "quietist" mode of creativity in which success attends those who open themselves to the inherent physiognomic appeal of the (naturally) found object. Conversely, the usage of the found object by Western artists such as Duchamp are less passive, their selection being presented as pure acts of will on behalf the artist. Duchamp claimed that his ready-mades possessed "the beauty of indifference", that is, the objects used in their creation were selected on the grounds that nobody could possibly imagine that there could be any particular reason for them to have been selected in preference over others. However, having "no reason" to select some "thing" as an object of ready-made art, is in itself a reason, since it is motivated by the need to avoid selecting anything for whose selection some reason might be proposed. Consequently, even the purportedly "arbitrary" ready-mades of the Dadaists, forced themselves on these artists "who responded to the appeal of their arbitrariness and anonymity, just as the Buddhist landscape artists responded to their mutely speaking boulders".

Western Collage consists in reassembling preexisting images in such a way as to form a new image answering a poetic need. Max Ernst defined it as "the chance encounter of two distant realities on an unsuitable level", a formula which

Figure 8. "Splitting the Atom to Kill Time" by ariel Brearly, aka. Kerry Wimpenny. Virtual Collage, Second Life® and digital post-processing, 2009

also finds resonance in Lautréamont's proposition: "Beautiful as the encounter of a sewing machine and an umbrella on a dissecting table, it gives us a remarkable method of triangulation that does not provide measures, but brings to the surface unrevealed mental images". Aragon states that collage "is more reminiscent of the operations of magic than of those of painting since it hinges on the artist's success in persuading us to recognize the connection of visual elements on the plane of poetry". Asked, if his collages were visible poetry Jean Arp replied "Yes, this is poetry made through plastic means".

It would be overly optimistic, if not indeed downright foolhardy, to view the bulk of the content created by the population of Second Life® in such a light as to do Jean Arp's words justice: While there is noteworthy output generated by a relatively small group of skilled builders, for the overwhelming part Second Life® creativity is devoid of the intrinsic attributes which would qualify it as meaningful to anyone outside of the person(s) engaged upon the task itself. However, when viewed in its entirety as a socio-cultural manifestation, if not indeed a way of life; it may again be foolhardy to disregard the cumulative creative endeavor in Second Life® as the mere doodlings of a largely amateur society. What seems to be at work is akin to what happens in a petri dish: While one cannot yet foresee the ultimate direction and outcome of the experiment heed should still be paid, not only in terms of what may materialize in the synthetic world itself, but in terms of how what materializes there may in the end be of relevance to artistic processes at large.

Creative Activity for Behavioral Change

A pertinent result of creative activity in a synthetic world, when considered in relationship to the avatar, is the behavioral change "the created" effectuates upon "the creator". Yee and Bailenson have reported upon the relevance of the physical attributes of the three dimensional avatar, finding that both the height and the attractiveness of an avatar in an online environment are significant predictors of the player's performance. However truly startling is also the finding that according to The Proteus Effect, not only does the appearance of the virtual body change how dyads interact with others in the online communities themselves; but this effect is indeed powerful enough to be carried through to subsequent face-to-face interactions amongst the physical handlers of the avatars participating in the experiment (Yee, Bailenson, 2007).

Ascott's Groundcourse, with its emphasis on behavioral change, utilized the creation and enactment of new personalities as an integral part of the educational process. A personality is a thing comprised of many layers of complexity, comprised of a gamut of attributes, ranging from intelligence and temper to genetic make-up. However, based upon the importance which Suler (1996) as well as Yee and Bailenson (2007) place upon the physical attributes of the avatar the author has focused her current area of interest to the creation of visual identities and indeed multiple visual identities all belonging to one human handler as the visual manifestations of the diverse facets of the human persona: A previous exercise based upon a concept borrowed from Robotics, "The Uncanny Valley", conjoined with Kristeva's definition of "The Abject", has been the subject of a prior publication (Ayiter, 2008). A second experiment, alpha.tribe - a virtual fashion enterprise jointly operated by 5 alt avatars belonging to the author, addresses the ability of splitting the (visual) creative identity through physically divergent manifestations of one human, taking cues from the literary tradition of noms-de-plume and heteronyms into a visual domain of creativity. A novel, third addition to the series in which visual identity will be investigated from a synthetically genealogical standpoint is currently being deliberated upon.

CONCLUSION

In "Exodus to the Virtual World" Castranova alerts his readers from the onset that the book is of a speculative nature. However, after this opening statement he continues on to list the scientific instruments by which he is constructing his model. Given the solidity of his assessment tools as well as his academic expertise in economics and public policy, it would not be too imprudent to regard his predictions as anything other that informed deliberations, which it might behoove his readers to take into serious consideration: Even if his cogitations come to bear fruit only partially, humankind may find themselves living in a vastly altered world, or indeed in multiple worlds, "synthetic" and "real" simultaneously.

We may find ourselves in a social milieu where the bulk of recreational time, if not indeed work hours, are spent in fantastical, frivolous, playful and fun activity; where economic demand and supply are shaped by parameters that are currently being forged in online synthetic virtual worlds. Returning full circle to the days of the pre-industrial revolution designers and artists may find themselves to not only be the conceivers but also the crafters and merchants of their own creative output; an output whose intrinsic descriptors, function and usage may be vastly altered to those of the present day.

While it is important to educate dedicated, professional artists and designers in a manner which would take into account the essential requirements of establishing viable creative practices in the synthetic worlds of today and tomorrow; art educational enterprises may also need to prepare themselves to disseminating learning to a social group, the emergence of which was already foreseen by Ascott in 1966; i.e., "a new, leisured class" that will be involved in creative pursuits, furthermore a class which falls outside of the boundaries of traditional art educational practice.

This text has thus attempted to raise some of the issues related to a need for an, at least partial, restructuring of existent art educational curricula; positioned within the context of educational strategies which seem suited to the task at hand as well as an examination of related material from Cyberpsychology and Presence Studies, Ludology and present day creative activity in the metaverse.

REFERENCES

Ascott, R., & Shanken, E. (Ed). (2003). *Telematic Embrace: Visionary Theories of Art, Technology, and Consciousness*, University of California Press. Berkeley, CA.

Ayiter, E. (2008), Syncretia: A Sojourn in the Uncanny Valley, *New Realities: Being Syncretic*, Springer, Vienna, AT

Castranova, E. (2007), Exodus to the Virtual World, Palgrave MacMillan, NY, NY.

Damer, B., & Judson, J. Dove, j., (1997) Avatars! Exploring and Building Virtual Worlds on the Internet, Peachpit Press Berkeley, CA, USA

Dewey, J. Art as Experience, 1934, (1980) Perigree Books, NY, NY.

Gell, A. (1998) *Art and Agency*, (pp: 30, 31), Oxford University Press. UK

Gredler, M. E. (2004). Games and Simulations and their Relationships To Learning. In Jonassen, D. H. (Ed.), *Handbook of Research on Educational Communications and Technology*. Mahwah, NJ: Lawrence Erlbaum Associates.

Hansen, M. (2006) Bodies in Code: Interfaces with Digital Media. Routledge. New York, Ny.

Hill, J. R., Wiley, D., Miller-Nelson, L., & Han, S. (2003). Exploring Research on Internet-based Learning: From Infrastructure to Interactions. (pp: 433 – 461) In Jonassen, D. H. (Ed), Harris, P. (Ed). *Handbook of Research on Educational Communications and Technology*. Lawrence Erlbaum Associates. Mahwah, NJ.

Huizinga, J. (1955). *Homo Ludens: A study of the play-element in culture*. Boston, USA: Beacon Press.

IJsselsteijn, W., deRidder, H., Freeman, J., & Avons, S. E. (2000). Presence: Concept, determinants and measurement. *Proceedings of the SPIE, Human Vision and Electronic Imaging*, San Jose, USA.

Joerissen, B., (2008) The Body is the Message. Avatare als visuelle Artikulationen, soziale Aktanten und hybride Akteure. *Paragrana, Volume: 17, Issue: 1*.

Kafai, Y. (Ed.). Resnick, M. (Ed). (1996) Constructionism in Practice: Designing, Thinking, and Learning in a Digital World, (pg: 11) Lawrence Erlbaum Associates. Mahwah, NJ.

Kirriemuir, J. Measuring" the Impact of Second Life for Educational Purposes. Retrieved on 29/03/2008. http://www.eduserv.org.uk/upload/foundation/sl/impactreport032008/impactreport.pdf/

Lagorio, C. (2007) The Ultimate Distance Learning, The New York Times. Retrieved on 13/06/2007 http://www.nytimes.com/2007/01/07/education/edlife/

Linden, T. (2009), The Second Life Economy - First Quarter 2009 in Detail, Retrieved on 09/26/2009 https://blogs.secondlife.com/community/features/blog/2009/04/16/the-second-life-economy--first-quarter-2009-in-detail

Lobard, M., & Ditton, T. (1997). At the heart of it all: The concept of presence. *Journal of Computer-Mediated Communication*, Indiana, USA.

Mantovani, G., & Riva, G. (1999) "Real" Presence: How Different Ontologies Generate Different Criteria for Presence, Telepresence, and Virtual Presence, *Presence: Teleoperators & Virtual Environments*; Vol. 8 Issue 5, p540, 11p.

McCullough, M. (1996). *Abstracting Craft: The Practiced Digital Hand*. Boston, MA: MIT Press.

Mcpherson, M., & Nunes, M. B. (2004). *Developing Innovation in Online Learning: An Action Research Framework* (pp. 46, 47, 54–60). London, UK: RoutledgeFalmer.

Mezirow, J. (1991). *Transformative Dimensions of Adult Learning*. Jossey Bass Higher and Adult Education Series.

Papert, S. (1980). *Mindstorms: Children, computers, and powerful ideas*. Brighton, England: Harvester Publications.

Papert, S. (1990). Introduction. In Harel, I. (Ed.), *Constructionist learning* (pp. 1–8). Cambridge, MA: Media Laboratory Publication.

Sheared, V., Sissel, P. A., & Cunningham, P. M. (2001). *Making Space: Merging Theory and Practice in Adult Education* (p. 250). Westport, CT: Bergin and Garvey.

Slater, M., Pertaub, D., & Steed, A. (1999). Public speaking in virtual reality: Facing an audience of avatars. [USA.]. *IEEE Computer Graphics and Applications*, 19.

Steffe, L. P. (Ed.). Gale, J. (Ed), (1995) Constructivism in Education. (pg: 351). Lawrence Erlbaum Associates. Hillsdale, NJ.

Suler, J. (2007). "The Psychology of Avatars and Graphical Space in Multimedia Chat Communities", Retrived on 05/05/2008 The Psychology of Cyberspace, www.rider.edu/suler/psycyber/basicfeat.html (article orig. pub. 1996)

Sutton-Smith, B. (2001). *The Ambiguity of Play*. London, UK: Harvard University Press.

Yee, N., & Bailenson, J. N. (2007). The Proteus Effect: The Effect of Transformed Self-Representation on Behavior. (271-290). *Human Communication Research*, 33.

KEY TERMS AND DEFINITIONS

Avatar: Originally Sanskrit word which denotes a computer user's representation of himself/herself or alter ego, whether in the form of a three-dimensional model used in computer games, a two-dimensional icon (picture) or a one-dimensional user-name used on Internet forums and other communities. The term "avatar" can also refer to the personality connected with the screen name, or handle, of an Internet user. This sense of the word was coined by Neal Stephenson in 1992 novel Snow Crash, who co-opted it from the Sanskrit word "avatāra", which is a concept similar to that of incarnation.

Creativity: A mental process involving the discovery of new ideas or concepts, or new associations of the existing ideas or concepts, fueled by the process of either conscious or unconscious insight.

Immersive: A three-dimensional image which appears to surround the user of a virtual environment.

Metaverse: A fictional virtual world, first described in Neal Stephenson's 1992 science fiction novel Snow Crash, where humans, as avatars, interact with each other and software agents, in a three-dimensional space that uses the metaphor of the real world. The word metaverse is a portmanteau of the prefix "meta" (meaning "beyond") and "universe".

Online: Being connected to a computer mediated telecomunication web, i.e.i the internet.

Play: a range of voluntary activities that are normally associated with pleasure and enjoyment. Play may consist of amusing, pretend or imaginary interpersonal and intrapersonal interactions or interplay. The rites of play are evident throughout nature and are perceived in people and animals, particularly in the cognitive development and socialization of those engaged in developmental processes and the young. Some play has clearly defined goals and when structured with rules is entitled a game, whereas, some play exhibits no such goals nor rules and is considered to be "unstructured" in the literature.

Presence: A theoretical concept describing the effect that people experience when they interact with a computer-mediated or computer-generated environment (Sheridan, 1994). Lombard and Ditton (1997) described presence as "an illusion that a mediated experience is not mediated" (Abstract). They explained that the conceptualization of presence borrows from multiple fields including communication, computer science, psychology, science, engineering, philosophy, and the arts. And a variety of computer applications and Web-based entertainment depend on the phenomenon to give people the sense of, as Sheridan called it, "being there".

Synthetic World: Term coined by economist Edward Castronova who has argued that "synthetic worlds" is a better term for 3 dimensionally embodies, interactive, immersive virtual worlds.

Virtual World: A genre of online community that often takes the form of a computer-based simulated environment, through which users can interact with one another and use and create objects[1]. Virtual worlds are intended for its users to inhabit and interact, and the term today has become largely synonymous with interactive 3D virtual environments, where the users take the form of avatars visible to others graphically.

Chapter 12
Kritical Art Works in Second Life

Dew Harrison
University of Wolverhampton, UK

Denise Doyle
University of Wolverhampton, UK

ABSTRACT

The main thrust of this chapter is to interrogate current creative practice as it develops in form and content within the new Virtual World platform of Second Life (SL). Kriti is an SL island for playful experimentation and research, it is becoming known for showing art works, not as a virtual gallery emulating a real life space with walls, ceiling and hung paintings, but as an island with installations, screened works, land-works. Works are built in SL only to be exhibited in SL, they are not imported copied images of real-world objects, they are to be visited by the SL community through avatars, delivering that odd, uncanny experience of distance and nearness at the same time. The first Kritical Works in SL show was curated for screening at ISEA2008 in Singapore, the artists were SL practitioners familiar with, but intent on pushing the creative possibilities of the platform. The second exhibition Kritical Works in SL II curated for ISEA2009 now includes real-world artists beginning to explore SL, and examples new interests in bridging the virtual world to real-world gallery spaces. A Kritical Works panel brought together curators and artists to discuss the second exhibition and further unpack the issues and challenges of creating art works in SL while addressing the role of the curator within this.

The chapter outlines the curation process for both of the exhibitions; presents the artists' ideas and works; and with the panel's insights, (some of whom are somewhat skeptical of an artist's intent where SL work is concerned) draws them together into a new reading of art in Second Life.

DOI: 10.4018/978-1-60960-077-8.ch012

VIRTUAL WORLDS, ART AND SL

Virtual Reality (VR) constitutes a wide variety of applications associated with its immersive and highly visual 3D environments. Michael Heim identifies seven different approaches to using VR: simulation, interaction, artificiality, immersion, telepresence, full-body immersion, and network communication. (Heim, 1993) Although there are many applications for this technology, mainly simulated conditions for emulating real-life learning experiences, as used in medical and military training, there are also artists exploring VR for its new aesthetic possibilities and immersive qualities. Although there has been extensive research in VR and virtual worlds within the humanities and sciences, there has been little research undertaken concerning the use of virtual worlds for creative practice. A particular feature of SL is the accessibility of the platform for building and customising spaces. Using the SL building tools to create objects and manipulate terrain, along with the application of the SL programming language, it is possible to have a high level of control when creating a virtual environment. This has enabled a growing community of artists to take advantage of both the capability and online accessibility of the platform, and of the relatively low cost incurred when compared to that of developing a unique online virtual world from scratch. Second Life (SL) is a relatively new virtual world launched on June 23 2003, developed by Linden Lab in California as an Internet community. Its *Residents* interact with each other through avatars to socialize, explore the plethora of islands on the *grid*, and to create and trade virtual property and services with each other *inWorld*. When launched there were barely 1,000 users on board, there are now over 16 million with an SL account. (Rymaszewski 2007:5) It was envisioned primarily as a commercial venture, and still serves as so, but anyone can hire/buy an island and build on it as fancied. There are currently over 500 art gallery islands, with a growing market economy, these mostly emulate real-world galleries with imported digital images on show. More interesting online spaces such as *Odyssey* (www.art-virtual.jimdo.com) have started to exhibit works made on the open platform itself with the SL 3D modeling toolset provided, these works are objects in constructed space that might invite avatars to walk, or fly, through them, around them, in the sky, on land or undersea, to enjoy sounds made in this spatialised audio environment – to experience the spatial nature of the works themselves.

Kriti offers a similar space to artists with experimental projects, but it has a research agenda attached too.

KRITICAL WORKS IN SL

The University of Wolverhampton purchased an island on the SL grid in July 2007 initially for the purpose of pedagogical and doctoral research in digital media. Since then Kriti island has developed to host a number of events, research projects and experimental works. Following the first inWorld exhibition on Kriti in 2008, and five years into the new SL platform itself, this online space rapidly assumed a sense of real presence for those involved and became a focus for collaboration, nationally to the UK and internationally. The *Kritical Works in SL* exhibition showed the work of ten artists in Second Life and as part of the Inter-Society of Electronic Art event (ISEA) in Singapore. The project aimed to bring together a range of art works from the SL community to explore whether common themes were emerging for creative practice on the platform: Where there perhaps certain characteristics of the virtual fabric of the SL space? Was there a possible maturing of the languages and spaces within SL? Was there a commonality of approach to creativity and aesthetic values? The works can speak for themselves but a clearer explanation of intent was required from the artists themselves towards clarifying the project questions. Each artist contributed a paragraph on their work and the different approaches to describing their work is very telling, some being poetic, some technical, some concerned with aesthetics, some with experimenting with the toolset. Five examples of the artist texts (in their own words) and short bios are given below alongside an image of their work.

The artists selected were already exploring the SL platform in some way in their creative practice, with varied interests in art, design, media arts, virtual environments and sound technology. The aim was to include a diversity of practice and to encourage responses from a range of backgrounds. Some contributors are very well known within SL but only through their virtual personas or counterparts. Their works in this exhibition were responses to one of the ISEA2008 themes, 'Reality Jam'. The confusion of real and virtual is hotly debated within the SL platform itself and forces us to re-evaluate our perceptions and registers of what is real. The ten pieces were created with a particular focus on the agency of the avatar, as bridging the two worlds, the real and the virtual, for the art viewer.

ARTISTS' WORKS

The artist's statements are taken from the exhibition catalogue.

Jacki Morie. Title of Work: Remembrance and Remains

Description of Work

This project creates a space of remembrance, contemplation and renewal for those affected by the Iraqi war. It is to serve as a gathering place where people can explore an Iraqi village in Second Life that is full of memories, sounds and images of the people who might have lived there. It contains a large remembrance wall where people can post photos, write about experiences, and tell their

Figure 1. Seven of the artists from the 2008 exhibition

stories. It is hoped that this artwork will, through sounds, videos and evocative imagery, give some sensibility of the spectrum of wartime experiences, good and bad.

Angrybeth Shortbread. Title of Work: Ping Space

Description of Work

Ping Space consists of a 40x40x40m cube with a void interior and small sound emitting objects. The Sound is a mixture of organic pink noise - ie. water/wind and binaural beats - sine wave tones of sound that range from 7 - 30 hz difference. The type of beats and other sound design within the void is controlled by an external source outside of Second Life, presently by a RL installation in Leeds, using a web interface (although this web interface will be accessible from the another virtual world called Twinity). Avatars flying around inside the void will also be sending data back out to the interface - effecting its presentation. Between these two spaces ping playful interaction - where each space's activity effect the other. The use of binaural sounds, explores the 'virtual sound' - an auditory experience created in the brain.

Robbie Dingo. Titles of Works: whisperBox, Machinima Works, Watch the World and Meteors

Description of Work: whisperBox

Folk music can be loosely defined as music which is passed on (typically from friend to friend or amongst a community) by performance rather than through a musical score. Whisperbox is an audio installation built in Second Life with this definition of Folk Music in mind. It appears as a circle of 7 speakers and listens for conversation from near-by avatars. If detected, it picks out certain words and/or letters from their chat and interprets these as pitches for short percussive tones which are distributed as looped patterns of notes and rests across the speakers. Additionally

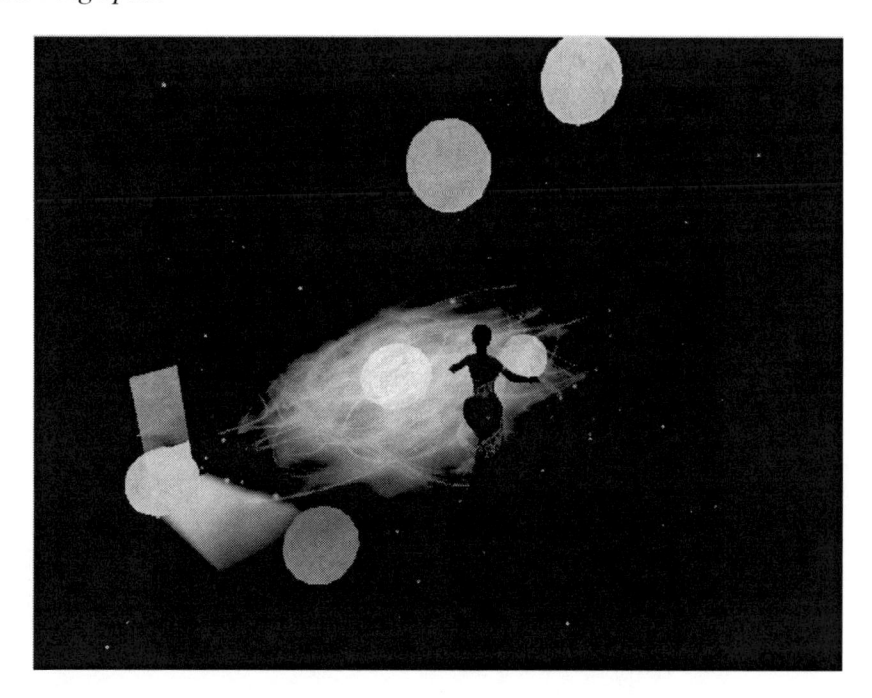

Figure 2. Inside Ping Space

it reiterates the music patterns converted from earlier conversations, thus perpetuating the music.

Description of Work: Watch the World

"Frameless heads on nameless walls, with eyes that watch the world and can't forget - like the strangers that you've met". [Don McLean]. Ever looked at your favorite painting and wished you could wander inside, to look at it from different perspectives? Spend a single day in one of mine, from early sunrise on a new day, to dusk when lights come on in cosy homes; through a peaceful night, till morning. This work is dedicated to the many weird and very wonderful strangers from around the globe I have met, but have never really met.

Description of Work: Meteors

Rob was invited to work with Kirsty Hawshaw, a lovely UK singer, as part of the new Millions of Us - Artist-in-Residence Program in Second Life. The resulting work is this short machinima created in Second Life set to a great song *Meteors* by Kirsty.

Cubist Scarborough. Title of Work: White Cubist Chair

Description of Work

This chair is one of a series of experiments that seek to subvert perspective and photo-realism in the 3D virtual world. It's a fully functional chair, complete with sit target, and is free to copy.

Wandering Fictions Story & Lime Galsworthy. Title of Work: Map to Grid

Description of Work

We are locatable, at least by definition if not by form. The architecture of the human body has already given way to the non-human. Working with the process of digital narrative and the non-human body as metadata, experiences of virtual worlds are described in new ways. One discovery: understanding our geographical position in this digital era is absolutely essential as we are able to live in, and embody, multiple realities. Our 'I's

Figure 3. White Cubist Chair

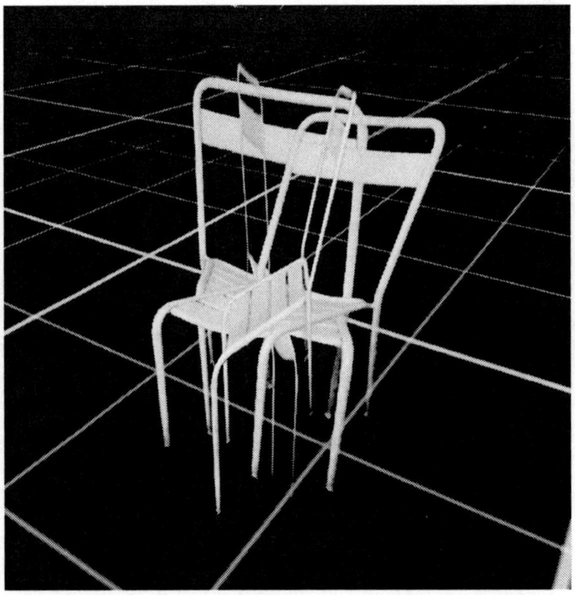

travel through multiple spaces and times. Dongdong and Wanderingfictions exchange stories and conversations. When and where they take place, maps out both a physical and virtual geography. Their states of mind, or physical bodies, are forced to confront both physical and virtual space. They are mapping the grid through the method of psychogeography. They are situated yet they still struggle to determine that exact position, our dwelling place. The two artists have already explored a conversational based narrative exploring their non-human bodies in *Embodied Narrative: the virtual nomad and the metadreamer* (Doyle & Kim, 2007). Working together for a second time, as Lime and Wanderingfictions they re-present their conversations as actual snapshots of maps created.

THE NON-DIGITAL ARTIST AND SL

In our current high-tech culture artists can engage with unorthodox media, tools and ideas, beyond the confines of Modernism. In the move beyond the media specificity of Greenbergian pure modernist art making, we have returned our interest from the flat play of the picture surface to re-engage with our desire to step beyond paint and canvas and into the painting. This has been an art investment since the birth of perspective but by employing new digital technology artists can now more fully accommodate these immersive yearnings through virtual worlds. Digital artists frequently use computers and new technologies to explore areas within and around art practice and cultural theory, but until recently the art world has preferred more conventional approaches to art concerns. *Kritical Works in SL II* includes established real-world artists who are now exploring virtual worlds as environments for practice.

David Em was the first fine artist to create navigable virtual worlds in the 1970s. His early work was done on mainframes at III, JPL and Caltech. Jeffrey Shaw explored the potential of VR for the fine arts in his *Legible City* (1989), *Virtual Museum* (1991), and *Golden Calf* (1994). Char Davies created immersive VR art pieces *Osmose* (1995) and *Ephémère* (1998) with a unique aesthetic of transparency. Maurice Benayoun's work introduced metaphorical, philosophical or political content, combining VR, network, generation and intelligent agents, his work includes *Is God Flat* (1994), *The Tunnel under the Atlantic* (1995), and *World Skin* (1997). Other pioneering artists working in VR include Luc Courchesne, Michael Atavar, Knowbotic Research, Rebecca Allen, Perry Hoberman, Brenda Laurel, Jacki Morie and the other artists in the Kriti exhibition. But what are the circumstances that have enabled this new interest in the virtual world of SL which is coming from the real-world/non-digital artists?

The 'digital art' historian Charlie Gere gives a convincing argument for current new media art practice as being an outcome of our present 'digital culture'. (Gere 2002) This culture has emerged from a synthesis of art, technology and science - the Cybernetic era of influential discourses, which includes information theory, general systems theory, molecular biology, artificial intelligence and structuralism. According to Gere, art practice reflecting these concerns began exploring questions of networking, telecommunications, information, and the use of generative techniques. By the mid-to-late1960s artists were incorporating technical electronic objects in their work and beginning to employ video and computers as new media. Christiane Paul suggests that contemporary artists are using new materials to engage with established art concerns, and that there are currently old concepts being explored within new artistic practices using emerging digital technologies, she states that "Some of the concepts explored in Digital Art date back almost a century, and many others have been previously addressed in various 'traditional' arts". (Paul 2003:5)

It is generally accepted that a direct line can be established between current art practice and the ideas of Marcel Duchamp, whether new technologies are involved or not, as Michael Rush declares in his book on new media practice "what branch of contemporary art, for example, would not claim Marcel Duchamp as a predecessor?" (Rush 1999:9). This view is upheld, but questioned, when considered in relation to computer-art by Lev Manovich who speaks of a distinction between Duchamp-land and Turing-land. Duchamp-land being the established art world and Turing-land being exemplified by ISEA, Ars Electronica and SIGGRAPH. Manovich asserts that the convergence of these two worlds will never happen where they have different agendas, with Turing-land being oriented towards state-of-the-art computer technology rather than content. He states that even though "Duchamp-land has finally discovered computers and begun to use them with its usual irony and sophistication" it will not accept practice from Turing-land, for it is only ever concerned with 'art' and not with "research into the new aesthetic possibilities of new media." (Manovich website) Dew Harrison continues to site Duchamp as the bridge between Conceptualism and computer-mediated practice initially through hypermedia technology (Harrison, 1997) and currently through AI behavioural methods of 'flocking' (Harrison, 2009). Art history books concerning 21st Century practice, now necessarily, include sections on digital and sci-art, in "art tomorrow" Edward Lucie-Smith (Lucie-Smith 2002) understands technology as linked to an enhanced representation in art, which therefore centres on video and, to a lesser extent, on digital still photography. He also determines a direct link between sci-art and Conceptual Art when he asserts that "The diagrams made by scientists in order to explain their ideas can be seen, and indeed have been seen, as direct forerunners of Conceptual Art". Margot Lovejoy (Lovejoy 1990) in her contribution to an Art Journal issue on computer art makes a strong case for the importance of photography as the basis for the use of electronic media in art practice. Manovich looks at any new digital media practice through the theory and history of the still and moving image "I draw upon the histories of art, photography, video, telecommunication, design, and, last but not least, the key cultural form of the twentieth century – cinema". (Manovich 2001) Frank Popper (Popper 1993) in his search for the roots of Electronic Art identifies seven different sources from which contemporary technological art has drawn its inspiration: Photography and Cinematography; Land Art; Light + Kinetic Art; Cybernetic Art; Installation Art; Performance; and Conceptual Art. All of these are arguments for the inclusion of the digital within the established art world of the non-digital and have been more clearly defined through the recent positionings of Jon Ippolito (Blais and Ippolito 2006) and Edward Shanken (Shanken 2009).

KRITICAL WORKS IN SL II

Most forms of the digital can be accommodated within virtual worlds, and SL is fast becoming the platform where Duchamp-land and Turing-land can meet. *Kritical Works in SL II* holds evidence of this in its exhibition of a work by the 1970s Conceptual artist Lynn Hershman. With a focus on artistic and inWorld collaboration a selection of artists were invited to explore the physical space of a real-world gallery and the virtual space of Kriti island, through their art works, four of the artists, including the curator, were also to be physically present at the ISEA conference this time. The exhibition includes three existing pieces, by real-world artists. Paul Sermon's *Liberate your Avatar*, Lynn Hershman-Leeson's *Dante Hotel* and Joseph DeLappe's small Gandhi figure from his *Tourists and Travelers* show in 2008. Sermon's work extends his existing telematic explorations and is re-presented in this group exhibition. The inclusion of Hershman's *Dante Hotel* from her L2 project and DeLappe's 8" Gandhi figure are intended to draw out the potential themes of emerging languages of artistic and creative practices in virtual worlds. The remaining five pieces were new and adapted works, and two projects have physical objects, as counterparts to their virtual exhibit, on display in the Golden Thread Gallery, Belfast.

ARTISTS' WORKS

The artists' statements are taken from the exhibition catalogue.

Lynn Hershman-Leeson

Title of the Work: The Dante Hotel Part 1 and Part 2, San Francisco, California

The Life Squared project sprung out of the desire to reanimate my archive, located in the Special Collections Library at Stanford University. The team at the Stanford Humanities lab, including Michael Shanks, Henry Lowood, Jeffrey Shanks, Henrik Bennetson, Henry Seligsman and Jeff Aldrich and the social network Second Life we re-animated two projects, Roberta and The Dante Hotel. So the project aims at nothing less than converting the archive into wholly new works that are created in a mixed reality architecture and environment. This means reshaping the archival experience as active, fragmented, exploratory, and personal.

The Dante Hotel in Second Life is an investigation of a simulated hotel room, in real life, in real time, that examines the context of its own location. It is set inside the user's social space of Second Life. It is an archaeological space through which people pass, leaving clues about their identities. As Hershman-Leeson put it in an interview about The Dante Hotel project, "once someone has occupied a hotel room, we can find out who they were by what they've left behind."

The original Dante Hotel project began in 1972, as a collaboration with Eleanor Coppola. Hershman and Coppola rented rooms in a run-down hotel located in the Italian neighborhood of San Francisco. The artists installed objects in the room, creating one of the first public art installations outside of a traditional gallery space in the United States. Hershman's room presented traces of a life - fragments or clues to an identity but also set specifically in the site. The work provided a strategic jumping off point in several respects: it opened themes that would continue as threads through Hershman's life as an artist; it began a life chronology reflected both in biography and the Stanford archives; it occupied a historical space in a specific time, which can be explored through historical and archaeological methods; and it reconfigured a public space as artistic space in ways that were stealthy and ambiguous.

Implementing the technologies of online game communities and pervasive media instigates a hybrid genre. Archives derived from past materials, but digitally relocated, become the content for a "meta-archive" that facilitates deeper analysis, investigation and exploration of the original work. Using emerging and pervasive technology as part of the structure is a pioneering method of engaging the archaeology of space, the plasticity of time and the multi-layered interpretations of embedded artefacts. There are profound philosophical implications when working within the territory of an animated social network. What does life extension mean in this world that defies gravity? Can/should avatars die? Converting the archive into a digital format of hybrid genre allows to dynamically revisit the past while simultaneously the past becomes content in this context. We replicated all installations participating in the distributed network exhibitions, Life to (i)nfinity, into in Second Life. Trespass, participation and site specificity were core elements of the re construction. The challenge was to rebuild an archive that relies on a fictional history as its spine. The recent version of *The Dante Hotel* is an investigation of a simulated hotel room, in real life, in real time, that examines the context of its own location. It is set inside the user's social space of Second Life.

Hershman's statement for the Second Life project incorporated the following description:

A new "bot" character will be created:

1) to use innovative technologies to investigate archives and develop new digital models for introducing new forms of active engagement with them.
2) to create a new context for the investigation of contemporary art.
3) to expand the audience for archives and contemporary art.
4) to instigate a hybrid genre through which to rework cultural archives.

In the 1972 version of *The Dante Hotel*, visitors would enter the Dante building, sign in at the desk, and receive keys to the rooms. Residents of this transient hotel became "curators" of the exhibition. The room, number forty-seven, re-created the ambience of presumed former inhabitants' stay based on materials gathered from the neighborhood, including books, eyeglasses, cosmetics, and clothing, all clues to their possible identity. A radio broadcast of local news in counterpoint to the sound of audiotaped breathing was installed under the bed. Pink and yellow light bulbs draped shadows over two life-sized wax cast women in bed. Above them was wallpaper made of repeated photographs of the room itself. The presence of these repeated photographs lends itself to the idea of replicated digital imagery that is available to be cloned and reused by Second Life's visitors.

In Life Squared, *L2,* visitors enter the hotel when they click on a blue box that signs them into the project, and then click on a red box that gives them a key to open the hotel door. The space is a remix now of original photographs, from the archive of *The Dante Hotel*, with virtual avatars trespassing, changing things and leaving their trail. Instead of a desk clerk, there is a "bot" guide named Dante, who guides visitors to the room. Further details can be found by visiting the link to the Stanford Humanities Lab Site: http://slurl.com/secondlife/NEWare/219/154/26/

In the real-world Dante Hotel piece, Eleanor Coppola kept her room open for one week. Coppola hired a friend, Tony Dingman, to live in the private space of room forty-three and be available to be watched whenever visitors came to the exhibit. Polaroid shots of the objects in the room and the

Figure 4. Dante Hotel in SL

subtle changes made through time were taped to the wall. Hershman's room was intended to stay open permanently, twenty-four hours a day, gathering dust and being perpetually reconstructed by the flux and changes that occurred naturally through viewers' interaction. The Second Life experience is basically the same, but reframed to the installation space of a screen. The floor of the Second Life building is designed on top of the actual floor plan retrieved from City Hall, an apt groundwork for the piece.

Paul Sermon. Title of the Work: Liberate your Avatar

Description of Work

An interactive public video art installation incorporating Second Life users in a real life environment. Located on All Saints Gardens, Oxford Road, Manchester, for the Urban Screens Festival, October 12th 2007 from 5pm to 6pm. http://www.paulsermon.org/liberate/

Liberate your Avatar brings together fifteen years of telepresence research by artist Paul Sermon with his current experiments and experiences in the online three-dimensional world of Second Life. In this project, commissioned by Lets Go Global for Urban Screens, Paul Sermon maps the actual 'first life' town square 'All Saints Gardens' on Oxford Road, with its Second Life counterpart, allowing both 'first' and 'second' life visitors to coexist and share the same park bench in a live interactive installation that focuses on the 'big screen' in All Saints Gardens as a portal between these two parallel worlds. Conceptually this project enters into a second life/first life feedback loop and by doing exposes the paradox in Second Life - have you ever noticed, there are no mirrors in this world? The project goes beyond this initial threshold and examines the history of 'All Saints Gardens', and relocates Emmeline Pankhurst within the Second Life 'All Saints Gardens' where she remains locked to the railings of the park, reminding us of the need to continually evaluate our role in this new online digital society. The project continues to play on the Manchester location as a focus of demonstration and protest in both worlds, identifying the 'big screen' as the mediator of change, and in its title echoing an analogy of Gil Scott-Heron's words 'the revolution will not be televised'.

The merged realities of 'All Saints Gardens' on Oxford Road, and its online three-dimensional counterpart in 'Second Life' was to, for the first time, allow 'first life' visitors and 'second life' avatars to coexist and share the same park bench in a live interactive public video installation.

This project looks specifically at the concepts of presence and performance in Second Life and

Figure 5. All Saints Gardens in SL

first life and attempts to bridge these two spaces through mixed reality techniques and interfaces. The project further examines the notion of telepresence in Second Life and first life spaces, the blurring between 'online' and 'offline' identities, and the signifiers and conditions that make us feel present in this world. This work questions how subjectivity is articulated in relation to embodiment and disembodiment. It explores the avatar in relation to its activating first life agent, focusing on the avatar's multiple identifications, such as gender roles, human/animal hybrids, and other archetypes, identifiable through visible codes and body forms in second life. The project aims to evaluate the diversity of personas and social life styles of the avatar.

Angrybeth Shortbread. Title of Work: Gestalt Cloud

Description of Work

The Gestalt Cloud is a SecondLife™ multi-user installation that invites online social play. When an avatar is within the Cloud's generating 10m3 space, it reacts by creating cubes engulfing the avatar, producing a low resolution 3D cloud that follows the user about. When additional avatars enter the generating space of the Gestalt Cloud, the overlap of each avatar's cloud becomes darker. When a group of four or more avatars are in close proximity, their combined cloud will be rewarded with precipitation, beginning a sequence of change to the space around them.

Denise Doyle. Title of Work: Meta-Dreamer

Description of Work

The project, *Meta-Dreamer*, is an installation that is presented in two spaces, one physical, the other, virtual. The installation created on Kriti is connected through virtual objects/avatars that have been materialized into the gallery space itself. Working with digital materialization expert, Turlif Vildbrandt, data has been extracted from *Second Life*, and digitally materialized objects created. By experimenting in digital processes and the type of material used, attempts have been made to represent jade, and clouded glass, amongst other textures.

Joseph deLappe. Title of Work: MGandhi 1

Description of Work

"MGandhi 1" is an 8" rapid prototyped 3D print of my Mahatma Gandhi avatar from Second Life. The figure is the first of series of three physical, sculptural reproductions of my avatar, MGandhi Chakrabarti (the other two include a 15" MGandhi 2, finished in genuine gold leaf, and a 17' tall monumental sculpture entirely constructed from cardboard and hot glue). The Gandhi sculptures was created as after a mixed-reality performative reenactment of Gandhi's famed 1930's Salt March which took place in Second Life. Gandhi's Salt March was undertaken to protest the British Salt Act of 1882. The reenactment in Second Life took place in 2008, starting on March 12 and ending on April 5. Over the course of 26 days MGandhi and I marched throughout second life using a specially converted, self-powered treadmill that translated my steps in the physical world to virtually control Gandhi's steps in Second Life. In total, we walked 240 miles - all the while interacting with residents in-world and with visitors to the gallery space at Eyebeam in New York City where the project was enacted. Post-reenactment, the smaller Gandhi figures were created through a process of avatar extraction, utilizing an open source application, Ogle and the rapid prototyping printer at Eyebeam. The monumental Gandhi sculpture was created using the same extracted 3D data that was processed in Blender and translated to papercraft using Pepakura Designer.

Figure 6. MGandhi 1 and Wanderingfictions Story – digitally materialized objects. Golden Thread Gallery (2009)

Jo Mills. Title of Work: Dysmorphia II

Description of Work

Jo has created an installation space within the Second Life virtual environment, taking inspiration from tales such as 'Through the Looking Glass'. Here, the spectator is invited to experience life beyond the mirror, a familiar yet unnerving space where nothing is as it seems. To create this, a space has been built in which objects can be seen both correctly and distorted, separated by a plane which the avatar can travel through to reach the other side. As well as recognisable objects, the space also includes photographic work created by Jo based around the theme of alternate realities.

Taey Kim. Title of Work: Strangers in the Neighbourhood

Description of Work

The story starts from this hypothesis: what if Na Hyeseok (1896 – 1948), a landscape painter who lived through the colonial period in Korea, and Mary Wollstonecraft (1759-1979), the first British feminist, had visited in my neighbourhood, Stoke Newington, part of the London borough of Hackney in England? This work is a snapshot of a ghost story, and a travelogue, but also an account of living as a woman in the hidden places in my neighbourhood. I started to travel with my bed to experience and to represent trans-historical characters and sleepwalkers. Now the story is migrated to Kriti Island in Second Life to exhibit the visual stories in a virtual space.

My imagination stems from the virtual encounter between these two women in trans-historical context in the contemporary set. I conducted my research with Dr Anna Birch who is a theatre Director, who also lives in Stoke Newington, and we wanted to build a very local story with these hypotheses. We exchanged letters as Hyeseok and Mary, discussing about interiors and exteriors, liberalism, virtuality, revolution and history across the time, nations and divergent cultures. I was searching for the nature of 'locality' via this fictional meeting. The conjunction of interior and exterior, private space and public space are also called into question. The forces of 'development' and 'redevelopment' are seen to threaten the cohesiveness and history of communities.

Jacki Morie. Title of Work: Traceroutes

Description of Work

Morie has created a series of artistic memory experiences in *Second Life*™. These can be experienced by visitors via their personal *Second Life* avatar that they log in with themselves (from remote locations) or by a custom avatar that is provided for gallery attendees who have no *Second Life* account. The basis of the artwork is experiences based on interesting, yet fictional, life memories a person might have. These memories start before birth and extend to one near the end of life. In 2007 she exhibited a virtual environment entitled *The Memory Stairs*. This artwork was a fully implemented virtual environment that could be experienced in stereo-vision via a head-mounted display, and navigated through with a game controller. It featured full surround sound and scents delivered at appropriate places in the experience by means of a bespoke scent collar. *Traceroutes* is work is based on that piece, but expands its visual and interactive scope within the online world space. The main interface wherein one connects with the memories is a custom-built tower structure on Kriti Island that contains a spiral path or stairway. The memory experiences are placed along this path. The space of the virtual world is more fluid than real space and I can experiment with the user's perception of that space within each of the memories. There are eight memories, expanding on the original four created for the 2007 *Memory Stairs* artwork.

DIALOGIC EXCHANGES FOR VIRTUAL CURATION

The curation process for the second Kritical Works exhibition was documented in order to present the issues involved to a panel of experts. The panel members were brought together specifically to discuss the changing and emergent understandings within contemporary curation as it opens to embrace creative practice in virtual worlds. Under the ISEA2009 theme 'Transformative Creativity – Participatory Practice' this panel presentation was to consider, debate and reflect upon the exhibition *Kritical Works in SL II* as presented online and in the Golden Thread gallery space, with respect to digital and real-world curatorial practice. The particular focus for this panel was to be on virtual worlds and the SL platform when understood as a creative space for exhibition and to address the following questions: How well does the 'idea' trans-locate across virtual and gallery spaces? To what extent does an inWorld exhibition anchor a virtual world into a physical environment? As phase two of the Kritical Works project, this exhibition will not only present the SL island artworks and trans-locate them into a real-world gallery space, it will also present the virtual artists invited, while documenting and recording the inWorld curatorial process itself. This periphery but valuable data was gathered to inform the interrogation of a new form of curation through panel discussion.

The panel brought together experts in the fields concerning online/real-world curation and collaborative practice, with artists developing creative projects in virtual world platforms. Prof Lizbeth Goodman (SMARTlab Digital Media Institute) provided Creating Collaborative Platforms. Prof Beryl Graham (CRUMB) offered current research into Online and New Media Curation and Kate Pryor-Williams, real-world curator (Wolverhampton Art Gallery) contributed the established curatorial perspective. Annabeth Robinson gave her experience as a virtual world artist exhibiting in both of the Kritical Works shows. Denise Doyle (University of Wolverhampton), the curator of *Kritical Works in SL and II*, provided an overview of the process of SL curation. Prof Dew Harrison (University of Wolverhampton) with experience of online curatorial practice chaired the panel.

The panel were asked to discuss the following considerations:

- What are the challenges of online and in-World curation?
- Are new virtual practices and virtual relationships transforming creative opportunities?

- How does the inWorld curatorial process differ from that undertaken by the real-world curator?
- Are virtual spaces and digital technologies enabling us to re-evaluate the relationship between curator and artist? Between artist, curator and the process of creative practice? Or do they force challenges to the established templates of creative practice and agency?

The panels responses are given below as requested from them, and again, in their own words to avoid mis-interpretation. Of the six panel members, three were exhibiting artists with one of those also being the SL curator, their positions as artists are given earlier in this chapter. The views of the remaining three panelists are therefore paramount in this section as two of them are online and real-world curators, both of these represent the more skeptical views on SL art.

An interesting point to note is that many artists deemed to have established a current critical reputation in the art world have not necessarily flocked to the virtual spaces. Where is Tracey Emin's virtual textile shop, where is Steve McQueen's virtual Venice Bienniale pavilion? Artists still need validation among the current real world art systems in order to be critically recognised. These structures and hierarchies are made up of a number of complex factors that give an artist and piece of artwork validation. The complex network includes reviews by critics, appearances at Biennials, international group and solo shows, publications, talks, presence at art fairs, gallery representation. The majority of these systems are still very much located in the real world, also linked to a real-world economic structure that defines artistic value. I would argue that unless a revolution is created in our attitudes we are still at a stage where in order to get validation many of these systems need to followed. Whether that is through virtual artists being critically reviewed, exhibited and written about in the real world environment, or through a virtual world replication of these current art world orders and structures.

This argument is reinforced in some ways by the project *Kritical Works in SL II*. Although the exhibition is presented in the virtual environment of Second Life, essentially it is still getting its critical peer review through those current art world systems. Our debate at this conference, endorsement in the academic world, presence in a publication and the application of a real world curatorial model for selection, confirms the virtual world's current reliance on real world structures. If this and other virtual world exhibitions do not go far enough to challenge current curatorial roles what could be an alternative curatorial process? It is clear that the virtual world presents us with some exciting new opportunities for creative expression, and leads us to challenge and question our existing structures of power elites dominating the real world art world. The public is becoming increasingly digitally literate and there seems to be less fear and more acceptance of the digital world being presented in the museum and gallery environment with passionate curators wishing to push boundaries and explore the cutting edge areas of production. When the virtual world and real world art projects meet there is a wonderful space and opportunity for new dialogues too occur. Although sometimes beset by challenges of display in a gallery environment what we are witnessing amongst audiences is a refreshing range of accepting attitudes, open to allowing this experimental art form into our art spaces. The virtual world has already bought about changes in the curatorial dialogue but how far the art world systems and structures will be directly changed is still an open point for debate as we watch the digital future unfold.

THE SL CURATOR'S PERSPECTIVE

Denise Doyle is the co-curator of *Kritical Works in SL* and sole curator for *Kritical Works in SL*

II, as such she presents below the description of her curatorial process for these exhibitions. Her avatar and virtual counterpart, Wanderingfictions Story, has been the link and the collaborator in the virtual space of the SL grid for this work.

Curating Kriti

The process of inWorld curation was both interesting and challenging. At least half of the contributors I have never met, or spoken to, in person. The curation has occurred almost entirely through avatar to avatar. Immersion was an essential ingredient in experiencing the works in the exhibition. The visitor was able to come and wander and experience the works within the space, much of what Second Life has to offer in terms of rethinking and reflecting upon our own reality is to do with our experience of space and scale from a third person (avatar) perspective. In terms of an exploration of creative practice there was, and still is, a growing interest amongst the international artistic community in the Second Life platform. The works exhibited contributed to the exploration of the potentials and limitations of the medium itself.

A number of meetings were held inWorld, and a blog space was set up to monitor the development of both projects. What seemed so striking about the first phase projects was the potential to further explore the relationship between the real and the virtual through an examination and exploration of the physical and virtual spaces. The aim of the second phase was then to explore this relationship between Kriti island and the physical gallery space at the Golden Thread Gallery, Belfast. The presentation of the documentation alongside the physical exhibits in the real-world gallery space, was to offer a point of reflection which would contribute an important understanding of the paradigm shifts occurring through digital technologies, and in particular, virtual worlds. This documentation of the curatorial process, from the planning stages of the virtual space through to the development of the virtual works themselves, including Machinima filming of activities and developments on the island, plans of the island through the curatorial process, and images of the artists and the artworks, was intended to be presented alongside the physical works in the gallery space itself. These peripheral 'documents' became an interesting part of both the creative and the curatorial processes, and

Figure 7. Wanderingfictions curating the 2008 exhibition

Kritical Art Works in Second Life

gave a playful addition to the thematic concerns of the exhibition itself, for example, the group photoshoot of the avatars inWorld. Unfortunately, due to the real-world gallery curator's generic plan for ISEA work, this was not fully realized and was only possible within SL itself.

InWorld communication resolved the issues of long-distance dialogues, even with the problematic time-zones involved where SL runs to California time, and proved to be an extremely effective way of curating the SL island virtual space. Paralleled with meeting artists in the real-word and taking them around gallery spaces for their work, the inWorld meetings on Kriti enabled the avatars to work to a particular site. Even more so, the meetings held in the latter stages of phase two build became points of collaboration, and of sharing skills and knowledge. In fact, I would suggest that the curation became a collaborative experience with a small number of the contributing artists. The default of SL is that, unless you specifically set your objects for others to share, no-one can move anything that you have created. This forces a different kind of collaboration as the space is effectively sculpted out, 'live' and in real time, as decisions could be made and realized instantly. The aim of this second phase was to give the whole island space over to Kritical Works in SL projects, and provide a full walk around experience. In light of this, the walkway from phase one was extended all the way around the island, where you could chose to walk, or fly, around the exhibits and related information.

The greatest problems were in the email conversations with the ISEA curators who were trying to meter out physical space to artists in the Belfast gallery. Due to their lack of SL experience, they had great difficulty in realizing space for the real-world objects that resulted from their virtual counterparts, they could only visualize them as flat images and texts documenting the actual works on the island. They were adamant in offering a computer terminal with wall-space for 2D images until a full explanation was given over the phone, then all was resolved with the following email text: "We really apologise, we are trying to make the best show for all those participating but within the capacity of what we can offer here in Belfast. We are delighted to show the three objects and we like it and the work. I can confirm that we will give you a three hour platform at the Waterfront, and I will be back to you personally on Monday, and will follow this up by phone to get it pinned down asap."

Figure 8. Six of the artists from the 2009 exhibition

CONCLUSION

Doyle presented her curation experience first and completed her talk with the view that "There is no doubt that virtual worlds and online gameplay are offering new creative forms for both the artist and the viewer/participator. Particular to the SL platform are the discursive game/non-game elements that perhaps make some audiences question the nature of the space itself. The fact that a number of the artworks presented specifically work with the potential of interaction from the avatar to trigger events, sounds or changes in the virtual environment, points to future potential that could be further explored." This led to a determined fission between the real-world and SL curators where 'audience' is concerned. Both Graham and Pryor-Williams understood curation to be 'audience centred' in that it is weighted towards the consideration of audience access to art works in a specific space/environment, whether virtual or real. Graham also sees curators as control freaks, uninterested in collaborative curation. Doyle's practice of SL curation is 'artist centred', a process of mapping sites for island installations in an avatar-to-avatar collaboration for creating events rather than spaces. This positions the audience as the artist's concern and is exampled by Angrybeth Shortbread's work where four avatars need to participate simultaneously to make her piece happen. But how does the artist orchestrate four avatars to view the work, where do you find the SL audiences? Do SL exhibitions rely only on other artist visits as audience? DeLappe's work is a performance where he walks through a plethora of SL islands meeting the public as he goes and is not reliant on other artists to act as audience for this. For Doyle, the audience was already set in place in that it was provided via ISEA with provision for the delegates to access Kriti island, her concern was therefore centred on supporting her artists.

There was also an issue of SL curation relying heavily on real-world curatorial practice and real 'art world' systems and structures, which suggests that new forms of curational activity need to be developed to more fully engage an audience with work on the grid. In this particular case, for instance, there was a catalogue published in paper form to accompany the exhibition, as in any physical gallery. However, *Kritical Works ii* had an investment in the artists' bridging of the virtual and the real through their art works, and the ISEA delegates were directed to the Golden Thread Gallery to visit the materialized objects from the exhibition as well as to access the virtual space. The materialized catalogue was in keeping with this curatorial 'bridging' decision and had a SL counterpart on the island which many considered as more appropriate and informative, when the space and layout of the works was seminal to viewing them.

Grahame argued that SL work is a matter of copying for artists as well as curators, in that it is representative of the real world, but results in a sanitized aesthetic, as exampled by Hershman-Leeson's Dante Hotel project, which in it's original real-world state was a place for down–and-outs, a scruffy dog-eared building. SL can't do 'messy' and is a heavily controlled public space with limitations and restrictions, it is therefore not capable of supporting such works as Robert Morris's "Assault Course" as shown at the Tate Gallery, April 1971. In defense of SL as an art platform, Goodman considered the SL curator as advantaged in being able to enable social responsibility and inclusion within in a social, and easily accessible, virtual world space. Finally, Pryor-Williams considered that as established art museums and galleries were investing in user generated content to further contextualize their exhibits, and also in there being a general move towards understanding the art museum as providing democratic exhibition spaces, then the inclusion of access to the social virtual world platform of SL would be a positive enhancement to the real-world curator's programme. This led us to the final conclusions that artist's will continue to explore the SL platform and push it's limitations to further their own practice. That SL curator's will develop new forms to accommodate artists working on the grid. That continued dialogue between the audience-centred real-world curator and the artist-centred SL curator will eliminate misunderstandings and bring new social art forms to our shared public spaces.

REFERENCES

Accessed March 2009.

Blais, J., & Ippolito, J. (2006). *At the Edge of Art*. London: Thames & Hudson.

Doyle, D. (2008). *Kritical Works in SL. Exhibition Catalogue*. Morrisville, North Carolina: Lulu Publishing.

Doyle, D. (2009). *Kritical Works in SL ii*. Exhibition Catalogue. Wolverhampton: CADRE publications.

Doyle, D & Kim, T. (2007). Embodied Narrative: the virtual nomad and the metadreamer. *International Journal of Performance Arts and Digital Media* 3: 2 & 3, pp.209-222, doi: 10.1386/padm.3.2&3.209/1.

Gere, C. (2002). *Digital Culture*. Reaktion Books.

Harrison, D. (1997). Hypermedia as Art System. In Drucker, J. (Ed.), *Digital Reflections: The Dialogue of Art and Technology. Art Journal, Fall, 56 (3)* (pp. 55–59).

Harrison, D. (2009). The Writing on the Wall. In *ISEA2009 15th International Symposium on Electronic Art*. Belfast: University of Ulster.

Heim, M. (1993). *The Metaphysics of Virtual Reality*. Oxford: Oxford University Press.

Lovejoy, M. (1990). Art, Technology and Postmodernism: Paradigms, Parallels, and Paradoxes. *Art Journal*. Fall, 257 ff.

Lucie-Smith, E. (2002). *art tomorrow*. Paris: Terrail.

Manovich, L. (2001). *The Language of New Media*. Cambridge, MA: The MIT Press.

Manovich, L. *The Death of Computer Art.* http://www-apparitions.ucsd.edu/~manovich/text/death.html

Paul, C. (2003). *Digital Art*. London: Thames & Hudson.

Popper, F. (1993). *Art of the Electronic Age*. London: Thames & Hudson.

Rush, M. (1999). *New Media in Late 20th-Century Art*. London: Thames & Hudson.

Rymaszewski, A. http://secondlife.com/whatis/economy_stats.php Accessed June 2009

Shanken, E. (2009). *Art and Electronic Media*. London: Phaidon Press.

KEY TERMS AND DEFINNITIONS

Bridging Virtual to Real Spaces: A form of practice now becoming apparent in the work of SL artists where they create real-world objects from code to be exhibited in real-world galleries.

Digital Culture: As defined by Charlie Gere in his book of the same name, situates creative digital art practice in virtual worlds.

Non-Digital Artists: How non-digital artists are now beginning to use virtual worlds and specifically the SL platform for their practice as exampled by those exhibiting in the Kritical Works shows.

Panel Debate: Refers to the discussions of a panel of experts brought in to ISEA 2009 to focus on the Kritical Works exhibition and virtual curation process.

SL Exhibition: Kritical Works in SL i and ii, two online art exhibitions in Second Life curated for ISEAs 2008 and 2009 as the focus of this paper.

Virtual Curation: The two exhibitions contributed understandings to a discussion on online or virtual curation where the curator might never meet the actual artists involved but work solely avatar-to-avatar.

Virtual Worlds: A general overview of creative practice in virtual worlds.

Chapter 13
Digital Media and the Quest for the Spiritual in Art

Ina Conradi Chavez
Nanyang Technological University, Singapore

ABSTRACT

'On earth, painters, sculptors, musicians, dream dreams of exquisite beauty, creating their visions by the powers of the mind, but when they seek to embody them in the coarse materials of earth they fall short of the mental creation. The marble is too resistant for perfect form, the pigments too muddy for perfect color. In heaven all they think is at once reproduced in form, for the rare and subtle matter of the heaven-world is mind staff, the medium in which the mind normally works when free from passion and it takes shape with every mental impulse. Each man, therefore, in a very real sense, makes his own heaven, and the beauty of his surroundings is indefinitely increased, according to the wealth and energy of his mind' (Besant) (Ringbom, 1986).

INTRODUCTION

Although the artist cannot recover the richness of inner visions in the coarse materials of the everyday world, digital art of today is given challenging role to continue the quest of the spiritual. Today's generative algorithm based painting offers limitless freedom in the choice of line, shape, and color as long as artist sees to it that representational forms are excluded. Algorithm provides artists with wealth of artistically exploitable ideas and images, representing the almost ephemeral forces and manifestations that could emerge spontaneously and involuntarily.

The intent of the following is to present a series of art works, my own artist practice pursued through the process of research and discovery, the outcomes of the work done at the School of Art Design and Media, Nanyang Technological University titled: *Digital Imaging in Singapore: The Integration of digital imagery with traditional art media and techniques for site specific architectural, urban and landscape settings of Singapore*. The work delves into an exploration

DOI: 10.4018/978-1-60960-077-8.ch013

of an innovative approach toward image creation methodologies, researching and integrating emotive and subjective abstract imagery in digital, traditional and non-traditional forms.

Made at the School of Art, Design and Media at Nanyang Technological University where part of the educational mission is to provide opportunities for faculty and students to pursue art as research, the art-work is created with the financial support from Academic Research Fund (AcRF). With a focus on exploring digitally generative systems and techniques for integrated image generation, painting methods with digital technologies and integrating traditional art methods and materials, the emphasis of art practice is on exploring the limitations of the digital medium, abstract painting and other forms of visual imagery, striving to achieve greater creative levels. With it artist's creative nature is manifested in a way and to an extend that before would be hardly realized.

WORKS: REASSESSING THE PAINTING

On Meaning of Abstraction in Art

'There is no aspect of life that does not reveal to us an infinity of the new and unexpected if we approach it with the knowledge that it is not exhausted by its visible aspect, that behind the visible there lies a whole world of the invisible, a world of comprehensible forces and relations beyond our present comprehension. The knowledge of the existence of the invisible world is the first key to it.' (Ouspensky, 1912)

In the following study of the main concern is the profound experience of abstract visual phenomenon in which the imagery has emotive impact rather than narrative meaning. Usually wordless and silent invisible experiences provoke a wide range of emotional responses that cannot always be put into words and are only possible to be expressed in art. (Zegher, 2005)

Abstract art is often misunderstood by the viewing public as being without meaning.

'Non-associative and abstract art challenges the viewers in a particular way: They are required to look with fresh eyes at pictures that are different. They have to discard old habits, such as the desire to recognize something. Abstract art does not imitate, it represents in a different way. Viewers are denied the satisfaction of re-encountering a known reality. One of abstract art's great discoveries is undoubtedly to have made reality's energetic side visible again. It helps us to comprehend that nature

Figure 1. Natural Systems: Decomposition of Blue', Detail, Digital Painting, 118 x 84cm (©Ina Conradi)

is just as invisible, immaterial and dynamic as it is tangible, concrete and static. The importance of the in-between is rediscovered. The abstract representation of reality is founded on the two-way flow of visual energies.' (1994) (Gottfried Boehm from Gooding, 2001, p. 91-92)

When groups of artists started moving away from representational art towards abstraction at around 1910, there was never a desertion of content in their paintings. Instead, they drew upon deeper layers of meaning to constitute the spiritual in their work. The subject and content of this non-representational form of art were found within the inner states of the artist and depicted through the paint medium to trigger an empathic response in the viewer. This power of abstract art in embodying messages and engaging the viewer is of particularly interest to this project (Khoo & Conradi, 2009).

From its' beginnings, the abstract work was meant to represent something almost vague and intangible, like a feeling or spirituality, a means of reawakening the experiences and emotions in art. The starting point for this study is comprised of image generation researching the diversity of stylistic methods of abstract art painting: painterly automatism with free creation of imaginary forms, action painting and color field painting with a highly articulated, psychologically based use of color. Through a course lasting a few months, diverse hybrid methodologies of spontaneous mark making were repurposed in creating still images and moving painterly abstractions. Short animated clips that use time and imaging, the process of painting using algorithm were generated. The properties, limitations, and site-specific scenarios associated with an image process guided art, while developing methods to extend the intertwined boundaries of image aesthetics, physical potential of animation and digital code. Here, painting is reassessed through the use of the new artist's palette employing algorithms to be constructed into lifelike installations of natural phenomena in motion. The practice launched a never-ending painting process; a pattern than can go on indefinitely, that is open ended, and suggestive of ceaseless coming into existence.

ARTIST PALETTE

Reassessment of Materials: Algorithm

Artists are selecting, highlighting, and thereby conferring form upon the formless and setting the seal of their style upon it using digital code. Today, sophisticated electronic techniques allow us to find unexpected formal aspects in the depths of material (Figure 2).

The newly developed artistic palette of algorithm brushes originally drew inspiration from science, nature and mystical thought. The resulting images question the primacy of representation and describe invisible forces and processes that lie behind nature and its ephemeral phenomena. The main goal was to crystallize that connection between natural arbitrary flux and spiritual existence. Following the rhythm of continuously moving, unfolding and floating geometries, an endless process of painting evolves. Algorithm is by definition open-ended, in a state of flux and suggestive of perpetual potentiality, seriality, repetition, merging into a blurred field. These works use these procedures, structures and processes, making insignificant, the traditional concept of integrated finished and framed paintings. The images started as simple drawings of elementary form: circle, square, triangle and elliptical structures. Light, color and formlessness are engaged in the repetition of certain geometric brushes and shapes marking the experimental and the infinite through reacquiring structures of linear pattern abstraction as a means to achieving the invisible and immaterial. Through machine 'crafted' occurrences, the marks whirled into complex dynamic patterns. Autonomous and in-

Figure 2. (a) Algorithm brushes: 'Decomposition Red' Digital Rendering. (b) Algorithm brushes: 'Decomposition Blue,' Digital Rendering (©Ina Conradi)

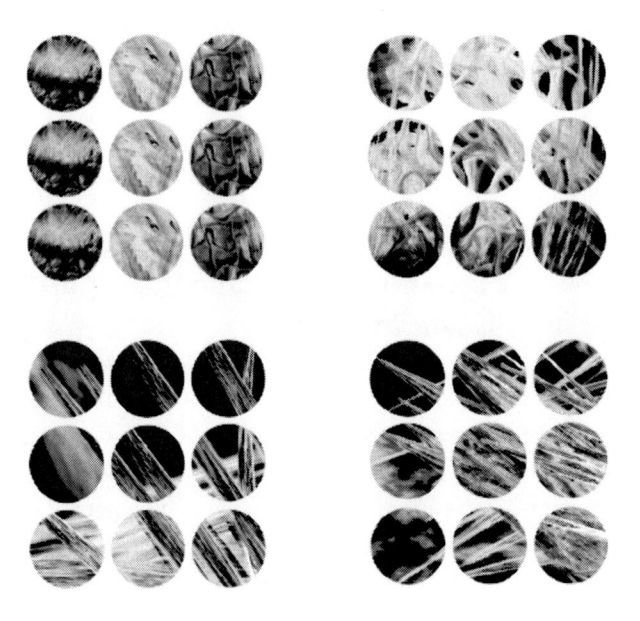

dependent expressions of algorithm had enabled development of spectacular pictures of chaotic geometry. Forces invoked were to be gathered and later articulated into painted digital stills or were to be moved in short animated clips.

The idea of using automatic techniques of algorithms as a means to generate imagery, includes expressive needs parallel to risks like the ones Jackson Pollock took as 'action painter' in abandoning traditional studio routines. The act of digital painting creates an equal collaboration between painter and machine that results in a set of rules and in an organizational model that once in motion produces something that is very often independent of the author. As organic algorithmic brushes are growing and blending smoothly with one another, the images are spontaneous, expanding and assuming multifarious shapes. Self-generating painting is conceived again as an action fraught with risk, but leading, when successful, to the thrilling discovery of original and revelatory image. The main significance is not only on techniques employed but rather on the artist's understanding of new digital means and radicality of the works that can be produced using algorithms. The technique is not merely mechanical method but is integral to aesthetic expression itself. Often symmetrically mirrored lavish kaleidoscopic imagery reflects characteristics of algorithms as comprehensible yet unpredictable descriptions of themselves. As such, it is in principle easily possible to articulate any number of the same work. Within a picture, the system of rules can also be expanded at will. Work becomes prototype of similar series of work. Given the arbitrary repeatability, this set of works requires, the question of originality arises. Nonetheless: even though we are familiar with the simple method behind the generation of this picture, there is always something mysterious and exciting about it. (Pamminger, 2008).

Paradoxically the commitment might be to the creation of imagery that is done through 'automatic', hallucinatory method, arrived at without conscious deliberation, and as the consequence of spontaneous arbitrary 'actions'. Its unpremeditated and unpredictable forms were derived, from the individual or collective consciousness, rather in the manner that free association or dreams had been demonstrated by psychoanalysis to bring into view things normally hidden from the conscious mind. (The Museum of Modern Art, New York, 1999) (Beyeler Museum AG, 2008)

ABSTRACTION IN VISIBLE

Abstraction in Visible is a series of visual explorations proposed for Singapore Cityscapes.

The field of abstract art investigations transcended from the artist's studio into an urban landscape to create new aesthetic encounters through conjuring visionary objects and familiar environments in public settings of Singapore.

When the digital print combines with traditional painted medium, of what importance is the significance of the surface? The proposed visualizations and materialized works delved into an innovative approach toward image creation methodologies, researching and integrating emotive and subjective abstract imagery in digital, traditional and non-traditional forms. Implementing the latest technologies in image creation works documented emotive spaces and impressions of Singapore to build illusions of three-dimensional creative vision; moving forward to using this imagery as a resource for more creative work that will be comprised of an exploration of manipulated surfaces and mixed media structures in site specific settings (Figures 3 and 4).

Contemporary painting practice is not taking advantage of integrating traditional painting methods of abstract painting with available digital prototyping and printing devices to create high impact artworks in public spaces. Usually the outcome of digital painting is dull digital print that lacks the physical presence interest and excitement of traditional art forms. To avoid that the aim was to customize unusual materials and transferring digital images to absorbent, non-absorbent and three-dimensional surfaces. Layering digital paintings within virtual and real urban public spaces originated extraordinary art visualizations extending beyond traditional canvases. Working on an architectural scale created an even greater physical presence of digitally generated images. Once imagery is combined with outsized and multidimensional surfaces, the creative possibilities become vast and influence applications of new emerging technologies. Parallel with digital visualizations and prototyping experimenting with a wide range of numerous uncoated flexible and rigid substrates was possible with the help of specialized large format UV curable flatbed inkjet printers, - UV VUTEk® QS2000 series and Mimaki JF Series, perfecting ultraviolet (UV) curable inks. A number of physical prototypes were test-printed using various printing substrates (rice paper, textiles & dye sublimation, textured canvas, metal, acrylic polycarbonate, Dibond).

Fascination with contemporary installation and architecture evoked a desire for renewing and

Figure 3. Public Art Proposal for Esplanade Exterior 'Natural Systems: Elixir of Redness' 14 panels 1100 x 227x 200 cm, variable, Tempered glass, exposed U-channel with *concealed*ground mounting, Singapore. (©Ina Conradi)

Figure 4. 'Elixir of Redness' is depending on existing architecture, counterbalancing it with the seamless mix of technology and image's painterly aesthetic. Grand in scale and opalescent it is consisting of modular transparent red voids that are changing its phenomena depending on the light and atmosphere. The light reflected of the transparent surface and present color creates new interference patterns thus giving form to the invisible. As elusive virtual "Elixir of Redness" enables us to see given space as infinite, dimensionless without form, and void. At the same time the differences between genuineness of location and intruding art are blurred as both of these are equally important in the process of conceptualizing art authenticity in desired locations. (©Ina Conradi)

Figure 5. Singapore City Scapes, 2008, Singapore Photography: Quek Jia Liang. (©Ina Conradi). The photographs used in these studies are not so much a representation of things as a presentation of time in space. The frozen cityscape's spatial reality frames and encloses the visual by exploring, unfolding and anchoring the art in existing space.

elevating the existing sites through various art proposals and virtual renderings. Art works triggered and evoked new emotional responses to given space, transforming urban and built, into a more personalized interaction to contemplate new space. These virtual proposals were not meant to be manifest of the artist's identity, or an idea or personal emotion but rather an effort to heal and at same time exalt the given urban life landscape. Glamorous and non glamorous, the places were transfigured into new visual experiences serving as harmonious meditative states of everyday metropolis life. Throughout the design process, digital prototyping was used for art visualizations

Digital Media and the Quest for the Spiritual in Art

in various outdoor and indoor architectural settings. The fusion of art and technology was visualized through the realistic reconstruction of 3Dimensional sets and photo space manipulation. The aim of purposing these artworks in a large scale constructs was to transform the existing space into new one, creating a sense of awe and to engulf the viewer in a totally new environment, with different imagery generating different feelings, emotions and thoughts through colors, form, material and physical construct of these artworks. Viewers were not limited to viewing the artwork from a distance, be it a canvas on a gallery wall or an image on the computer screen. Rather, the viewer interacted with the manipulated environment, which become the art itself.

Actual construction of large scale physical structures can prove to be extremely costly, hence the importance of digital pre-visualizations as a precursor to actual construction. From which, the flexibility of the digital medium allowed to accurately explore multiple versions and variations of space, artwork, lighting combinations. Other than large scale structures, the digital imagery were repurposed and conceptualized in different possible mediums. The results were realistic visualizations that depict the extent and possibilities in which such digitally created imageries can be dynamically employed within the existing environments around Singapore and repurposed in different digital mediums with the intention of using technology to push traditional artwork beyond its norm into a sea of possibilities (Figure 6). (Tan & Conradi)

The below pieces were prototyped in relation to the exhibit site architecture. Taking full advantage of the computer's ability to create motion and points of view that are not available by any other means a playful point of immersion created aesthetically engaging and inspiring experiences in the designing process. The paintings became autonomous and independent expressions and experienced physically in relationship to the viewer's movements through the space. The exhibit space in Kyoto Textile Center measured just over 12 sq meters. More than a dozen of pre-visualizations were made to explore the possibilities and creative ways of integrating existing large-scale artworks with the constraints of the limited space. In a series of digital planning and prototyping for this art installation in Japan, both 3D flythrough animation and the visualization of art objects were utilized to model, to augment or to engage synthetic image with the given architectural space. As a result architectural space could easily be transformed and the painting transitioned from the 2D canvas surface into an architecturally integrated moving image. This new painting would to expend into space and onto a light translucent fabric. Whether spilling out on the floor or hanging in space, by moving beyond restrictions of frontal picture frame its compelling immediacy challenges the ways that audiences look at art. Digital prototyping again underscored continuing investigation into convergence of digital technol-

Figure 6. Exhibit Proposal Digital Rendering and pre-visualization; translucent fabric hangings 20 sq m, International Contemporary Textile Art Centre & Gallery Gallery EX Kyoto, Japan, 2009 (©Ina Conradi)

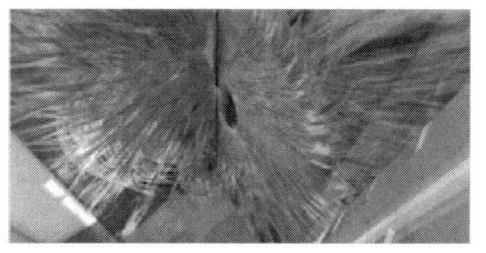

ogy and traditional painting methods. The analysis of painting through the associated technologies devised to augment the execution of future creative designs and visualizations. The pieces were prototyped in relation to the exhibit space and architecture. Taking full advantage of the computer's ability to create motion and points of view that are not available by any other means, a playful point of immersion created aesthetically engaging and inspiring experiences in the designing process. The paintings became autonomous and independent expressions and experienced physically in relationship to the viewer's movements through the space.

EXHIBITIONS

About Luminous Abstractions and Sacred Geometry

Several ideas are common to the world views of most myths: the universe is a single, living substance; mind and matter also are one; all things evolve in dialectical oppositions, thus the universe comprises paired opposites (male- female, light-dark, vertical-horizontal, positive - negative); everything corresponds to a universal analogy, with things above as they are below; imagination is real and self-realization can come by illumination, accident or an educed state: the epiphany is often suggested by heat, fire, or light (Figure 7). The ideas that underlie mystical beliefs are often augmented by illustrations that, because of the ineffable nature of ideas discussed, were abstract or emphasized with the use of symbols. (Senior, 1959) (Tuchman, 1986)

Sacred geometry is inextricably linked with various myths and traditions. Thus sacred geometry treats not only the proportions of the geometrical figures obtained in classical manner, but the harmonic relations of the parts of human beings with one another; the structure of plants and animals; the forms of crystals and natural objects, all of which are manifestations of the universal continuum. Since the earliest times, geometry has been inseparable from magic. Even the most archaic rock-scribing are geometrical in form. The complexities and abstract truth expressed by geometrical form could only be explained as reflections of the inner most truths of the world's being. Complex concepts could be transmitted from one to another by means of individual geometrical symbols or a combination of them. The fusion of symbol and meaning in Hinduism's esoteric ideas is expressed through the use of basic but sacred geometric figures in architecture. This influenced the artwork planning for exhibit in Post-Museum.

Figure 7. Paired Opposites (©Ina Conradi)

Figure 8. Natural Systems: Primal Atom, Ten Energy Whirls, Detail, Digital Painting, 118 x 84cm (©Ina Conradi)

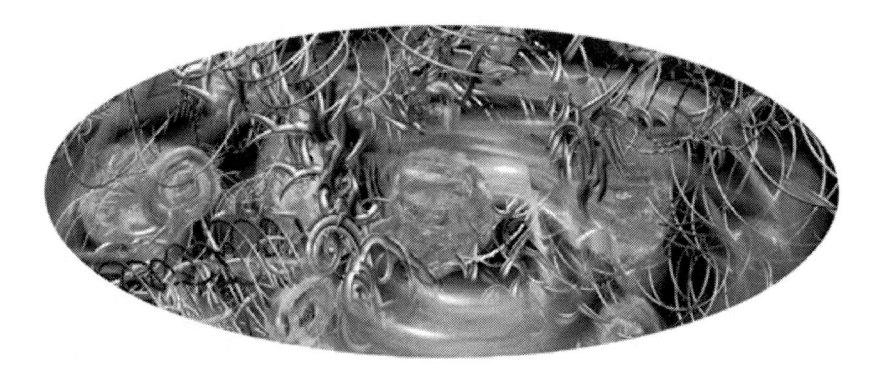

Sri Veeramakaliamman Hindu Temple has integrated conceptualizing expressions of natural systems and behaviors in its sacred architecture and organic abstract geometry of its interior patterns (Figure 8).

INTERNAL EXTERNAL

Exhibit Installation in Post Museum, Singapore 18/2/-3/8 2009

'...The dreamwork is not simply more careless, more irrational, more forgetful and more incomplete than waking thought; it is completely different from it qualitatively and for that reason not immediately comparable with it. It does not think, calculate or judge, in any way at all; it restricts itself to giving things a new form...the dream above all has to evade censorship, and with that end in view the dream work makes use of displacement of physical intensities to the point of transvaluation of all physical values. The thoughts have to be reproduced exclusively or predominately in the material of visual and acoustic memory traces...' (Freud, 1955)

'Internal External' is a multilayered environment combining video projection, 3D computer animation and space modifiable backlit digital still image. The traditional concept of painting as a surface and space integrated and limited by the frame is argued and contrasted with ambivalence between the mark and its support. Open-ended unframed painted surface maintains separate existence and is only bound by the gallery walls. The exhibition space is transformed into an intricate spatial environment consisting of tall backlit compositions that are aspiring to escape the frame and create a unified visual field (Figure 9).

Using dreams as a conceptual motif, and adopting simulated natural phenomena such as flow and growth, the work explores organic and natural design leading the viewer thorough the aesthetic impression of a dream like state sensibilities, feelings and experiences that go beyond language. Inundated with light, color and motion, works are visually translating invisible realities, the ephemeral and ever changing appearances of nature in flux. The overlapping and mirrored layers of complex imagery are engaging in the repetition of marking the void: experiential, spiritual and infinite. Exploring phenomenal nature of the space between the observer and the picture plane, between the viewer and the viewed, new emotive and tangible contemplative space awareness is created. Enhanced by light and augmented through an increased scale, the works could be viewed as a vista from the very entrance into the exhibit space or even further from the street

Figure 9. 'Internal External Digital Rendering The Center of Creation, Blue Conjuctio, Elixir of Gold and Gold Conjunctio', Digitally generated still images on UV VUTEk® QS2000 backlit polycarbonate 100 sq meters; 11 backlit panels, each panel 2.27x 1.00 x 0.008 m, variable Art Installation Post-Museum, Singapore, 2009. (©Ina Conradi)

through the open door of museum space; but it is at the point of entering into and moving across the field of luminous colors and patterns that the distance collapses as a kind of limitless depth and engulfs the viewer. From the visible external matrix drawn are invisible relational dynamics between things that could only be realized in abstract form, experiences that are wordless and silent, but nevertheless of verifiable and tangible pleasures. Installation thus fleshes out an experience of bliss, beautiful and consoling, perception as satisfaction, experience takes shape between object and viewer (Figure 10).

CONCLUSION

Drawing on the ideas of the merging of digital painting, avant-garde film-making and sound cultures, resulting work aims to craft immersive, interactive and 3D animated installations using digital images, seeking innovative convergence of art and technology to transform spaces into novel experiences. Through an integrated design and fabrication practice, created are experimental, built environments that enhance and celebrate the potential for social interaction through sensation and physical engagement. To achieve these results, novel image making methodologies were

Figure 10. There are two main visual parts to the installation: one- the re-constructed museum interior using tall backlit compositions depicting hallucinogenic dream space, the other- the projected experimental animation at the back of the exhibition space mirroring the map of that dream universe. (©Ina Conradi)

employed, experimental animation was developed and applied towards architectural installation in unorthodox ways. (Conradi & Vasudevan, 2010)

REFERENCES

Besant, A. (1899). *The Ancient Wisdom: An outline of Theosophical Teachings, 2nd edition, 146.* London: Theosophical Publishing Society.

Beyeler Museum, A. G. (2008). *Action Painting, Jackson Pollock.* Catalogue. Beyeler Museum AG.

Braque, G. (1917). *Reflections on Art.*

Conradi, I., & Vasudevan, J. (2010). *Internal External, Ina Conradi.* (J. Vasudevan, Ed.) Sinagpore: Ina Conradi.

Eco, U. (2004). On Beauty, A History of a Western Idea. In *The Contemporary Re-assessment of Material* (p. 438). Secker & Warburg.

Freud, S. (1955). *The interpretation of Dreams*, Standard edition, Vol. V, 1900 reprint. London: Hogarth press.

Georges Poulet, b. (1961). *The Matamorphoses of the Circle.* Baltimore: Johns Hopkins Press.

Gooding, M. (2001). Abstraction: An Introduction. In Gooding, M. (Ed.), *Movements in Modern: Abstract Art* (pp. 6–7). Cambridge: Cambridge University Press.

Khoo, Y. H., & Conradi, I. (2009). *Painting Using Experimentall Animation.* Singapore: NTU URECA Proceedings.

Pamminger, W. *(2008).* 3delyxe, Transdisciplinary Approaches to Design. In *Serial Machine, Generative Systems* (p. 180). Frame Publishers.

Ringbom, S. (1970). *The Sounding Cosmos: A study of the Spiritualism of Kandinsky and the Genesis of Abstract Painting.* Acta Academiae Aboensis.

Ringbom, S. (1986). *Transcending the Visible: The Generation of the Abstract Pioneers.* In M. Tuchman, *The Spiritual in Art: Abstract Painting 1890-1985* (p. 137). Los Angeles: Abbeville Press Publishers.

Senior, J. (1959). The Way Down and Out: The Occult in Symbolist Literature. In Tuchman, M. (Ed.), *The Spiritual in Art* (pp. 39–41). New York: Greenwood Press.

Tan, C. Q., & Conradi, I. *Digital Imaging in Singapore.* Proceedings of the URECA@NTU 2008-09.

The Museum of Modern Art. New York. (1999). *Jackson Pollock, New Approcahes.* New York: The Museum of Modern Art, New York.

Tuchman, M. (1986). *Hidden Meaning in Abstract Art.* In M. Tuchman, *The Spiritual in Art: Abstract Painting 1890-1985* (p. 19). Los Angeles: Abbeville Press Publishers.

Zegher, C. d. (2005). *3 x abstraction: New Methods of Drawing, Hilma Klint, Emma Kunz, Agnes Martin 1862-1944.* The Drawing Center and Yale University Press.

KEY TERMS AND DEFINITIONS

Abstract: 1. disassociated from any specific instance b: difficult to understand: abstruse; 2. intellectual and affective artistic content that depends solely on intrinsic form rather than on narrative content or pictorial representation.

Abstract Art: Imagery that is expressing personal, spiritual, or metaphysical through non-representational forms.

Algorithm: A step-by-step problem-solving procedure, especially an established, recursive computational procedure for solving a problem in a finite number of steps.

Chapter 14
Plastika [Totipotenta]

Catherine Nyeki
Plurimedia Artist, "plasticienne", France

ABSTRACT

Over the last ten years, my work has developed into a biotope abounding in diversity, a sort of virtual "vivarium" composed of interactive pieces, musical creations, drawings and writings in numerous sketchbooks, films and installations. A digital work of art is by definition non-uniqueness; it is potentially transformable at any moment. It is both fascinating and intriguing to have access to such a "plastic material". My artistic research, which has flexible boundaries and echoes that of today's nanosciences, genetics and cell biology, has gradually led me to improve certain personal concepts like the sensitive microscope, tactile laboratory, imaginary incubator, parallel botany, nano evolutions, visible and audible strata, multilingual semantic zoo, bud cell, virtual tissue, body graft... and the list goes on. My contribution's title, Plastika [Totipotenta], is taken from my last solo exhibition in Paris that brought together recent works related to plasticity, questioning a sort of constantly evolving "modeling clay"-type thinking. This chapter is an invitation to question the various levels of plasticity concepts applied to some of my latest works that have been inspired by current biotechnology and my recent collaboration with a cell biologist.

INTRODUCTION

The conflicting ethical debates on stem cells and genetic manipulation that invade our screens and newspapers today cannot possibly leave anyone indifferent. How can a specific stem cell restore a part of a body in one case, then have the opposite effect in another? What is the program? If there is indeed one. We have seen for several years now that questions on relevance and uncertainties are increasingly being raised in the so-called "hard" sciences. My last privileged experience in 2009 with Michel Gho, a researcher at the French CNRS laboratory of the Department of Developmental

DOI: 10.4018/978-1-60960-077-8.ch014

Biology (University of Paris VI) was humanly very rich. The recent theoretical hypothesis concerning the role of chance in the evolution of cells opens up astonishing new perspectives on living plasticity. This discovery has a disturbing resonance with my artworks that have been, up until now, totally disconnected from the scientific community. My works, which belong to a virtual world but are increasingly anchored in reality, have evolved in a way that is similar to developing stem cells. This creative process has led me towards hybrid conceptual territories, like "malleable cellular buds" that possess both hereditary and chance factors. In the following presentation of several of my works, we will see how this mutation came about through the description of interconnections that are not always plainly visible. *Why Plastika [Totipotenta]?* Plastika as in plasticity. In biology, it is the ability of an organism to adapt to a given environment. Plasticity can be observed in the behavior of the brain. Plasticity generally means the ability to be shaped or formed. It differs from "elasticity", which refers to the ability of something to change temporarily and revert back to its original form. The ancient Greek word "plassein" means to shape, to model or to give birth. In plastic art, as the contemporary poet, essayist and researcher in neuroscience Marc-Williams Debono writes in his book, *L'ère des plasticiens: de nouveaux hommes de science face à la poésie du monde*:

"Plastic artists would like to experiment with reality without having any a priori knowledge. Just as neuroplasticity enables the brain to organize itself according to new experiences, the artist is the architect of his/her very own evolution… This is how the chaotic brain plans the ordered future of man, as well as his creative potential. Thus, man has taken a step further in evolution, which calls for both dynamic interaction between the environment and the genes necessary for brain development, and generic memories that recapitulate and reorientate the progressive stages in hominization."

[Totipotenta] refers to totipotent stem cells. Totipotency is the ability of a single cell to divide and produce all the differentiated cells in an. Totipotent cells formed during sexual and asexual reproduction include spores and zygotes. In some organisms, cells can differentiate and regain totipotency. For example, a plant cutting or callus can be used to grow an entire plant. Etymologically, totipotency means "full authority", indicating that these cells can theoretically be differentiated into cell types that may constitute any part of the body (e.g. epithelial cells, nerves, bones, etc.). Plasticity is a strange "nodal concept" covering several disciplines of the life sciences. Ours is an era of tremendous progress in the field of biology, in which advances in genetics, biochemistry, embryology, cell and evolutionary biology have been made parallel to one another. The recent progress achieved in molecular biology and genetics has begun to reveal correlations between countless observations that were previously unidentifiable. Embryonic development, genetics, the physiology of cells and organs and the evolution of species tend to offer an overall theoretical view according to Dominique Lambert (PhD in Physics and Philosophy) and René Rezsöhazy (PhD in Biology) in their essay on the surprising plasticity of living things entitled "Comment les pattes viennent au serpent" (How snakes come to have legs):

"Our reading of the latest findings in biology does indeed highlight, in counterpoint to the relative robustness and autonomy of the biological organism, the extent of its vulnerability, its heteronomy and its capacity to be determined not only and essentially by an internal program, but by an environment that is likely to 'in-form' and deform it significantly […] plasticity can be considered as an expression of existence in the essential systems of living things, or of a completely original, dynamic and constitutive connection established

between a certain robustness (i.e. maintaining consistency, structural invariance, etc.) and a kind of malleability (ability to be deformed or 'in-formed').

The evolution of theoretical biology shows the gradual emergence of formalisms of which the characteristic is to describe life in terms of multiple "stationary" states, locally stable with transitions between these states. This process reminds me of a particular specificity of 3D tools, which I currently use in my digital creations, called interpolation. This calculation method enables the creation of a succession of forms and surface mappings that are mathematically transformed from one state to another. Such modeling often generates "inter-forms" that are otherwise impossible to obtain. When I work, I have the extraordinary sensation of holding an incredible piece of "modeling clay" in my hand that can be sculpted into an infinity of forms.

The idea of plasticity as a necessity for living beings could find its origin in their molecular components, such as proteins. As D. Lambert and R. Rezsöhazy explain at the gene expression level, it could result from dialogue between cells, organism development and evolution. We cannot understand life merely from a physicochemical point of view. Life is a story of shaped species, molded by selective pressures imposed by variety and a changing environment. Developmental genes probably constituted the first demonstration of biological plasticity as a property of systems that may be deformed, while maintaining a structural unit.

"Today, the advances in science allow us to affirm that complex systems, especially biological organisms, have the same capabilities that we find in sculpture, in that they participate in the creation of the forms from which they originate, instead of being mere passive subjects. This implies that the container (the form) and the content (the object or organism) are co-determined. There is no domination of one over the other, but cooperation".

Moreover, as the biologist Jean-Jacques Kupiek explained recently, proteins are able to react with a multitude of other proteins. The consequence of this is that the chance factor is reintroduced to the core of the creation of life. What characterizes the living is a great plasticity and adaptability, and a real potential for diversity. The notion of balance replaces the notion of a program.

In parallel, living processes have inspired a certain number of digital artists such as William Latham in his famous film "Evolution of Form" (1990) based on a complex algorithmic program, Karl Sims in his interactive installation "Genetic Images" (1993) composed of visitor sensors, Christa Sommerer and Laurent Mignonneau in the "A-Volve" creature maker live installation (1994), Casey Reas in his "TI" piece (2004) composed of autonomous aggregate living surfaces, Antoine Schmitt in "Facade life" (2008), in which artificial animals respond to architectural borders such as doors and windows. What all these works have in common is the use of generative programs. Certain pieces create amazing plasticity and unpredictable forms that may or may not be manipulated by the spectators. Some are truly very high-tech performances with deep meanings, but we do often find a plethora of "software art" based on empty content. Right from its early beginnings, digital art has always borne a resemblance to biological processes mainly due to its virtual "re-generative" potential. On the other hand the bioart movement, *"an art that involves the manipulation of life processes, and that invariably brings biomaterials to inert forms or specific behaviors, uses tools and biotechnological processes in unusual or subversive ways, and/or invents or transforms living organisms with or without social or environmental integration",* as explained by the artist Eduardo Kac, raises some other issues. For instance, could manipulations of real live animals or body transplants become a work of art? This approach describes another kind of "plasticity process" that specifically makes use of living material. Generally, such productions are carried out in direct collaboration with scientific laboratories and must often be destroyed and disposed of after the exhibitions because of their intrinsic limited lifespan and also for sanitary reasons. Although my personal approach is different from this, my research does also deal with biological life issues. Whether on screen, in music, films or installations, my work has always been conceived as a malleable evolving organism. My approach is, paradoxically, both intuitive and extremely structured. Many of my drawings and writings, noted down in numerous sketchbooks, constitute an invisible but essential part of my work that are rarely shown in my exhibitions. This somehow allows me to be a chronological observer of my own "plasticity process". It is particularly exciting to discover strange loops that provoke curious seeds of thought when I re-read certain passages.

The *Cellula* series is an example of an outcome of this process. This creation was based on an original handsketched "stem drawing" developed from a population containing a wide variety of forms, all linked together as a single organic entity. Named using the root word [cellula], many "daughter drawings" have been generated on computer, based on such sketches. These drawings gradually evolved into "totipotent stem cells" that ended up invading the walls of the Paris Griesmard & Tamer Art Gallery in printed form in 2008. With two-word names that express their heritage in relation to the original "mother drawings", [cellula] *arbora*, [cellula] *coda*, [cellula] *expenda*, [cellula] *transgena*, [cellula] *vira*, etc., they represented a sort of "plastic anticipation" of the genetic manipulations suggested by recent developments in biotechnology.

From very early on in my first work in 2001, I started to consider the computer screen not just as a mere surface but also as an innovative "malleable space". As I mentioned previously, digital artwork is by nature formable and transformable

Figure 1. "CELLULA" "External Memories" exhibition © Nyeki 2008

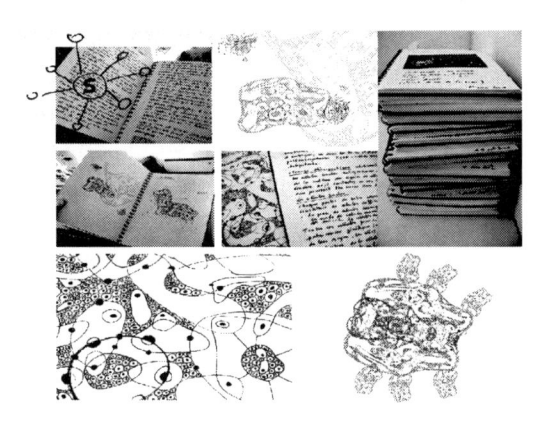

at all times. The particular flexibility of virtuality creates a sort of common "mold" that is shared between the "creation" process and the "presentation". This constant back and forth dual specificity had a strong influence on me. The concept of malleability and plasticity has been present since the very beginning. Without being subjected to the physical laws of tangible materials, the constant becoming of virtual art deeply questioned my relationship with this paradoxical "non-material" expression that can be modified infinitely in space and time, like some sort of special "modeling clay".

VIRTUAL ORGANIC PLASTICINE

My background as a scenographer dealing with the live arts was what led me to seek human interaction with the computer. Very quickly, I came up with the idea that I wanted the viewer him- or herself to take control of the programmed creation, rather than just leave the decision-making to the program. Pure algorithmic works are often disappointing because they generate mostly brilliant effects with great plasticity potential but are unfortunately void of meaning. My work combines natural intuitive actions with computation. This is how I ended up developing the concept of the "hand theater". In

2001, I jointly published a first interactive artwork entitled "*Micros Univers*" with Marc Denjean, an artist and programmer. It was based on a virtual interactive choreography of abstract 3D sculptures animated as microorganisms that were almost lifelike. This artwork enabled an experimental perception where *the very act of manipulating the image was, in a way, akin to "animating" it*. Thus, very intuitively, viewers could discover how simple elements could produce a myriad of complex, pleasant, or strange lively forms, just by displacing them, via overlap operations and symmetry. Based on the principle of metamorphosis and manipulations in real time, "*Micros Univers*" and particularly "*Mµ herbarium*" developed a virtual organic plasticine that revealed new visible and audible strata through "drilling touch" actions. "*More than any other part of the human body, the hand generates and orchestrates gestures. It reveals tactile senses and becomes like a light projector, the paintbrush of our eyes*" - extract from the Mµ herbarium libretto.

Following Micros Universe, we developed a more highly evolved project. We noticed in our first work that certain phobics frequently experienced surprising gut reactions to certain forms, especially those that were hairy. Therefore, we decided to investigate these tactile and visual paradoxical sensations triggered by virtual objects in greater detail, in order to systematize the phenomenon. Thus, we developed, with the help of *Mµ herbarium*, a *sensitive microscope, tactile laboratory and imaginary incubator*. This exhibit invited the "spect-actor" to browse through and manipulate an entire live population of 3000 manipulable entities, classified into families and according to imaginary physical laws. The keyword Mu is linked to the contraction of the expression "Micros Universe", freely mutated like a semantic transplantation in a polysemic "plasticity" with amazing convergence in different languages:

µ: prefix of the word "micro" [in Greek]
Mu: mutation, movement, metamorphosis
Mue: molt or slough off skin, change or breaking of voice [in French]
Mû: lost continent, myth of Atlantis
Mü: artwork, synthetic, artificial [in Hungarian]
Mu 無: nothing, vacuum, infinite [in Japanese]
Mu 木: tree, wood [in Chinese]

The interactive nature of Mµ herbarium allows one to handle plant-animals in an infinity of ways, much like a virtual modeling clay! "Active plasticity" is the key feature of the process. This project develops a sort of metaplasticity, "a plasticity of plasticity", a self-construction of sight. Endless interactive possibilities have multiplied from this experimental perception. Furthermore, this is an on-going experience that has been carrying on regularly even until today. Since 2005, we have had a lot of interesting feedback from the public. Published as a limited series, it has been shown in many art galleries, international festivals such as Ars Electronica (Linz, Kiev, Shanghai), and also cultural centers and museums in France and abroad. We have carried out many activities and interesting workshops with people from all walks of life, including people with Down syndrome. Today, this artwork, which began from pure artistic research, has multiple applications connected to real life. The origin of plasticity in creation is not a closed-ended issue; there are many ways of creating something. What we have tried to develop in this work is to give anyone the possibility to generate and memorize a multitude

Figure 2. "Micros Universe" © *Nyeki/Denjean 2001*

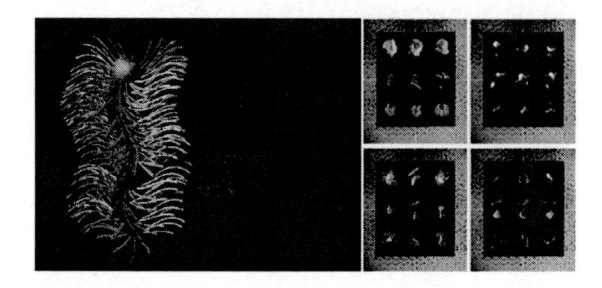

of combinatory forms and compositions through simple manipulation in a matter of seconds. This somehow provides access to a "plastic co-creative process". Every action is unique and not exactly reproducible, and therefore each manipulation may result in new interpretations with dynamic real-time revelations at any moment. The possibilities are infinite and exponential. Different structures and methodologies were developed for this artwork:

Observation
- In order to observe and meet the "inhabitants" of Mµ herbarium, we imagined a "virtual Petri dish" in which a hundred mobile and sound-producing hybrid beings (consisting of "body-nodes", "arm-branches", "firefly trees" and "vegetable clocks") evolve, much in the same way as organic cells. By means of this "sensory microscope", the spectator becomes a "laboratory assistant" of his/her own imagination, playing with test tubes of "senses", indiscreetly looking beneath the skins of animated "plant-animals", drilling into new invisible landscapes.

Classification
- We divided one hundred inhabitants on a classification chart into 10 families whose names were established according to their character: "Aquatic", "Bulbous", "Cottony", "Hollow", "Thorny", "Granular", "Luminous", "Hairy", "Stem-like and Spotty". Each inhabitant has three ranges of 10 variations, which altogether represent a group of 3000 organisms to be discovered and transformed. Each variation shows a succession of metamorphoses according to their nature: soft, prickly, luminous or aggressive, etc.

Nano Evolution
- Each variation highlights a succession of 30 metamorphoses with the same "plant-animal". The keyboard, which acts like a piano, allows one to discover other potential developments. However, this musical and visual universe is not fixed. The visitors can enjoy new rhythms by making use of pre-composed musical scores that can be reworked and saved. They can act as interpreters of their own manipulations and play surprising new "nano evolutions".

Alphabet
- Mµ herbarium consists of "sculpture networks" constructed using an alphabet of modular forms possessing tactile and audio properties. By considering the grafts done by gardeners to obtain new plants as a starting point, the assembly of various modules of "body-nodes" and "arm-branches" allows us to create a dozen families comprising inhabitants whose movements and breathing suggest hybrid organisms on the border between the animal and vegetable kingdoms.

Tactilo-Audio-Optical Grammar
- By analogy to a painter and his/her pallet, we created a "tactile pallet" containing a spectrum of a wide range of emotions and sensations such as anguish, repulsion, fear, aggressiveness, softness, lightness, fluidity and humor. Thus, this pallet enabled us to personalize the characters and behaviors of the "inhabitants". Based on the paradoxes of movement, matter and sound, we have created an "archipelago of senses"- tactile, musical and visual; a specific "tactilo-audio-optical" grammar.

Musical Creation
- Mµ herbarium incorporates original musical creations. While my Hungarian roots are often the source of my musical inspiration, I also introduce many other extravagantly funny and mysterious sound mixtures. Each "nano-looping-sound" is created in order to be rough, prickly, smooth, hard, soft, fluid, etc., much like a material. Then, like a sculptor, I use these materials to build "sound bodies" onto which I project modulations of my own voice, thus generating "MUtations"

Figure 3. Mµ herbarium © Nyeki – Denjean 2005

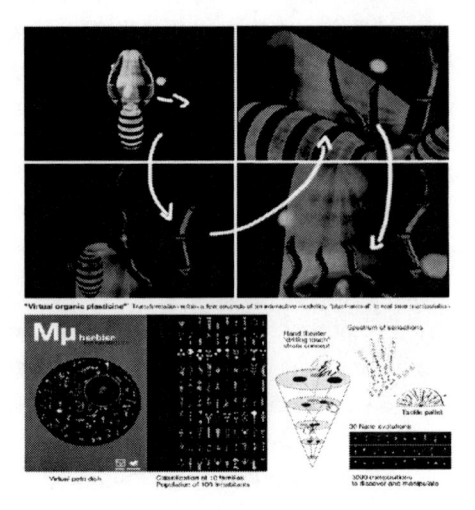

of textures, deformations and emerging sounds that are close to onomatopoeia and imaginary languages.

FERTILE EVOLUTION

Since 2005, my work has developed several outgrowths based on "Mµ herbarium" that are fairly different from previous accomplishments, especially in films and installations. Just as stem cells may transform into various specialized cells possessing different functions, the transplantation of primitive elements allowed me to generate a number of pieces in different dimensions and materials. The film *"Main Verte"*, created in 2006, would not have existed without the extraction of basic elements picked from several families created in Mµ herbarium. *Main Verte*, literally *"Green Hand"* in French, means having green fingers. In this day and age of transgenic plants, this work was an invitation to discover an artificial "organic" culture. Consisting of real soil, sand and rocks collected from various regions in France, it can be likened to a clock that never stops; branches derived from a variety of tree species are successively placed on the four sides of the screen by a pair by hands. This series features a multitude of virtual spores and grafts. In a simulacrum of life, *"Green Hand"* triggers an "internal sap" flow consisting of artificial "Klons" (combinations of an individual plant or animal and any offspring from them by asexual multiplication), which carry sounds that approximate the human voice. In this way, the viewer gets the strange sensation of witnessing some curious biological mechanism. Poorly differentiated plant cells enable regeneration that is the basis of vegetative propagation. This extraordinary ability is due to totipotency. The fact that plants are fixed organizations, with no possibility of moving from one place to another, requires them to be highly adaptable to their environment. This leads them to maintain a high phenotypic plasticity and therefore a high adaptive ability to preserve life. This is achieved through fragmentation, natural cuttings and layering. *"Green Hand"* was a musing on the creation of life. Eventually, through the agglomeration of "video stem cells", I somehow established another type of plasticity process. Using various superpositions of moving autonomous cells, I was able to produce a living tissue that was in the process of being formed; this procedure allowed me to create a sort of evolving virtual skin. Mixing real and virtual elements gradually led to the creation of a new hybrid world.

Several other works then followed: Mµ_tamorphe, Lemna Luciola, H2-Zoo, Globulus ... these were all linked to Mµ herbarium. Like a gardener, I planted new seeds and let them grow. I felt the necessity to transpose my virtual work onto other media, and a fundamental need to materialize a part of my creation in some other way.

PLASTIC PLASTICITY

In October 2008, I was invited to create an installation on four large ponds of the Science Museum in Pudong, Shanghai, during the e-Arts Festival. It was a typical mutation of my virtual work in

Figure 4. "Green Hand" © *Nyeki 2006*

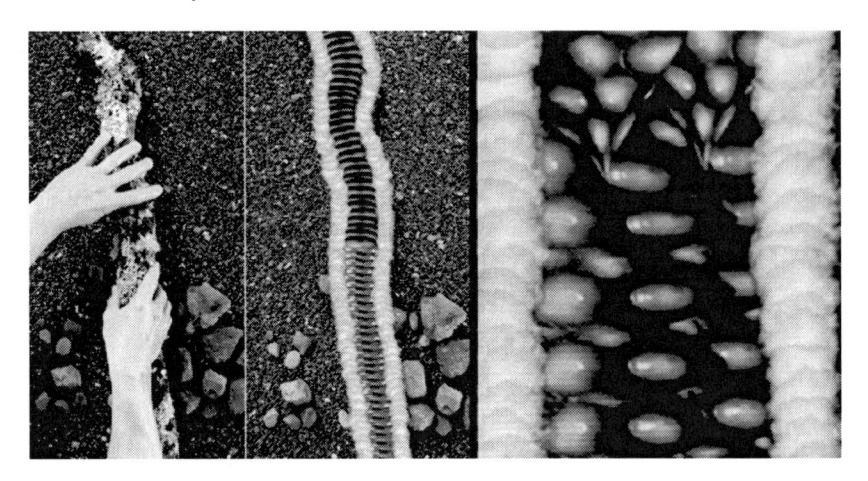

another dimension, environment and material. Using electronic light sensors in water, I developed a behavioral work comprising about 400 floating elements that reacted to the surrounding light. *Lemna Luciola*, literally "firefly lenses", was conceived in two versions to reveal a double vision of the same exhibit during the day and at night. In this day and age of genetic manipulation, new biomaterials, cell and molecular biology, this digital installation presented viewers with a particular point of view. The work consisted of a settlement of several hundred artificial "plasticized" lotuses made from encapsulated prints extracted from the "luminous" family of Mµ herbarium. In daylight, the abstract fireflies seemed to reflect an internal light. The water surface was dotted with several lenses, which constituted groups of small "islands" with natural behaviors. The lenses would float and move slightly according to the wind and the flow of water, small droplets would form clusters under the encapsulated lenses, natural microorganisms colonized the material, and fishes swam freely in all directions under these aquatic plants. At nightfall, all the luminous spots lighted up, appearing like constellations that completely transformed the water surface. Depending on the vantage point around the pond, the public would discover different configurations. This piece of art aimed at creating an experimental perception using artificial luminescent plants and called to mind experiments in bioluminescence. It is a "poetical-ethical" staging of contemporary research in biotechnology. *Lemna Luciola* mutated from a virtual world in a waterproof plasticized installation as a "plastic plasticity".

ELASTIC DIGITAL SOUNDS

Parallel to my visual works, I also carry out "sound plasticity" research. It represents a significant complementary part of my work. The border between sounds and music is a very fine one in what I do. I often gather sounds to compose music. Whether in interactive pieces, films, or, most recently, in an installation named Aqua [activa], I develop the same logical structures as I do in my visual approach in direct relation to biological processes. The way I compose is very similar to the "video stem cell" method that was briefly described above. I gather many sounds from musical instruments, noises that I create, or extracts of my own voice; then, I often proceed by cutting micro samples that I build up on to compose on multiple tracks. The result is very inventive. Superposition, coupled with melodic lines, allows me to create infinite original combinations of sounds and plastically organic tissues. As a result, I have developed a "sample cell" sound concept. I often need to write my sound scores using graphical notations. My sound work, being elastic and flexible, represents a sort of malleable territory, based on a lot of improvisations, which I then painstakingly organize and compose. My audio creations simply could not exist without such manipulation and precise remodeling.

VIDEO STEM CELLS

A few years ago, I discovered a process that could be likened to mechanisms for stem cell division and differentiation. As I mentioned in the film "Green Hand", I have developed the concept of "video stem cells". It consists of extracting small basic organic forms taken from real video sources. Let me elaborate on a particular methodology that I now systemize. This method involves extracting any part of a film with a moving or dynamic mask system. Without going into detail, this meticulous cutting up allows me to produce collections of micro moving modules. The *H2-Zoo* film was based on this very principle. "*H2-Zoo*" *Aquatico organic bestiary* develops multiple plasticity scales of observation ranging from micro, macro, nano, cellular, molecular to the gene level. My postulate for some years now is that cells, molecules and genes emit sounds when they undergo division. In French, when read aloud, *H2-Zoo* may take on two different meanings, one of which is H_2O, the chemical symbol for water. Based on real extractions with virtual encapsulated elements, the viewers were invited to discover a "primitive molecular zoo". As water is fundamental to any living organism, this film is a poetical simulation of hybrid virtual and real "cellular buds" that metamorphose and utter a mysterious proto-language. In a recent exhibition, this film was shown as an installation. This artwork has no front, top or bottom view: it can be seen from all sides. Projected on a white salt cellular form screen surface, people can turn about and sit, as they would around a real pond. The immersion in a pseudo-water representation is extremely captivating.

As an extension of this experience, I began to further my research in relation with the sciences. In conjunction with the bicentennial of the birth of Charles Darwin, I had the opportunity in 2009 to collaborate with a Paris-based researcher, Michel Gho, who is specialized in Developmental Biology, like I mentioned at the beginning of this chapter. This constitutes a first step for me and is very promising in terms of reflection and creation. This turning point in my work corresponds to a desire to better understand real life evolutionary processes. This is the reason why I felt the need to widen my knowledge with the help of a scientific point of view. So far, it has indeed provided me with some enlightenment, especially with regard to the concept of the mother cell (cell from which another cell of an organism, usually of a different sort, develops), and the daughter cell (cell resulting from the division of a single parent cell). Over a three-month period involving much study and many interviews and meetings with researchers, I was able to develop a new film concept with "*Motoo Viridis*", which was based on the creation of a "mother film" and two "daughter films". As a conclusion, I shall provide some details about this project. Motoo is derived from the name of the Japanese biologist, Motoo Kimura (1924-1994), and Viridis means green in Latin, or that which still has sap. As the author of the neutral theory of molecular evolution, the biologist and geneticist Motoo Kimura sparked controversy in the Darwinian world, saying that at the molecular level, the majority of genetic changes are neutral with respect to natural selection. Without denying the role of natural selection as a key factor in evolution, Kimura's theory makes genetic drift, an ultimately random changes, a primary factor in evolution. The chance factor multiplies the possibilities of variations and subsequently, life retains only those that are viable. A recent book with the contributions of renowned scientists published in February 2009 corroborates this theory:

"Le hasard au cœur de la cellule. Probabilités, déterminisme, génétique", coordinated by Jean-Jacques Kupiec:"Biology is currently undergoing a revolution. Living beings are not governed by an omnipotent genetic program. It has now clearly been shown that the chance factor lies at the heart of the functioning of genes and cells in organisms and plays a role that is still largely underexploited. While biology has always been dominated by "deterministic" or even finalistic theories since ancient times, the experimental results obtained in recent years herald a radical change of perspective. The new biology, by way of its probabilistic nature, will render obsolete the very idea of a program and genetic determinism (a concept that has been generally qualified as the "all genetics" thesis), an idea that was forged after what has come to be known as the "central dogma of molecular biology" (Francis Crick, 1958). But the new biology should not be construed as a denial of previously acquired knowledge in molecular biology. On the contrary, it constitutes an extension of the physicochemical perception of life. Additionally, it will inevitably have profound philosophical consequences. The fascinating question that it opens up consists in understanding how living things are created from molecular randomness".

Through pure intuition and prior to the publication of the book, I had already begun to work on a virtual cell plasticity concept including random and chance factors. *"Motoo Viridis"* develops an experimental perception by analogy with cell division processes, namely *mitosis* (process of cell division that results in the production of two daughter cells from a single parent cell; the daughter cells are identical to one another and to the original parent cell) and *meiosis* (type of cell division by which germ cells, i.e. eggs and sperm, are produced). Meiosis is a process of reductional division in which the number of chromosomes per cell is halved. I am currently creating a previously unreleased work.

FLEXIBLE BIOMEDIA

In this way, the film becomes an installation based on these notions. An interpretation of a poetical evocation of the theory of Motoo Kimura, *"Motoo Viridis"* presented three cell films where virtual seeds were mixed with real cells extracted from real microscope image sources using my "video stem cells" method. The cells divided, evolved

Figure 5. (Left) H2-Zoo ©Nyeki 2008 - (Right) Motoo Viridis creation Video plasticity research ©Nyeki09

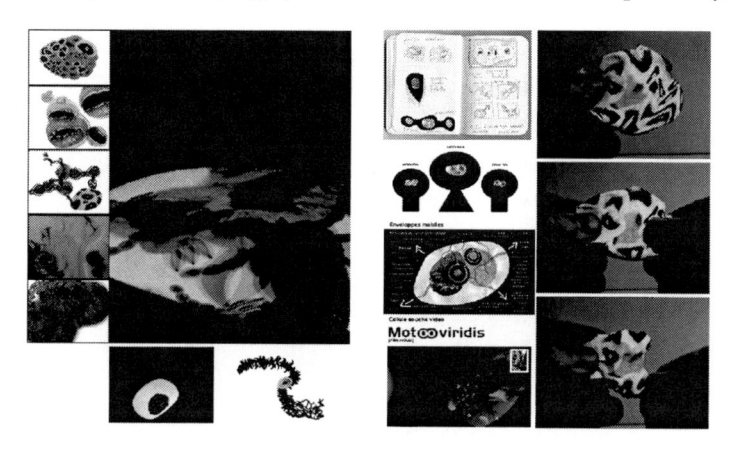

and generated living organic tissue close to that found in real life. Instead of the usual normalized rectangular screen, I developed moving envelopes that were relatively close to reality. The presentation in the exhibition proposed a projection on a large flat screen [mother cell] and two smaller flat screens [daughter cells] covered by a rigid black mask to hide the materiality and rectangular form of the plasma screens. In real life, "daughter cells" may be identical or asymmetrical, so the "daughter films" that were presented separately were directly replicated from the "mother film". To further highlight the parallel between artistic practice and scientific phenomena, the three movies were "randomly" presented by reversed symmetry and in asynchronous time. In addition, the corresponding sound research and development was important too. Based on the assumption that a cell emits sounds when it undergoes division, I composed a specific acoustic space from real materials, noises, voices and also speeches[1], all compressed into audio "sample cells" with a program originally developed by IRCAM, the French Institute of Acoustical and Musical Research and Coordination. The researchers at the IRCAM are the ones who created the ever fascinating castra voice in the 1994 Gérard Corbiau film, *Farinelli*, using a sort of sound-morphing process based on the voices of real men and women. I composed melodic lines for the "mother film" and more chaotic sounds for the "daughter films" that suggest random codification messages.

In summary, Plastika [Totipotenta] is a totally imaginary virtual world that, while being diametrically opposite to scientific methodologies, nonetheless seems to share certain amazing concepts with the latter. In the future, I am thinking of developing a "hybrid reality" that will include dynamic plasticity. I imagine myself creating a sort of "biomedia" material, much like a soft dough that mixes virtuality and reality, composed of "over-plasticities", through the development of malleable dynamic surfaces.

HYBRID WORLD

This overview of my work raises some issues on cell plasticity, the flexible frontier between the real and virtual worlds, simulation and simulacra paradox, and the origin of life. The role of the viewer and the human ability to interact dynamically with virtual creations are also involved. Aesthetic concepts have evolved in a way that is similar to that in which a plant grows: plasticity, totipotency, stem cell videos and virtual organic plasticine are some of the concepts that have proliferated freely. A singular conceptual garden has been developed, with many territories that still lie in fallow, at least, for the time being. Like a gardener, one can only expect one's plantations to become fertile after a certain period of time. My recent collaboration with various researchers opens up new perspectives. These experiences were more than a simple fertilizer. Video plasticine research will be extended to virtual body grafts known as xenografts and xenotransplants. My latest film, entitled INDY_Gene, features an imaginary longevity gene inspired by the Gene INDY discovered in 2000, and is a continuation of my work in progress. *"Life imitates Art more than Art imitates Life"* Oscar Wilde (Decadence of Lies)

REFERENCES

Debono, M.-W. (1996). *L'Ere des plasticiens: De nouveaux hommes de science face à la poésie du monde*. Coll. Sciences et Spiritualité, Editions Aubin.

KacE. (2009) from http://www.ekac.org/

Kupiec, Jean-Jacques [coordinator] (2009) *Le hasard au cœur de la cellule, Probabilités, déterminisme, génétique,* Syllepse Edition

Lambert, D., & Rezsöhazy, R. (2004). *Comment les pattes viennent au serpent Essai sur l'étonnante plasticité du vivant, Nouvelle bibliothèque scientifique*. Flammarion.

LathamW. (2001) from http://www.doc.gold.ac.uk/~mas01whl/

NyekiC. (2009) from http://www.nyeki.com/

ReasC. (2009) from http://www.reas.com/

SchmittA. (2009) from http://www.gratin.org/as/

SimsK. (2001) from http://www.karlsims.com/

SommererC.MignonneauL. (2009) from http://www.interface.ufg.ac.at/christa-laurent/

KEY TERMS AND DEFINITIONS

Artificial Graft: Virtual extension on plants and/or bodies.

Hybrid Harvest: Mixing of real and virtual worlds.

Plasticity: Ability to be formed and transformed.

Totipotency: Ability of a single cell to divide and produce all the differentiated cells in an organism.

Video Stem Cell: Cellular form of extraction from films to create virtual living tissue.

Virtual Plasticine: Manipulation of a moving entity in real time, on different scales and in different directions.

Virtual Tissue: Agglomeration of virtual moving video stem cells.

ENDNOTE

[1] I accelerated and, in particular, manipulated the recording of an interview with Mathias Mericzkay, a researcher at an INSERM laboratory, with whom I also worked.

Chapter 15
Spatial Design and Physical Interface in Virtual Worlds

Hidenori Watanave
Tokyo Metropolitan University, Japan

ABSTRACT

We would like to propose a new spatial model, the "Contents Oriented Space", conforming to the physical senses experienced in the 3D virtual worlds, as well as providing an appealing spatial experience, and present a design methodology making use of this new model. There are three necessary conditions for such "Contents Oriented Space": (1) The contents are visible from the outside; (2) The contents are directly accessible; (3) By being directly accessible, the contents become "spatial". By applying such a spatial model, it is possible to realize the architectural space in the 3D virtual worlds, conforming to the physical senses experienced in such environment, at the same time providing an attractive spatial experience. It is a new design methodology, able to be widely applied in the architectural space design for the 3D virtual worlds in general. The experimental use of the proposed methodology in the physical interface expanding this design methodology is also currently on going.

1 INTRODUCTION

In this chapter, the new spatial model, the "Contents Oriented Space", conforming to the physical senses experienced in the 3D virtual worlds, as well as proving an appealing spatial experience unique to the virtual worlds, and the design methodology making use of such model will be proposed through looking at specific examples. In recent years, 3D virtual worlds services have started to be popularized, as represented by Second Life (Linden Lab, 2003). In these services, 3D form making tools have been released, and space has, in a way, become "open source". Here, it is possible to freely create objects such as architectural buildings. Using this characteristic, many 3D virtual worlds architectural buildings modeling

DOI: 10.4018/978-1-60960-077-8.ch015

the real world space have been created. However, these models are not conforming to the physical senses experienced in the 3D virtual worlds, and have not been so attractive to the general users. Although few in number, there exist unique cases that conform to the physical senses experienced in the 3D virtual worlds, created by nameless/anonymous "residents". Nevertheless, these are not created under a common design methodology, therefore, the degree of each completed work will depend on the ability of each user; and it lacks the possibility of reproduction. Also, it is not possible to say they provide the appealing spatial experience unique to the virtual worlds, equivalent to that experienced in architecture in the real world.

The "Contents Oriented Space" is a new architectural space model in the 3D virtual worlds designed by the authors, based on the aforementioned cases. The three necessary conditions of the "Contents Oriented Space" are as follows: 1. Contents are visible from the outside; 2. Contents are directly accessible; 3. By being directly accessible, the contents become "spatial". By applying this spatial model, it is possible to conform to the physical senses experienced in the 3D virtual worlds, as well as realize the architectural space in such environment, providing an attractive spatial experience.

It is a new design methodology able to be widely applied in the general architectural space modeling in the 3D virtual worlds. The author has applied the "Contents Oriented Space" in the past, designing a large number of architectural/art works for the 3D virtual worlds as *"Archidemo"* (Watanave,2007). Hereafter, the existing design methods in the architectural space in the 3D virtual worlds will be examined in section 2; the "Contents Oriented Space" spatial model proposed by the authors and the design methodology making use of this model will be discussed in the section 3; the specific application of the design methodology and its results will be stated through reviewing past works dealt by the authors in section 4; the experimental use of physical interfaces currently underway will be introduced in section 5; and the conclusion of this chapter will be stated in section 6.

2 DESIGN METHODS' REVIEW

In this chapter, the existing architectural spatial design methods in the 3D virtual worlds will be examined by reviewing specific cases.

2.1 Imitation of Real World Architectural Buildings

There is a long history in visionary (unbuilt) architecture (Burden,1999). These works had the charm that cannot be achieved in real world. But they were works only in the thought, and not able to be experienced interactively. And also, there are plenty of previous work concerning the design of architecture of the real world that uses the virtual reality technology (Bridges,Charitos,1998) (Whyte,2002), or real world architectural buildings imitated in 3D virtual worlds(Maher,Simoff,Gu,Lau,2000) (Brouchard,2006) in recent years. In this case, the measurements and forms of a real world building are faithfully reproduced, and on the surface, it seems to be same as the building in actual existence. However, in experiencing the space through the 3D virtual world body, or the avatar, the interior of the building feels smaller than in reality. It becomes a stressful space as you hit the walls or get stuck in the spatial gaps. For the users used to navigating freely in space, not being restricted to the collision checking in the 3D virtual worlds, the walls and ceilings imitating the real world building become obstacles preventing movement, resulting in a stressful experience. Furthermore, there is a fundamental difference in the physical senses, where the real world is experienced through the entire "body", the 3D virtual worlds is experienced through an interface such as the personal computer.

2.2 Case of an Architectural Space Characteristic of 3D Virtual Worlds

In this section, examples of unique architectural space, conforming to the physical senses experienced in the 3D virtual worlds and which do not exist in the real world, will be shown. Further, these cases do not follow any common design guideline, and are individually created by each user (the "resident" of the 3D virtual worlds). Upper-side of Figure 1 shows a case of a foot bath in the 3D virtual worlds. Although it is not possible to reproduce a real-life foot bath in the 3D virtual worlds, where there is no sense of temperature, this 3D virtual world version of a foot bath is busy with many avatars. Since it is not an indoor foot bath, but an outdoor one, the group of avatars is easily seen and observed from afar.

The avatars that arrive can directly access the foot bath without being blocked by any wall or roof. The visibility and the accessibility from the outside is both high, and it has become a place convenient for avatars to gather. Left-down side of Figure 1 is an example of an architectural space which is accessed through the search word "Freebies (a slang in Second Life, Freeware)". No wall or roof exists here, and all the contents (freeware) are revealed to the outside. The imitation of the real-life spatial element is completely eliminated, and the architectural space is constructed by means of the contents alone. Similar to the foot bath, the contents within the architectural space is easily observed from the outside, and they can be accessed directly without being obstructed by wall or roof. It is a highly convenient architectural space for the users hoping to attain the freeware. Right-down side of Figure 1 is an example of an architectural space which is accessed through the search word "Door". Similar to the Freebies, no roof or wall exists, and the contents, in this case the doors, are revealed just as they are to the outside. Its visibility from the exterior and accessibility are both high, and it is an architectural space composed solely by the contents. These cases realize the spatial placement and access methods unique to the virtual worlds through the use of real world architectural elements. As a result, a site that is "popular" as well as "populated" is produced. In the next chapter, a new architectural methodology will be discussed, based

Figure 1. Cases of an architectural space

on these cases conforming to the 3D virtual worlds physical senses.

3 PROPOSAL OF A DESIGN METHODOLOGY

3.1 The Need for a New Design Methodology

Presently, no common methodology designing architectural space exists, that is appropriate for the 3D virtual worlds physical senses. In the case of section 2.2, space is created based on the ability of an individual user, and it is a "spontaneously generated" architectural space, so to speak. However, in the following few years, further popularization of 3D virtual worlds and its use in broader spectrum of fields is predicted(Nomura Sougou Research Lab,2007) (Itmedia,2009); and there are actual reports of increase in the number of its users. In the near future, the need for a common architectural space design methodology for the 3D virtual worlds to realize an appealing user experience is a must. Since the spontaneously generated example given in section 2.2 conforms to the physical senses experienced in the 3D virtual worlds, it is possible to come up with a new design methodology based on such a case. In the next section, as a clue to proposing the above mentioned methodology, past cases of architectural space design methodology in the real world will be discussed.

3.2 Architectural Design Methodology in the Real World

Architects, including Kazuhiro Kojima, proposed the "Space Block"(Hiiro,Kojima,Iwasaki,1996) (Inoue,Kojima,Hiiro,Watanave,1999), which modeled the architectural space that generated spontaneously. They have applied it to a new architectural space design and have successfully realized the architectural space that inherits its original quality. Although this "Space Block" is a result of sampling and layering various elements from the real world, it is not intended to reproduce the original space. The "Space Block" is utmost a "Space Model" as a means to achieve a goal, and different design techniques in order to realize a model in the actual design process have been explored. Of course, the design methodology that the "Space Block" is applied to aims at designing architectural space for the real world environment, therefore, it is not possible to use it in the 3D virtual worlds directly. However, the technique to apply to the newly designed architectural space in general, by modeling the cases of "spontaneously generated" architectural space, is effective even in the 3D virtual worlds.

By referring to this technique, it must be possible to create a new design methodology which can be widely applied to the architectural space in the 3D virtual worlds in general. At first, in the next section, architectural space of the real world environment and the case of section 2.2 will be each made into a model to be discussed.

3.3 The Architectural Space Model of the Real World

Figure 2 shows the architectural space of the real world and the 3D virtual worlds as a model.

The architectural space of the real world is made up of physical elements such as the roof/wall. At this time, imagine the flow within the architectural space for a person to arrive at the contents. Assuming the contents to be books in this case, the books within the bookstore will be observed through the glass window (dotted line arrow). In the general sense, it is obviously not possible to "penetrate through" the window in order to move. The person will have to follow a roundabout route, go through the entrance in entering the bookstore, and arrive at the books (solid line arrow). At this point, the continuous change in the scenery, the sequence, brought about by following the route, is the "spatial experience"

Figure 2. Model of architectural space in the real world

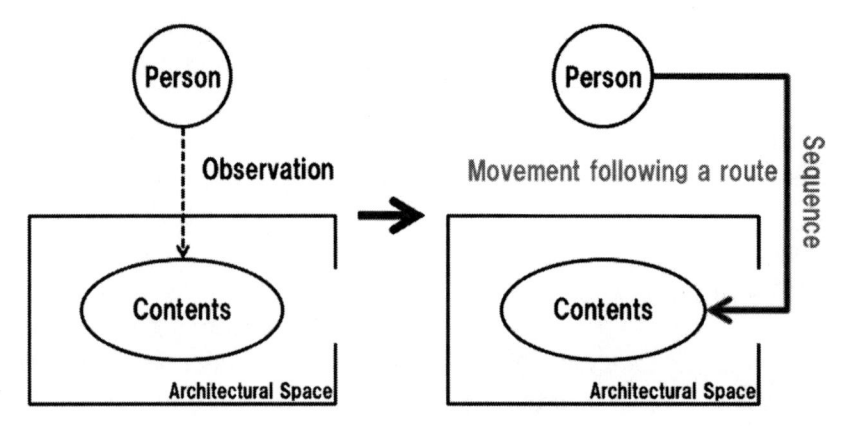

perceived by a person; and it is one of the attractive elements the real world architectural space possesses. However, as already stated in section 2.1, it is not possible to recreate such "spatial experience" just by imitating and reproducing the real-life architectural space within the 3D virtual worlds. Next, the case in section 2.2 is shown as a model in Figure 3.

In this model, the user/avatar observing the contents will directly access it following the navigation process (instantaneous transportation by a click). Even with the placement of a 3D object as a pretended architectural space in the surrounding area, it is possible to "penetrate" and arrive at the contents, if the collision check is off. This point conforms to the physical senses experienced in the 3D virtual worlds. However, unlike the model in Figure 3, the continuous change in the scenery, the sequence, does not exist; therefore, no "spatial experience" is generated.

In order to generate the appealing spatial experience "unique to the 3D virtual worlds" that is equivalent to that of the real world, it is necessary to produce the continuous change in scenery following a navigation process. In the next section, the behavior of Windows GUI will be discussed as a clue to this possibility.

3.4 "Making Contents into Space"

Figure 4 is a screen shot of Windows GUI behavior. The content is shown after double-clicking the folder within the window.

Figure 3. Architectural space model unique to the 3D virtual world

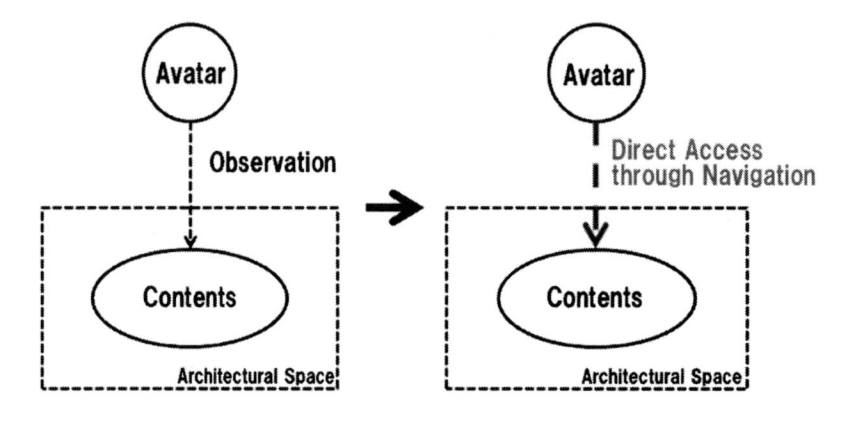

Figure 4. "Contents made into space" in the Windows GUI

This behavior is reinterpreted in accordance with the model in Figure 3. When the user directly accesses (=double-clicking) the observed content (=folder icon), the substance of the content (=folder) is reached. The architectural space (=window) itself does not change, but the contents (=folder) is developed into a new architectural space (new window) where the scenery (= appearance) continuously changes. That is to say, through the user's direct accessing, the "contents are made into space", and the "change of scenery" are generated at the same time.

With this chain of events as a clue, the authors developed an architectural space model for the 3D virtual worlds, the "Contents Oriented Space", which conforms to the physical senses experienced in the 3D virtual worlds, and that which provides a more appealing "spatial experience". The next section will explain about this model.

3.5 The Model of the "Contents Oriented Space"

Figure 5 is a spatial model brought about by projecting the "contents made into space", stated in section 3.4, in the 3D virtual worlds.

In this model, the "contents is made into space" through the direct access of the avatar, and the contents understood to be just objects up until then, changes into the space enveloping the avatar. The change of scenery generated here is understood by the user to be the "spatial experience".

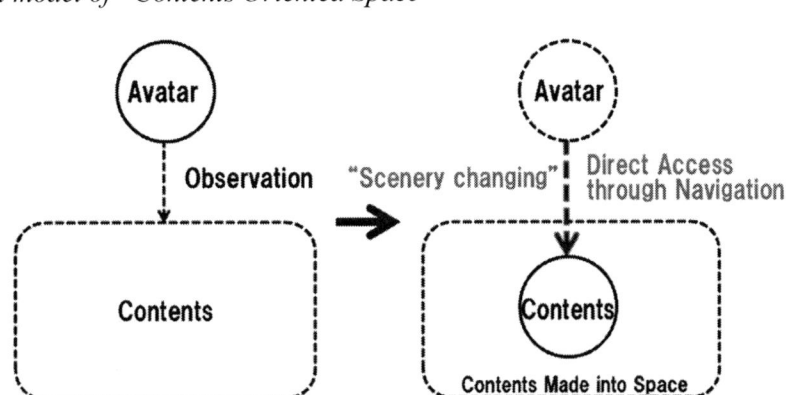

Figure 5. Spatial model of "Contents Oriented Space"

This model is based on the case stated in section 2.2, and conforms to the physical senses experienced in the 3D virtual worlds. Furthermore, it is also based on the concept of the "contents made into space" stated in section 3.4, and it provides the "spatial experience" equivalent to the architectural space of the real world. It is not necessary to state that this "spatial experience" is non-existent in the real world, but is unique to the 3D virtual worlds.

The authors named this spatial model the "Contents Oriented Space". The necessary conditions to realize this "Contents Oriented Space" is as follows:

- The Contents are visible from the outside;
- The Contents are directly accessible;
- By being directly accessible, the contents become "spatial".

This "Contents Oriented Space", similar to the Space Block stated in section 3.2, is utmost a "spatial model", and is a means to achieve a goal for an appealing architectural space design in the 3D virtual worlds. Accordingly, unlimited number of design techniques, applying the spatial model of the "Contents Oriented Space" and fulfilling the above mentioned conditions, are feasible. These design techniques accumulated will be established as the design methodology possible to be applied widely in the general architectural space design in the 3D virtual worlds.

In past group of works, the authors have tried various design techniques, where the spatial model of the "Contents Oriented Space" was applied. These works not only received high appraisal from users of the 3D virtual worlds, but won prizes and were selected for international festivals and conferences(Watanave,2007). In the next chapter, specific architectural space design techniques for the 3D virtual worlds the authors used to apply and realize the "Contents Oriented Space", and its result will be discussed in reference to appropriate examples.

4 THE APPLICATION OF THE DESIGN METHODOLOGY AND ITS RESULT

In this section, the authors will discuss the results of applying the design methodology making use of the "Contents Oriented Space", while referring to past pieces of work. As stated in the previous chapter, the authors have accumulated and investigated these design techniques in expectation of the future, and aims to establish the common "architectural space design methodology for the 3D virtual worlds" in the end.

4.1 "Archidemo"

"Archidemo" was held as an art collaboration with the aim to explore the possibility of architectural/city planning design in the 3D virtual worlds, with the cooperation of *Nikkei Architecture (Nikkei BP)*, from the period of August 2007 to January 2008. During this period, total of 24 researchers, creators, and students participated. This section will present the collaboration with the science fiction novelist, *Hirotaka Tobi*, the *"Numbers Beach"*, as an example of a work applying the "Contents Oriented Space" concept. This work uses the theme from *Tobi*'s work, *"Grand Vacance"*.

The entire space of this *"Costa Del Numero"* is made up of an all-sky periphery panoramic image of a business district, covered by a texture-mapped giant cube (64m square). Within its interior, several small cubes (16m square) are floating. The small cubes have a group of all-sky periphery panoramic images of a resort hotel textured-mapped; and furthermore, since the outward polygon is made transparent, the panoramic image (contents) within is visible from the outside [left-side of Figure 6]. When the user clicks on the small cubes, the user is teleported (directly accessing) into each interior space [right-side of Figure 6]. In this instance, the small cubes, recognized as the contents up until then, unfolds into an all-sky periphery panoramic

Figure 6. "Costa Del Numero"

image (contents made into space) surrounding the avatar. At this stage, the user will be virtually experiencing the architectural space of the resort hotel. By repeating the clicking process, it is possible to move between different contents one after another to experience its space.

By applying the spatial model of the "contents oriented space" in such a way, it was possible to achieve both a stress-free spatial movement and a dynamic spatial experience.

4.2 3D Image Database of Oscar Niemeyer

The year 2008 was the centennial year of the start of the Japan- Brazil diplomatic relationship, and the *"Japan-Brazil Exchange Year Project"* (*Niappaku 2008*) was carried out by the Foreign Ministry and the Japan-Brazil Exchange Year Executive Committee. The author created the *"3D Image Database of Oscar Niemeyer"* as part of its official project, the "Japan-Brazil Architectural/ City Planning Internet Archives". This project entailed reconstructing a group of images within the internet relating to Oscar Niemeyer, the representing architect of Brazil, as an image database possible to be operated and viewed intuitively.

This work was exhibited at the "File-Electronic Language International Festival Sao Paulo 2008" held in Sao Paolo. In the first state, group of cubes (2m square), with images related to Oscar Niemeyer texture-mapped, float in space [left-side of Figure 7]. As you can see from the figure, the image (contents) of each cube can be viewed from the outside.

When the user clicks on the cube of choice, the image box (6m x 8m) held within the cube appears, and approaches the avatar quickly (within 1 second), and it surrounds the avatar [right-side of Figure 7]. In this piece of work, by the contents approaching (direct access) the avatar, the concept of the "contents made into space" is achieved. Furthermore, when the image box is clicked on, the website where the contents appear will pop up and be shown.

In this way, by applying the space model of the "Contents Oriented Space", it was possible to realize both a stress-free interface for the viewing

Figure 7. 3D image database of Oscar Niemeyer

of image/web contents, and an appealing spatial experience.

4.3 SIGGRAPH Archive in Second Life

"SIGGRAPH Archive in Second Life" is the past group of venue photographs of ACM SIGGRAPH Emerging Technology/Art Gallery, reconstructed into an image database space which is possible to be operated and viewed intuitively. This piece of work was exhibited at the Art Gallery of "SIGGRAPH Asia 2008" held in Singapore in the year 2008.

In the first state, the plates with different years are floating in space. When the plate of choice is clicked on, a group of image boxes (6m x 8m), with the venue photographs of each year texture-mapped within, spurts out into space and unfolds extensively [left-side of Figure 8].

Since the outward polygon alone is made transparent, the photographs (contents) within are visible from the outside. When the image box is clicked on, the user will be teleported (directly accessing) into each interior space [right-side of Figure 8]. In this instance, the image box, recognized as the contents up until then, covers the entire field of view, and unfolds around the avatar as a form recognizable as space (contents made into space). In this way, by applying the space model of the "Contents Oriented Space", it was possible to realize both a stress-free interface for the viewing of the photo contents, and appealing spatial experience.

5 EXPERIMENTS USING PHYSICAL INTERFACES

Presently, the keyboard + mouse are the standard for the actual operating interface in the virtual worlds. However, if we are considering the design of contents that conforms to the "physical senses experienced in the 3D virtual worlds, as stated in section 2, the keyboard +mouse is not necessarily the most suitable interface. As a preceding example of research of non-keyboard + mouse internet interface, there is the research on the Brain Machine Interface by Ushiba and his fellow researchers(Ushiba,2009); but because of the need for skill, it is not easily experienced by many users, and also it is very costly.

In order to resolve these points, we are carrying out experiments using an interface other than the keyboard + mouse, through which it will be closer in experiencing the normal physical sensation, and moreover, possible to be handled easily by anyone. By using the Wii Remote as the interface in the *"Laval VRchive"* in Laval Virtual ReVoluti on(Watanave,2009,April), held in the year 2009, it was possible to achieve a more intuitive operating sensation. Also, through the eye tracking device, the *"Tobii Eye Tracker"* made by *Tobii Company*,

Figure 8. SIGGRAPH archive in Second Life

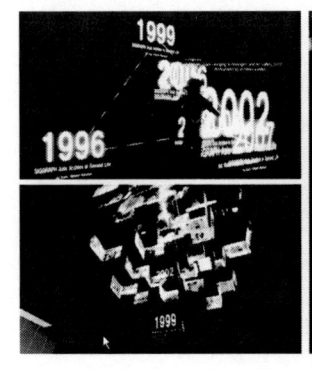

testing of the avatar operating through the use of the sight and "blinking" is underway. We would like to discuss these projects in another occasion.

6 CONCLUSION

Through the activities starting with the group of projects shown in section 4, the design methodology of the "Contents Oriented Space" has begun to be substantial. From seeing the facts that the past group of works have won prizes and were recognized in international festivals and conferences, as well as receiving constant appraisal, it is possible to consider our design methodology to have validity. Although the spatial model, "Contents Oriented Space", discussed in this article was developed with the most suitable 3D virtual worlds platforms available at present, the Second Life, as the basis, it is possible to apply in other 3D virtual worlds in general that have similar characteristics. From now on, we plan to further enhance the design methodology by means of the actual use of the physical interface in experiments, as stated in the previous chapters, as well as carry out activities in platforms other than Second Life.

ACKNOWLEDGMENT

I would like to thank Shigeru Owada, Koichiro Hajiri, Akihiko Shirai, and Yumiko Nakadate each for their cooperation in writing this article.

REFERENCES

Bridges, A., & Charitos, D. (1998). On architectural design in virtual environments. *Design Studies, 2*(18), 143–154.

Brouchard, J. (2006). *The Archi*, from http://archsl.wordpress.com/

Burden, E. (1999). *Visionary Architecture: Unbuilt Works of the Imagination*. USA: McGraw-Hill Professional.

Hiiro, M., Kojima, K., Iwasaki, K. (1996, August). *The Concept of the Space Block and its Achievable Media: Design Method Using the Space Block*, Japan Association of Architecture Academic Lecture Series, Vol. 1996(19960730) (pp. 463-464). E-1 Architectural Plan I.1996, Japan Assciation of Architecture.

Inoue, C., Kojima, K., Hiiro, M., Watanave, H. (1999, August). *Construction of the Space Block Database, Space Block as the Interactive Spatial Design Tool, Part 1*, Japan Association of Architecture Academic Lecture Series, Vol. 1999(19990730) (pp. 531-532). E-1 Architectural Plan I, Japan Assciation of Architecture.

Itmedia (2009, May). *Second Life Will Not End: Increasing number of users, Growing Economy*, from http://www.itmedia.co.jp/news/articles/0905/18/news037.html

Linden Lab. (2003). *Second Life*, from http://www.secondlife.com/

Maher, M. L., Simoff, S., Gu, N., & Lau, K. H. (2000). *Designing virtual architecture*, Proceedings of the Fifth Conference on Computer Aided Architectural Design Research in Asia (pp. 481-490), Singapore.

Nomura Sougou Research Lab, Inc. (2007, July). *IT Road Map*, from http://www.nri.co.jp/news/2007/070525.html

Ushiba, J. (2009, February). *Operation through the Characters of the Virtual World Second Life*, The Brain Science to Open the Future (No. 5) Technology Challenging the Interaction between the Brain and the Environment (2), Bio-industry 26 (2) (299), (pp. 89~93), CMC Publishing.

Watanave, H. (2007, July). *Archidemo*, from http://archidemo.blogspot.com/

Watanave, H. (2008, November). *"Contents Oriented Space"* ACM International Conference Proceeding Series; Vol. 352. Proceedings of the 2008 International Conference on Advances in Computer Entertainment Technology. Posters. (pp. 418-418), ACM.

Watanave, H. (2008, August). *"3D image database of Oscar Niemeyer"- NIPPAKU 100 (Memorial of Japan and Brazil exchange year) Official Project,* Artwork exhibited in FILE - Electronic Language International Festival in Sao Paulo 2008. SESI' Cultural Centre. Sao Paulo. Brazil.

Watanave, H. (2008, December). *"SIGGRAPH Asia Archive in Second Life"*. Artwork exhibited in SIGGRAPH ASIA 2008 Art Gallery (curated works). Suntec Singapore International Convention & Exhibition Centre. Singapore.

Watanave, H. (2009, April). *"Laval VRchive"*, Laval Virtual France Virtual Reality International Converence 2009 Proceedings (pp. 397-398), Laval Virtual.

Watanave, H. (2009, April), *"Archidemo - Architecture in Metaverse"*, Artwork exhibited in 9th edition of FILE - Electronic Language International Festival in Rio de Janeiro, in Media Art category, 2009

Whyte, J. (2002). *Virtual Reality and the Built Environment.* USA: Architectural Press.

KEY TERMS AND DEFINITIONS

Second Life(SL): Is a virtual world developed by Linden Lab that launched on June 23, 2003, and is accessible on the Internet. A free client program called the King Bee enables its users, called Residents, to interact with each other through avatars. Residents can explore, meet other residents, socialize, participate in individual and group activities, and create and trade virtual property and services with one another, or travel throughout the world (which residents refer to as "the grid"). Second Life is for people aged 18 and over, while Teen Second Life is for people aged 13 to 17.

Virtual Worlds: A virtual world is a genre of online community that often takes the form of a computer-based simulated environment, through which users can interact with one another and use and create objects. Virtual Worlds are intended for its users to inhabit and interact, and the term today has become synonymous with interactive 3D virtual environments, where the users take the form of avatars visible to others graphically. These avatars are usually depicted as textual, twodimensional, or three-dimensional graphical representations, although other forms are possible (auditory and touch sensations for example). Some, but not all, virtual worlds allow for multiple users.

Chapter 16
The City of Abadyl

Michael Johansson
PRAMnet, Sweden & China

ABSTRACT

In the City of Abadyl we try to explore a complex digital space in a setting that invites to participation. We provide a detailed and complex, yet open world that can be utilized in order to generate scenarios for the temporary co-creators of Abadyl, who would then interact in an optional environment and in the end producing new artefacts. Abadyl is a database that contains all the gathered information in different file formats, it is a storage facility for all of the physical artefacts, it is a website used for communication and documentation, it is a map for navigating the City. This combination of interactive situations and artefactual production we called "fieldasy". We are concentrated on developing collaboration in the production of new media and it's artefacts. We try to produce artwork that actually incorporates surprising visual and technical proposals that are unusual, enriching and engaging. By building prototypes and iterate it over time and amongst the co-creators, it let us explore this area in a fruitful way, moving between artistic intentions/screen writing, artefacts and digital generated expressions and script/code writing. Here the virtual object can challenge the physical with qualities that is very hard to achieve in the physical world, and in that conflict, new expressions can be developed. Today Hybrid creations have become a method for working with cultural production not only with different elements of form, but as blending identities of the creators as well. In our prototype work we focus especially on interactive installations and stage design; we realize that the digital design process both demand new forms of conceptualization and prototyping activities to support the design of the expression of the final artwork itself – and maybe in the long run propose a updated and appropriate design theory in this field.

DOI: 10.4018/978-1-60960-077-8.ch016

Copyright © 2011, IGI Global. Copying or distributing in print or electronic forms without written permission of IGI Global is prohibited.

INTRODUCTION

In my profession I have always had a very close relation with the material I have used for investigating and execute my ideas. For all my works, no matter what material I have chosen for its execution, it is of foremost importance to be in a constant dialogue with them. There should be no difference between the critical mass of once idea and the final outcome of the artwork itself; it should be interwoven unity both in theory and in its form. Since 1984 the computer has played a major roll doing just that. The strategy I have developed is how you as an single artist can set-up a long-term working environment where you want to explore methods for both artistic work practice and technical research&development and in the end produce artworks of relevance?

My approach is to establish an environment which can facilitate artistic work practice in a complex production environment such as the one of digital media, supported by invited artists, researchers and students. In other words, create a kind of gravity around a single project that other easily can transfer their knowledge into without having to provide all of its content from start. Today the city of Abadyl has involved over 120 people, Exhibited internationally as well as in Sweden and China where I recently live and been part of several external research projects in areas of textile, computer science, interaction design, cultural studies, theatre and film. In the chapter I will show how we set up complex and open framework and used a city as a metaphor, to serve as a playground for the ongoing projects and the invited co-creators.

The city of Abadyl is virtual and physical, it is a database that contains all the gathered information in different file formats, it is a storage facility for all of the physical artefacts, it is a website used for communication and documentation, it is a map for navigating, it is a collection of stories extracted by the co-creators, it is a board game that can be played in three different languages; Swedish, English and Mandarin, it is a collection of music and films recorded in Abadyl. The idea in using different form of representation and materials is to achieve a continuously iterative process driv-

Figure 1. A map of the city of Abadyl 1999-2009 Mobela-C version

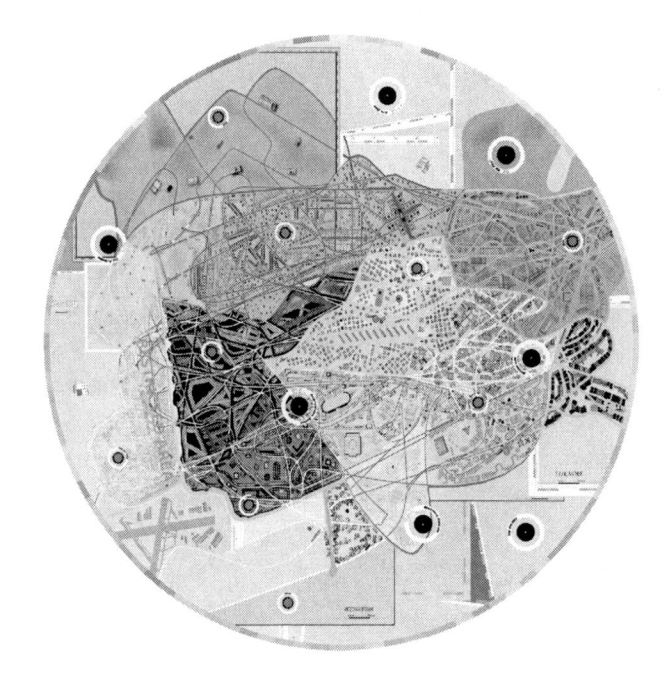

ing the different parts of the overall art project forward, developing relevant design theory and support innovation in area of technology and new media. In the chapter I put forward a design theoretical model and discuss the use of prototypes as a vital design material in my investigations. The intention is not so far the establishment of a so called great narrative based and extracted from the City. Inspiration has instead been retrieved from the art of novel writing and its practice in constructing worlds. In "Postscript to the Name of the Rose", Umberto Eco writes on the generative logic he has adopted, a logic both limiting and expanding creativity. The fundamental parameters guide what can and what cannot be included in a fictional but historically plausible universe. In *The City of Abadyl* we have chosen to focus more on the generative itself in this logic; that is to say, it is not about parameters resulting in a watertight consistent universe, but the main interest is in *what can be generated* from a number of predetermined parameters. However since the project have been running now for ten years and all of the input has reached a level of complexity and detail where a lot of things can happen and be put into play. In the chapter we give some examples of what has been generated and how I can conduct my work together with the co-creators.

BACKGROUND

Four things are worth mentioning in relation to this projects development over the years it has been running. They are fully described in different papers and articles through the years, Fieldasy 2004 and are you programmed to speak 2006. First of all it builds upon a former art project called "from an uncertain point in the Cartesian space". Second the experience we had with looking and talking to people visiting the studio and how they made up their own stories based on our unorganized research and reference material in "My computer is 36 m2". Third the development of the framework and set of rules in "maps, scales and objects" 2005. Forth the method "fieldasy" for working within the framework and how one can extract stories and artefacts from it.

From an Uncertain Point in the Cartesian Space

In 1997 I got a grant from the art council in Sweden together with a colleague of mine, to make a project called "from an uncertain point in the Cartesian space". In that project we went around the world in an investigation of a series of locations all having in common their state of being established by recognizable senders – dictatorships, religious and political ideologies, different kinds of utopias realised, or at least regarding themselves as realised. The idea materialised to somehow be able to destabilise these implemented utopias without destroying their utopian qualities altogether, their boldly thrown out suggestions of *something else*. To save these utopias from themselves by reprogramming them, introducing a constant distortion in their implementation. We then chose to digitally reconstruct these locations, "erect" them as 3D-models. Partly because this in fact is and is not implementing, but most of all because it rendered us the possibility of hands-on experimentation with these architectonic manifestations, joining them and exposing them to practical philosophy (or for that matter, some kind of *living*). And through the utopias (always pointing too toward the prevalent) and the virtual tools a way of engaging in dialogue with the world, examining its possibilities as well as those of the tools, without replacing presence with another as determining presence.

The Computer is 36 m2

At a certain point in the end of collecting all of the reference material - which was mostly directly digitized into a computer - my colleague said to me:

Figure 2. From an uncertain point in the Cartesian space together with artist Mikael Ericsson, the 36m2 database, Halmstad, Sweden 1999

-I do not know what is inside of that computer any more, let's print everything out.

So we did. It covered a 4 to 9 meters big wall and it occupied us for four days going through every image, sound, 3dmodel, video sequences and animation: But in the end we didn't create that hoped for overview of the project. But the overall structure there on the wall was very inviting for the people that came and went in the studio. They stopped, starting to look at the images, reading the text and by moving in parallel they started creating there own stories and told us things about our material that we didn't know. Unfortunately we didn't follow that direction at the time, but this way of inviting the audience, guests and colleagues was to become the basis for how I later set-up the framework for the city of Abadyl. Instead we developed new stories and expressions in relation to the actual place itself, this was later overlaid back onto the site itself in forms of projections.

The project was shown both as a gallery exhibition and on Internet in 1999-2000.

Maps, Scales and Objects; A Framework for Abadyl; 16, 7 and 100

Creating a world and communicating it among others is not easy. The task is very complex and demanding and the risk of being disappointed on what you will achieve is evident. But since we in the "from an uncertain point in the Cartesian space" project already had generated 2000 low-resolution and 550 high-resolution models of buildings, interiors, objects and exteriors, we saw a unique possibility to get a good start at least. The basic idea was to establish a space where to practice a critique of art, culture and society, through an investigation of philosophy and criticism in a dynamic material in a mixed reality space. So to be able to create such an infrastructure, that could host the already developed models I came up with the idea to set

Figure 3. From an uncertain point in the Cartesian space, Galleri Skanes konst Malmo Sweden 1999

The City of Abadyl

up a closed but yet complex space, an area with blurred boarders to the surrounded unknown. So to support this we looked for metaphors that could host this delimited space. The analogy to racing tracks was obvious; with roads that just looped themselves through their environment; Together with five architect students I had to find, write and draw a set of characteristics for every part of the city to be able to see which of our former material that could be transferred to the different parts of the city. A sort of simplified pattern language (Alexander, Ishikawa, Silverstein, 1977) had to be developed.

7 scale system that we introduced to handle the event space of the city it described the different levels on which objects and events can occur.

1.Environment, 2.Building, 3.Room, 4.Furniture, 5.Tool, 6.Interface and 7.Idea

16 The locations constituting the starting point are sixteen in number, and were initially held apart. But later they were merged into a connected city. As joining infrastructure the series of sixteen Formula One tracks was chosen, piled on top of each other to a joint figure. Hence aiming to expose architecture and cityscape to extreme strain (= life, unpredictability).formula one tracks became our point of departure, they were put on top of each other shaping an interesting ornamentation of roads just waiting for be driven on. The city then divides in sixteen parts with the idea to host their own separate ideology, architecture, fashion, lifestyles etc. 100 objects to represent the city or the world, was matched Peter Greenaways (Greenway, 1992) idea of an encyclopaedic approach. In the various quarters of the city 100 objects would be found, each devised in a way that constituted an elegant description and was manifesting the aspects of their very own district. These objects is helping us shape the differences on a broader range of levels in every part of the city and serves as a series of obstacles for the explorations. The purpose is to interfere with both the already built objects and the activities that are going to occur here later on. In this way we happened to examine our original objects once again, when we were depicting them in every detail. To be very thorough in investigating its form, visualises other layers in the object and creates conditions for both deconstruction and new narratives. To force out its meaning by depicting/modelling – to sort of philosophize from the street level and up. We used the different programming proposal cross-programming, trans-programming and dis-programming by Bernhard Tschumi (Tschumi,

Figure 4. The score for the framework based on the numbers 7, 16 and 100 - the seven scales, the sixteen formula one tracks and the hundred objects

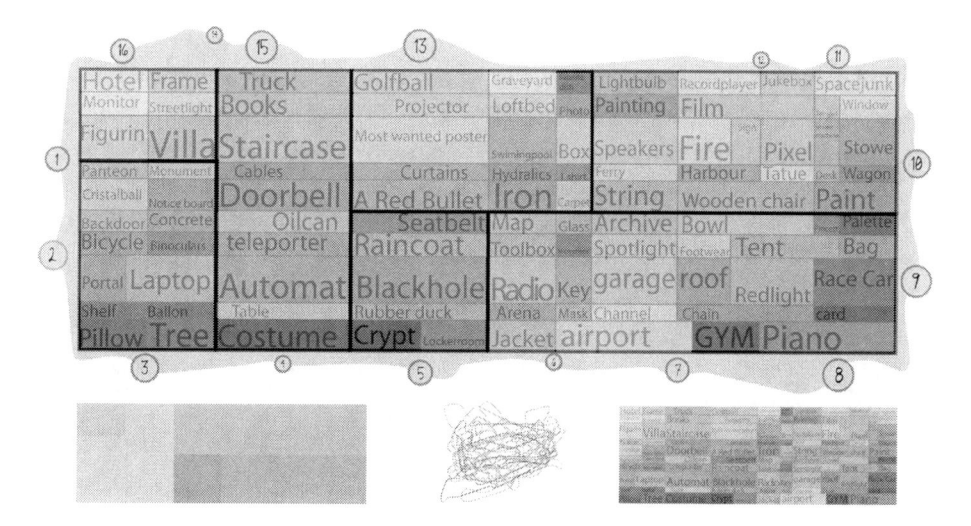

1994). This was implemented on a conceptual level for the 3d modellers. It described a series of Boolean operations with a different design formula for each part of the city. We also included a chart with profiles or silhouette of each city part, which supported both the modelling and the distribution of the architectural models in the city. The main architectural expression was then combined with the textures from other location than its origin, to create a both familiar and strange atmosphere in the different parts of the city.

Fieldasy

Fieldasy was the method we developed in 2003 – 2004 and was presented as an exhibition in Malmo Sweden at the Gallery Skanes Konst and later as a research paper for the Pixel Raiders conference. Fieldasy attempts to understand and redefine our world in a situation where information is lacking. This lack of information is used as a resource. By providing ambiguous fragments as a starting point, that put no constraints on imagination. For the exhibition we built a a structure that invited the audience to connect and navigate in-between the artefacts that was named Mobela_a. Like in the case of the 36 m2 computer, the viewer side-steps from shelf to shelf arranging their own stories of the objects, the shelves and body movements create disruptions like rhythms. Jaleh Mansoor refers how Kurt Schwitters in the first meeting with Hans Richter walked up to him and introduced himself as "I'm a painter and I nail my pictures together" (Mansoor, 2002). Painting and nailing seems to belong to different domains, but was integrated in his Merzbau, a gigantic project giving physical form to an assemblage of objects and spatial configurations. While side-stepping the furniture in the fieldasy01 exhibition, the body framed the viewing in an a laborious way – nailing, and at the same time performed an act of imaging, in combining the objects into visual stories – painting.

Digital art has come to focus, and rightly so, on the interactive meeting point between viewer and the art object, the way the viewer becomes an agent of change and participator by using interactive technologies. This is of course necessary to explore one of the intrinsic characteristics of new media. But in doing so there's also a focus on singular moments extracted from time. Other definitions might give way to how the metaphors and models could integrate how art can evolve over very large temporal spans. There's slowness on the border of inertia in the way the city of Abadyl develops. Involving many co-creators in the process develops nodes of expressions that may have meaning for the co-creators themselves in their work. That meaning gears into other levels and speed, when put together with other nodes from other actors. The exhibition situation is one such moment, when the speed of the artefacts implodes. They enter into a realm where they can be contemplated and juxtaposed into new stories for the viewers. The term 'floating work of art', with references to Eco maybe better depicts this openness, not only in directions of the narrative but also the character of work-in-progress (Dinkla, 2002). Fieldasy01, the exhibition, does not increase speed going from object to signs and

Figure 5. Fieldasy01 exhibition view

representations. It stops time to put forth material objects that are generative rather that produced. Generative in the sense that they have nor original nor final form. They are sprung out of a chain of association that generates new forms, which in its turn re-combines into new stories by the side-stepping act of the viewer.

STAGING THE CO-CREATION OF ABADYL

In the early 90`s I tried to combine different images and objects in the computer and then translated them into physical objects again. Here the role of technology was not only that of the tool. Because electronic media had also raised consciousness of an incidental flux in our culture where cultural production combines fragments, dislocates them and re-combines again. The cut-up of Burroughs or the game of Exquisite corps by the surrealists where no longer weird for ordinary people. The concept of sampling takes it's older relative the montage some giant steps further. Collage or montage is a kind of juxtaposition where you disrupt elements to put them in new combinations, the sampling technique works on a more genetic level. Since all media objects share the same foundation (Manovich, 2001) they can establish another kind of interpolation. Here the virtual object can challenge the physical with qualities that is very hard to achieve in the physical world, and in that conflict new expressions can fruitfully be developed. Today Hybrid creations have become a method for working with cultural production not only with different elements of form, but as blending identities of the creators as well. While the major part of research on interactive narratives has been aimed toward the exploration of interactivity in user experience of finished art works, we also aim at also exploring the perspective of collaboration in production of new media. Content dealing with narrative structures and its expressions often overlook the actual complexity of writing and prototyping the content fitted for the chosen media. Our long-term aim with research in new media is to formulate a set up of narrative tools or frameworks that can be used in a creative, collaborative process.

The Role of Scenarios, Prototypes, Perspectives and Iterations in Abadyl

Scenarios. The role of scenarios in design has been that of writing narrative descriptions of use. Other cultural domains have generated more speculative methods for collaboration. Originating from the idea of autonomous writing the surrealists borrowed methods from academic disciplines such as sociology, anthropology and psychology to elaborate methods in the form of games for exploring the mechanism of imagination and intensifying collaborative experience. They subverted academic modes of inquiry to undermine rationality and invented playful procedures to release collaborative creativity (Gooding, 1991). The role of procedures and systematic strategies, while still being playful makes a creative constraint. Research on creativity points to processes, which not stems from a vacuum in the individual mind, but that they are a result of serious and known strategies. This applies to many aspects of artistic work. Changing a constraint might be at the core of creative thinking (Boden, 1997). Other researchers stress the process of association, how one item by acts of creative association creates a new item (Brown, 1989). The scenarios acted very much as constraints, but also as a first generator in a chain of associative artistic work producing the artefacts. How do we go about exploring a complex digital space? Of course, we could let people walk the streets of Abadyl in for example a game engine, but we so far choose to go in another direction. We used the framework of Abadyl, to stage different events and spaces in the form of written scenarios. These are handed over to the invited temporary citizens and co-creators. They can then act in relation to the scenario, in and by themselves chosen environment. That in the end helped them produce the artefacts. Our scenarios try to bring field studies and fantasy together, to slowly create a discreet dynamic tension and/or displacement between person, objects, time, places and events that are not usually - if ever – associated into new and surprising conjunctions. By using scenarios we are able to provide detailed and specific data, which the co-creator can use as background material for their action. Hopefully the co-creator themselves imports qualities into the world, which do not and cannot stem from the City of Abadyl itself.

Example: An interview with one of the participating artist in the fieldasy exhibition revealed that; "Imagination was tickled by the knowledge of being part of a networked mapping I didn't know in detail. The scenario got me going, but I felt no repressing obligation towards it and also felt more liberated that in the situations of my own work where I'm the responsible and potential object for critique"

Prototypes. So how do we use digital materials as a "design material"? We know how to use computers and software as tools when directing and conceptualize traditional productions, but we still have a lot to learn when it comes to seeing digital material as a design or artistic material in itself. However it is not so strange, as digital design is not as mature as traditional design, and digital material have characteristics that differ a great deal from those with which we are accustomed. Digital materials are usually more complex and flexible, less transparent and tangible. But above all, it is more cumbersome to learn and produce than most realize. We have to point out the need for a more profound relationship to digital materials and tools. We believe that increased complexity in creative development calls for both disciplinary depth and integrative skills. So there is a demand for a deeper challenge between the virtual and the physical objects, a will to explore their incompatibilities rather than merging them together into one.

The answer we have come up with is to use prototypes as our main surface of communicating and testing our ideas and concept. And by letting them evolve in different media and materials the final hybrid will host an interesting comprehension perspectives. This kind of work actually incorporates surprising visual and technical proposals. Building prototypes let us explore this area by iterating between artistic intentions/screenwriting and digital generated expressions and script/code writing.

Example: Do-Fi-Do is a application developed in the period 2005 - 2007, together with two former interaction design students of mine at Malmö University, Magnus Wallon and Johan Salo. The idea was to use RFID tags and readers in different ways to quickly prototype interaction events in different settings; in a building, in a room, on a table, in a scale model etc. The application created works with different media formats, images, sounds, video and flash (swf) files. In record mode you drop the desired file in a dropbox in your computer and than connect it to a RFID tag by moving a card over the RFID reader. The set-up processes for connecting media files to a more physical and spacial format as RFID makes it quick and easy. In Play mode one start to explore media clips, distributed in a chosen environment. When accessing the clips they are played in real-time with no latency and you can mix and shift between them with no interruptions. A single computer can handle a series of readers simultaneously through usb or RS-232. After years of doing prototyping activities with it together with students we made a version that so stable that it was used in a series of commercial work as well as in two of my artworks. Code & Tom 2008. Based on a film by Thore Soneson, you follow a journalist Tom Hasper in his attempt to find a female group of graffiti artist. were the audience/user reveal a series of video clips by rearrange a

Figure 6. RFID prototyping, RFID cards, do-fi-do application, Code and Tom prototype, F1 (flash based story engine) and f1 prototype (RFID & scale model story engine)

series of everyday objects on a table. Recently it was used in a two reader set-up in the sound based work Mobela_c. Here the user could extract a series of descriptive stories about the different parts of the city in English and Mandarin. The voices a explored by moving two RFID readers over a map with mark out points with the titles of the subject of exploration.

Perspectives. In our prototype work with the artefacts, we focus especially on interactive installations; we realize that the digital design process both demand new forms of screen writing and prototyping activities to support the design of the expression of the final artwork itself. We have tried to create three perspectives that put traditional roles of production aside. Each perspective will all consists of people with different expertise, sharing the idea of creating a common production and explore a new format for thinking (concept development), testing (prototyping) and finalize a work of art. Here the actual development will be conducted simultaneously in all three perspective at the sometime, with constant iterations in-between them. The perspective are as follows:

Expression – is everything that has an actuation and presence in the play/performance/installation. Here all the traditional media producers, put forward examples that can be tested and tried out in different ways. It is about calibrating and creating expressions in relation to the other exploration clusters, involving; actors, groups, audience, film, sound, stage design, images etc.

Play mechanics – is about matching the story and the audience/players development through spacial interaction, actuation and sensory input, designing the underlying structure that support the possible outcomes of all scenes in the play. Here scriptwriting, dramaturgy, stage and interaction design propose and test different set-ups. This will be an layout of time and space to establish a dialogue with the other two exploration clusters involving; spacial and dramatic models, dramatic tension, sensors, actuators, spacial functions, interaction methods, interaction sequences, deliverance, believability.

Dynamic behaviour – is about matching every aspect of programming with time, space interaction and narrative progression. This will be an computer model and the program core which opens up and establish relevant in and outputs to the other two exploration clusters involving; processes, computer models, game play- architecture, state machines, sensor calibration, computer-vision, narrative progression, discrete mathematics

The overall goal with this set-up or approach is to look at how we can build a narrative and interactive framework, and create a testing and production environment. We also see a need to develop a format for creating and sharing knowledge in-between different professions and stakeholders in this process – from tests through rehearsals and prototypes into the finished productions. The three perspectives is our way of setting up the experiments, and invite people to participate in and contribute to them. Hopefully this will create a meaningful dialogue between all of the projects co-creators. Here are no examples yet and this work lays ahead and is described in the part. Later in this chapter called: *Journey to Abadyl(PRAMnet,2009)*

Iterations. Is commonly used in different design settings and environments. It is part of a work practice in game-(Zimmerman, 2003), software- (Larman, 2003), interaction- (Ramsey, 2009), and industrial-design (Roland, 2008) etc. In the area of art and especially in relation to new media it can be used a bit differently, where the shortcomings and quirks of the prototypes themselves will be used as major features and qualities in new artworks. When trying to follow the artistic intention or concept, the prototypes itself can produce qualities not known beforehand. But it has to reach a certain state of a complexity, a complexity max (Hoberg 2006) to be able explored thoroughly. Therefore a production environment is crucial to facilitate practice based research in the area of art and new media. To many art projects are just illustrations of technology, and to many design projects unreflected uses convention and qualities from art. We strive to facilitate a openness towards what happens in-between the design cycles. The idea with our iterations is to develop and put forward qualities that are hidden from the concepts point of view and by doing so - have the concept redeveloped itself through the results and experiences we achieve by the iteration, going from concept to detail over and over again. We try to explore what is being shadowed by the concept itself. Therefore we do not formulate any detailed specification in relation to the concept beforehand that we have to match other than some dramatic or situated qualities we are after for a certain production. But at the same time it is important for us to be able to control the in and outputs of the prototype and how it works in relation the overall IT structure of the project.

Example: Iterations is the actual point of departure for Abadyl, it tries to make the co-creators iterate around a world with a very low density, constructed by the maps, scales and objects described earlier. By using file formats and a programmable structure, all of the digital material is connected through a series of dynamic links that will be updated as soon as some of the original material is edited or new files are created and put into the overall projects. Therefore it took five years to fully get every object in place for the 100 objects matrix. Since it from the beginning contained a lot of objects that did not match the criteria we set up halfway through the project - that all of the 100 objects should be made by human hand. So as we exchanged them, the story routes, the layout of the images on the bone china got updated with new obstacles and therefore had to be revised and new stories and images created. It is when you follow an intention to it´s smallest detail and reflect on the overall concepts execution influenced it you are able to see if it works as intended or if you have to go back to the "drawing board" and implement some changes or revise the whole concept from scratch based on your new insights of the details behaviour. The most important software packages we use in this is set-up is Autodesk Softimage, Quidam, Form-z, Adobe CS, Python, Mind manager, Open office and Extensis portfolio.

Figure 7. Mobela_a (collection space for the 100 objects), b (the database) and c (the voices of the constructors)

The City of Abadyl

INVITING SOMEONE FOR A WALK ON THE STREETS OF ABADYL

To be able to host even more complex and specific projects in Abadyl there was need for try to describe the city planning and architectural theories we used and misused during the design of both the different city parts and the architecture of the city. In 2003 four Interaction design students from the Malmo university art and communication department used the Abadyl framework to create a computer game. During their work they wrote a series description of the different city parts. To start to explore different methods of handling the future events in the project they looked at The GURPS (Generic Universal Role Playing System). They set up rules on every level in the city. The case that we choose to test their approach was to make an interface and a database for the creation of characters in Abadyl, a system that provided an integrated set of tools for the "minds" and "bodies" of the future citizens. They also wrote a detailed description of all of the sixteen city parts used as background material in their staging of the characters and their life's. In 2005-2008 two short movies was made in Abadyl called "In Search of the Militant Code" together with Thore Soneson.

In 2008 these were redeveloped by Jette Lund and Kathrine Nielsen for the prestudy of Journey to Abadyl. By doing this work they raised a lot of questions about how the city was constructed, what kind of underlying ideas that made each part look the way it did – their was a need for the co-creators to have some more philosophical points of departure, so I wrote the Seven voices of the constructors of Abadyl. In this work I paired some of the writers that inspired me over the years into shaping the framework of the city. The idea was to take the authors in small groups for a walk in Abadyl and listen to what they had to say and the one´s how finally made it was: Alexander / Venturi, de Certeau / Calvino, Wallenstein / de Botton, Taylor / Mankiewicz, Damasio / Cambell, Baudrillard / Koolhaas, Auster / Manovich, Davis / Hoffmeyer and Sorkin / Söderberg. Here are some quotes from the seven constructors of Abadyl:

Pin point (a text that delft with the mapping of space and time in real and fictional spaces)

You see, writings and especially literature has us experience space and place in a myriad ways that have little to do with real space.

So there is not much correlation between writing about activities in space and then try to recreate them in a factual way. It is like when you try to be on two different locations at the same time, it is the logistics of hell.

Sometimes mapping these worlds can be a waste of time and miss the point of creating maps. Literature of all kinds has so many things to tell us about space and place, but the things it has to communicate are not necessarily of the sort that lends itself to cartographic representation.

In a book people can talk for ten minutes going down a stairway, which in real life takes about 25 seconds. The planners (a text about constructing spaces of surprise)

Most of all I have been frequently occupied with architecture and buildings as objects. In other words my notion of theory has not being there to explain or analyse these objects, but to face architecture – the theory and the built environment in frontal collision, like trains on the same track, in opposing directions. Facing the building, the

Figure 8. Stills from the "in search of the militant code" 1 & 2 with Thore Soneson 2005, 2008

theory should explode, it has to explode, it has to blast, blow up.

Same strategy I assume in relation to other theories, intellectual suppositions, ideas. I regard them also as objects, try to lash out to them – more or less as like when you try to hit and extract certain particle in a atom smasher.

The walkers (a text about the Flaneur)

To provoke the perspectives imagined by the urbanists, city planners, painters and cartographers. Establish a blind practice that are foreign to the geometrical or geographical space of both visual and theoretical construction

The art and act of walking is to the urban system what the speech act is to language. Where bodies follow the thick and thin of an urban text.. Walking is like writing and never be able to read it. And by walking the streets of a city one shapes and shares the narration that creates humanity.

FROM CONVENTION TO THE UNKNOWN AND BACK AGAIN

Journey to Abadyl

In the recent prestudy of "Journey to Abadyl" (PR-AMnet 2008) made by a multidisciplinary group of artists, production companies and researchers agreed on to focus on some common areas of interest. We have decided to try to build a series of prototypes that are going to be developed to fully functional stage technology. The first part is called the Traora Jukebox and is set in one of Abadyls older parts. We will build a prototype that will be explored in a series of four interactions and three technologies, RFID, Blue tooth and computer vision. We will try to put forward knowledge for handling production complexity as well as research methodology, and prototype development, using scenarios, prototypes, perspectives and iteration.

Based on scenarios our model of exploration starts in a unfamiliar territory, were all of the co-creators try to put forward, experiment and establishing iterations between the perspectives. Providing knowledge to the different stakeholders and influence the future development and research. By using prototypes as our surface for communication and proof of concept we will try to established useful conventions that are mature enough to be part a production. The models for this work is fully described (Lowend, 2009).

At the moment 2009 - 2011 we are hosting three more major activities inside Abadyl. The first is Conversation China, the artefacts is about getting to know new people from different cultures and backgrounds at a dinner party. The idea is to produce a complete 150-piece Porcelain Dinner Set that invites to conversation among the attendees around the table. To do this we will use Joseph Campbell book The Hero with a Thousand Faces as the main source of inspiration, and put that in relation to the specific constraints that The city of Abadyl consists of.. The second is Strip Hit Music where we build a music and text machine and make four pieces of music in two different musical genres which in the end will be made into two machinima short films. The third one is 48 - Us and Them were we will build for room sized buildings based on the architecture of Abadyl were we invite the audience by sound explore same of the inhabitants of Abadyl.

CONCLUSION

Abadyl is a proposed city, a fantasy, a set of codes and models, a library of artefacts and prototypes and foremost it is it´s co-creators. It has become a large database of material that is interlinked through the architecture of a city, regardless of its incompatibilities. That space has proven itself as a continuously evolving platform for staging both immediate and long-term projects. By establishing a multidisciplinary common ground for a art practice, interaction design and technology development, through an investigation of philosophy and criticism in a dynamic material has so far been successful. It creates a open-ended way of working with art and design based on scenarios extracted from the framework and rule-set of the city. By using the scenarios the co-creators produce beforehand unknown artefacts in the area of art, design and new media. The Scenarios relate to the overall project of the city but is loosely defined as to allow the creation of art works, design and innovation, that enriching the database. The artefacts themselves are autonomous from the mother project in the sense that they can be exhibited, used and distributed by themselves and appear in other contexts. They also act as generators while they generate new and unforeseen processes which extend into new and likewise unforeseen contexts. The co-creators disseminate their knowledge into the platform, but they also extract something which can inform their own future practice. The most obvious quality of letting the participants work with physical objects in relation to fiction and a digital generated presence is that it is much easier to communicate their result in a group. To bring objects and talk about them is very convincing especially if they are products of fiction. Choosing exhibitions, workshops and a website as format for the internal and external communication of the overall project seems very fit. It makes it attractive for people to continuously involve themselves in the further evolution of The city of Abadyl.

REFERENCES

Alexander, C., & Ishikawa, S. (1977). *A pattern language: towns, buildings, construction*. New York: Oxford University Press.

Boden, M. (1997). *The Creative Mind: Myths and Mechanism*. London: Weidenfeld &Nicholson.

Brown, R. (1989). *Handbook of Creativity* (Glover, B., Eds.). London, New York: Plenum Press.

Calvino, I. (1997). *De osynliga städerna (invisible cities)* (Swedish edition). Stockholm: Bonniers Förlag.

Dinkla, S. (2002). The Art of Narrative. In Rieser, M., & Zapp, A. (Eds.), *New Screen Media*. London: Brittish Film Institute.

Gooding & Brootchie. (1991). *Surrealist games*. London: Redstone Press.

Greenaway, P. and Akademie der Bildenden Künste in Wien.(1992) *Hundert Objekte zeigen die welt = Hundred objects to represent the world: 100*. Verlag erd Hatje, Stuttgart.

Hoberg, C. (2006) *Komplexitets Max*, Dialoger Förlag, Stockholm, Sweden Jackson, Steve, (2002) *GURPS basic set third edition*, Steve Jackson games, USA.

Jaques, P. A., & Viccari, R. M. (2006). Considering students' emotions in computer-mediated learning environments. In Ma, Z. (Ed.), *Web-based intelligent e-learning systems: Technologies and applications* (pp. 122–138). Hershey, PA: Information Science Publishing.

Larman, C., & Basili Victor, R. (2003). Iterative and Incremental Development: A Brief History. *IEEE Computer*, *36*(6), 47–56.

Manovich, L. (2001). *The language of new media*. Cambridge, Mass: MIT Press.

Mansoor, J. (2003) Kurt Schwitters' Merzbau: The Desiring House, in Invisible Culture, Issue 4, Retrieved October 12, 2009 http://www.rochester.edu/in_visible_culture/Issue4-IVC/Mansoor.html

Montie Roland. (2008) Product Design Speak 101: Linear versus Iterative Design. Retrieved October 12, 2009, from http://carolinanewswire.com/news/News.cgi?database=pipeline.db&command=viewone&id=8

PRAMnet. (2009) Journey to Abadyl. Retrieved October 12, 2009, from http://www.pramnet.org/PRAMnet_JourneyToAbadyl_web/JOURNEY_TO_ABADYL.pdf

Ramsey, A. (2009) Three Reasons to Start Designing Iteratively Retrieved October 12, 2009, from http://www.andersramsay.com/2009/03/01/three-reasons-to-start-designing-iteratively

Tschumi, B. (1994). *Architecture and disjunction*. Cambridge, Mass, London: MIT Press.

Zimmerman, E. (2003) Play as Research: The Iterative Design Process. Retrieved October 12, 2009, from http://www.ericzimmerman.com/texts/Iterative_Design.htm

ADDITIONAL READING

Akerman, J. R., & Karrow, R. W. (2007). *Maps: finding our place in the world*. Chicago: University of Chicago Press.

Alexander, C., & Ishikawa, S. (1977). *A pattern language: towns, buildings, construction*. New York: Oxford University Press.

Allen, S. (1999). *Points + lines: diagrams and projects for the city*. New York: Princeton Architectural Press.

Baudrillard, J. (1990). *Fatal strategies*. London: Pluto.

Baudrillard, J., & Zurbrugg, N. (1997). *Jean Baudrillard: art and artefact*. London; Thousand Oaks, Calif.: SAGE Publications.

Bolter, J. D., & Gromala, D. (2003). *Windows and mirrors: interaction design, digital art, and the myth of transparency*. Cambridge, Mass.: MIT Press.

Calvino, I. (1974). *Invisible cities*. New York: Harcourt Brace Jovanovich.

Certeau, M. d. (1984). *The practice of everyday life*. Berkeley: University of California Press.

Certeau, M. d., & Giard, L. (1998). The practice of everyday life: *Vol. 2. Living and cooking*. Minneapolis: University of Minnesota Press.

Damasio, A. R. (1994). *Descartes' error: emotion, reason, and the human brain*. New York: G.P. Putnam.

Eaton, R. (2002). *Ideal cities: utopianism and the (un)built environment*. New York, N.Y.: Thames & Hudson.

Eco, U., & Alexanderson, E. (1985). *PS till Rosens namn*. Stockholm: Bromberg.

Engeli, M. and Eidgenössische Technische Hochschule Zürich. (2001). *Bits and spaces: architecture and computing for physical, virtual, hybrid realms: 33 projects by Architecture and CAAD, ETH Zurich. Basel*. Boston, Birkhauser.

Lindgren, M., & Bandhold, H. (2009). *Scenario planning: the link between future and strategy. Basingstoke, Hampshire*. New York, NY: Palgrave Macmillan.

Manovich, L. (2001). *The language of new media*. Cambridge, Mass.: MIT Press.

Marx, C. (2007). *Writing for animation, comics & games*. Amsterdam, Boston: Focal Press.

McCullough, M. (1996). *Abstracting craft: the practiced digital hand*. Cambridge, Mass.: MIT Press.

McCullough, M. (2004). *Digital ground: architecture, pervasive computing, and environmental knowing*. Cambridge, Mass.: MIT Press.

Mitchell, W. J. (1992). *The reconfigured eye: visual truth in the post-photographic era*. Cambridge, Mass.: MIT Press.

Office for Metropolitan Architecture. and R. Koolhaas (2004). Content. Köln, Taschen.

Shaw, J., & Weibel, P. (2003). *Future cinema: the cinematic imaginary after film*. Cambridge, Mass.; London: MIT Press.

Silver, M. (2006). *Programming cultures: art and architecture in the age of software*. London: Wiley-Academy.

Spiller, N. (2000). *Maverick deviations: Neil Spiller: architectural works (1985-1998)*. Chichester, West Sussex: Wiley-Academy.

Taylor, M. C. (2001). *The moment of complexity: emerging network culture*. Chicago: University of Chicago Press.

Taylor, M. C., & Tansey, M. (1999). *The picture in question: Mark Tansey and the ends of representation*. Chicago: University of Chicago Press.

Tschumi, B. (1994). *Architecture and disjunction*. Cambridge, Mass.: MIT Press.

Wardrip-Fruin, N., & Harrigan, P. (2004). *First person: new media as story, performance, and game*. Cambridge, Mass.: MIT Press.

KEY TERMS AND DEFINITIONS

Abadyl: Is the title of a painting by Michael Johansson made in 1987 for the exhibition populärminne (Popular memory) it is the combination of the word abstrakt (abstract) stad (city) and akryl (acrylic paint), it now lends it´s name to the project presented here. Years later when internet searching became available I found out that it also is a small mission in the USA.

Co-Creation: The aim to use co-creation is to break the common position of the single artist and instead put forward a positive friction between Abadyl and the invited co-creators perspectives and professions. To create a temporarily and mutual relationship where we can interact, evolve, reflect and innovate in a multidisciplinary setting that blurs the boundaries between art, research and innovation.

Fieldasy: Tries to bring field studies and fantasy together, to slowly create a discreet dynamic tension and/or displacement between persons, things, times, places, and events that are not usually - if ever – associated into new and surprising conjunctions.

Iterations: We strive to facilitate a openness towards what happens in-between the design cycles. To try develop and put forward qualities that are hidden from the concepts point of view and by doing so - have the concept redeveloped itself through the results and experiences we achieve by the iteration,

New Media: All media objects that could be variably, dynamic, meta-tagged, manipulated, modular, algorithmic, cloned, copied, coded, trans-, cross- and dis coded or procedural.

Perspective: Common method to see the world through some one else´s eyes, through a philosophical statement or system. To help us rediscover the ordinary and the world that surround us.

Prototypes: Is our approach to explore the profound relationship between digital materials, tools and artefacts.

Scenarios: Helps us to provide detailed and specific data, which the co-creator can use as background material for their action and artefactual production.

Chapter 17
Social and Citizenship Competencies in a Multiuser Virtual Game

Germán Mauricio Mejía
Universidad de Caldas, Colombia

Felipe César Londoño
Universidad de Caldas, Colombia

Paula Andrea Escandón
Universidad de Caldas, Colombia

ABSTRACT

Colombia is a country that is growing in technological and economical areas, but cultural diversity, armed conflict, and everyday violence are forces that lessen the progress. The Ministry of Education proposed in 2004 standards for citizenship competencies that intend to teach children and youths abilities to live peacefully and respect the others. These standards have encouraged multiple initiatives and innovation to achieve goal in social and citizenship competencies. A research group of Universidad de Caldas created and evaluated a serious virtual game to support this policy. The game is a multiuser virtual game called Civia that shows a metaphor of collective challenge. Peaceful interaction, participation and respect are values required for survival in the game. Players can take individual decisions that affect positively or negatively the collective status, but an overall positive balance is needed to maintain collective resources. It was expected that collective auto regulation led to the learning of patterns and competencies to live peacefully together. Currently, there is no consensus in the research community about what and how video games can take learning outcomes and behavioral effects. An evaluation of the game shows positive results; however, some concerns about the complexity of everyday life about social interaction and learning transferability arise. The authors discuss findings according to proposed theories and models about effects of video games in education and behavior.

DOI: 10.4018/978-1-60960-077-8.ch017

PROBLEM ANALYSIS

Context in Conflict

The *information age* has been leading to a historical transformation process. The *new* economy, centered in technological and corporate innovations, is modifying the social processes in cities. Cities that have gone into this order have been key spaces for economic progress but they have also been challenged to achieve social balance and cultural identity. Factors related with armed conflicts and economical inequality, particularly present in developing countries, reduce the likelihood of a smooth entry into the new economy order. This is the case of Colombia, a South American country marked by contradictions. Economic and technological development and cultural diversity contrast with scant improvements in social and conflict matters. According to the Colombian Department of Statistics, the country has 45.088.439 inhabitants (Departamento Administrativo Nacional de Estadística [DANE], 2009a) and a high poverty index (46%) (DANE, 2009b), yet media and information technology show high penetration rates. In 2008, 83.8% of households owned a cell phone, 88.5% of households owned a television, 46.4% of people used computers, 37.5% of people older than five years accessed the Internet, and 40.91% of children between five and eleven years old had played video games in the previous month when they were asked (DANE, 2009c; DANE, 2008). Although the literacy index (91.6%) and school attendance for the range of age between 5 and 24 (65.5%) (DANE, 2005) are high, the book reading mean in the last twelve months was only two. These facts put digital media as a source for social change strategies because its use is high and seems to be increasing.

Apart from this, the Colombian armed conflict has integrated and disintegrated broader civil population groups in the war front among the Colombian army, paramilitary groups, and insurgent guerrillas. Their fighting, which is part of an irregular war, occurs mostly in national countryside territories like rural areas, indigenous reservations and natural parks. Although armed violence occurs mostly in the countryside, the consequences of a violent environment are also visible in cities. The phenomena of social displacement and non-military demobilization affect everyday life and the development process in the cities. Urban centers show everyday effects of new interactions with immigrants from the countryside, violent media exposure, and domestic violence conflicts. This has led the government and public and private institutions to carry out initiatives for strengthening peaceful social interaction among children, youths, and adults.

Social and Citizenship Competencies

Social competence is an educational concept derived from the challenges given by the interaction of individuals in communities. Semrud-Clikeman (2007) posited that "social competence is an ability to take another perspective concerning a situation and to learn from past experience and apply that learning to the ever-changing social landscape" (p. 1). She added that this ability is closely related with emotional competence and its lack can lead to aggressive behavior. In violent contexts, social competence becomes even more essential in social education and social change processes. The concept of social competence is also related to citizenship competencies, which mean both claiming human rights and fulfilling responsibilities. Antanas Mockus, a former Bogotá mayor, contended that someone is a citizen when we he or she respects minimal responsibilities, respects other rights, and engenders basic trust (Mockus, 2004). In Colombia these competencies have become the subject of national research and policies because violence has affected social interaction for more than five decades. Chaux (2009), for instance, surveyed previous research about the effects on children of the violent political conflict in Colombia and concluded that there are three consequences. First, the environment of violence affects children's cognitive and emotional competencies. Second, the resulting competencies stimulate aggressive behaviors. And third, educational initiatives targeting these competencies can promote peaceful interaction, even in a violent context.

Aware of this, in 2004 the Ministry of Education introduced the National Program of Citizenship Competencies (Ministerio de Educación Nacional [MEN], 2004a) as part of a national policy. This program promotes standards for active learning of social and citizenship competencies in elementary and secondary schools, which is a constant challenge due to the violent conditions of the country. The standards define citizenship competencies as "a set of knowledge, attitudes, and cognitive, emotional, and communicative abilities that, properly articulated with each other, make the citizen willing to act in constructive and just ways for the society" (trans, p. 8). This definition emphasizes in the education of the mind and emotions to foster critical analysis of knowledge and abilities. Citizenship competencies center on knowing how to do and realize actions, which requires a contextual application of concepts in diverse circumstances. MEN determines particular standards for different school levels to develop gradually specific abilities in three group types: peaceful social interaction, democratic participation, and respect and diversity. These groups take in four types of competencies: cognitive, emotional, communicative and integrative. Chaux (2009), who led the standards teamwork, explained that the proposal was based on previous research of how a violent environment affects cognitive and emotional aggression. He clarified that due to the autonomy of schools, the Ministry of Education cannot force schools to apply these standards. Instead, it constantly encourages schools with incentives to promote them. He added that the standards are clear about what is needed but did not explain how it can be achieved. This fact has led to multiple formal and informal initiatives, which promote innovative ways to educate children in social and citizenship competencies.

Behavioral Effects of Video Games

The power of entertainment and immersion of video games are the central impact that can be considered the basis of their success in reaching players. The concept of immersion can be understood as a rich stimuli environment or a high cognitive attention trance. There are many devices that increase the amount of stimuli to immerse a person in a multisensory environment; for example, 3D animation, sound, virtual glasses or vibrating controllers. Video games use this type of immersion with multimedia elements; but their most powerful tool for immersion is the cognitive participation. Murray (1997) argued that digital objects are a participative media that combine both representation and presence. It means that digital users, and consequently video game players, have the experience of being active participants in the representations that are presented. This combination is not available in other media like literature or theater. Players are present in the narrative, belong to it and modify it, which creates a powerful immersion environment.

In addition to this active engagement aspect of video games, this project investigated the antisocial and prosocial consequences of video games, which are central for their use in increasing social and citizenship competencies. Experimental psychology researchers have focused on the effects of violence present in media and video games (e.g. Bartholow, Sestir, and Davis, 2005; Singer & Singer, 2005; Anderson, 2004; Bushman & Anderson, 2002). They reported that exposure to violent video games correlates with short and long-term aggressive behavior. Anderson and Bushman (2002) proposed a unified theory of human aggression called the General Aggression Model (GAM). This model states that negative influences can increase aggressive behaviors. In particular, repeatedly exposing children to media violence causes long-term personality results. Further, when they become adults, attempts to reduce aggression may be ineffective. Conversely, other authors have discounted these aggressive results. For instance, Gee (2007) argued that there is not evidence in current research that violent games result in real life aggressive behavior. He explained that video games and other media like movies and books have short-term effects on people but a link between public statistics of violence and the use of violent games and media has not yet been demonstrated. In addition, Prensky (2006) contended that video games can have more benefits than threats. He said that so far research has not identified a general negative influence of video games because children interact with a lot of more complex factors in life.

Recently, there has been an interest in studying whether prosocial games can teach prosocial behavior. Greitemeyer and Osswald (2009) reported two experiments in which subjects reduced aggressive responses after playing prosocial and neutral videogames. As well, Gentile et al. (2009) presented three studies from different countries where they found prosocial behavior effects after subjects played prosocial games. However, if the arguments used above by Prensky (2006) about violent behaviors are extrapolated to prosocial behaviors, the complex interaction of children in real life might also affect prosocial behaviors, which suggests a low validity of these findings in regards to prosocial games. This discussion seems far from a consensus because of the lack of evidence of long-term effects of both antisocial and prosocial content. Violence might be seen as a "necessary evil" for game motivation because the market has widely used this type of content. Nevertheless, there are a lot of neutral or prosocial games that indeed have shown high motivation. Przybylski, Ryan, and Rigby (2009) did a study on this issue and reported that violence is not essential for motivation and enjoyment. They found that competence and autonomy carry out such role and violence add little to desire to play. This result suggests that non-violent games may also lead to enjoyment and while at the same time avoiding eventual consequences of aggressive behavior.

Although there is not a general agreement especially on the long-term effects of video games, Buckley and Anderson (2006) proposed an extension of the General Aggression Model that includes all types of inputs and outputs, which they called the General Learning Model (GLM). They developed the model "to explain how video games teach and influence behavior" (p. 369). This model asserts that learning is a result of interaction between personal and situational variables. In the process, affects, cognition and arousal interact and reinforce themselves to influence personal state and, hence, the way people appraise situations and make decisions. Individuals experience constant cycles depending on game content that continuously shape knowledge structures and might lead to personality changes. Whereas this model explains real effects of video games on learning and behaviors, the question of how much these changes affect overall social behavior in a community is still unanswered. Despite that, the model provides a useful psychological framework that can be used to build initiatives and projects using games for social change.

Educational Effects of Video Games

The educational potential of games has been studied in different fields, from the inquiry of humans as *homo ludens* (Huizinga, 1955) to imagery production through game imaginations (Flusser, 2000). Traditional games have been widely used in education; toys, educational materials, playful activities and games are common learning aids especially for preschool children (Goldstein, Buckingham, & Brougère, 2004; Hartman & Brougère, 2004). Digital technology has created a new theoretical debate about games because of their new interrelations with physical spaces, immersive properties and cultural exchange potential. Flusser (2000) argued that telematics and computers provide emergent opportunities for imagery because they build iconographies from algorithmic code. In spite of this, video games are relatively a new tool for educators and many do not know how to use them (Kerr, 2006). Prensky (2001) was one of the first to advocate for education centered on learners using digital games. Goldstein et al. (2004) argued that the participation alone in games implies a know-how processing about the game rules, which is already a learning outcome. They added that digital games also promote exploration, enhance discovery and become effective for knowledge acquisition.

Certainly educational researchers agree that players learn when they are engaged and challenged in a video game; however, they differ in how learning is transferred to and replicated in life (Kerr, 2006). In Prensky's humanistic approach, the relaxation and motivation of games allow learners to be active and process knowledge easily with less effort (2001). Gee (2003) gave another explanation from a cognitive perspective. He contended that players exceed active learning and experience a critical learning, which is a conscious cognitive process that allows knowledge to be transferable to other contexts. But Kerr (2006) noticed that there is not evidence of such transferability. From a sociocultural perspective, Linderoth, Lindström, and Alexandersson (2004) observed that rules of game play, rather than aesthetic representations, are a more significant element of interest for game players in game interaction. This last particular perspective suggests that a prosocial game designer should carefully consider the game core mechanics to obtain learning outcomes of social competence.

Later, Gee (2008) explained that five conditions must be fulfilled to obtain long-term learning from games: (1) the presence of specific goals, (2) opportunities to interpret and reflect on game experiences, (3) rapid feedback to avoid interference with learning, (4) opportunities to apply previous game experiences, and (5) discussion with others of game experiences. This advance of the cognitive perspective appears to be consistent and useful. Gee has studied mostly outcomes of commercial games where learning is a derivation of the content not designed for learning. This raise doubts in games for learning because when Gee's conditions are applied to video games with learning goals, they might break the immersion trances because they demand a high level of consciousness about the rules and goals. In consequence, games with specific learning goals need further analysis because in some cases designers might want to hide learning objectives.

Civia Project

The use of videogames for education and social change has become popular since the beginning of this century. As discussed in the previous section, psychological and educational research supports the premise that video games with adequate content and frequency of use can lead to social competence and prosocial behaviors. These outcomes are in urgent demand in the world and especially in developing countries that depend on human development to reach social progress. In Colombia the use of information technology is increasing at an accelerated rate (DANE, 2009c; DANE, 2008), which makes the scene fertile to apply educational strategies using digital technology. Schools are not the only place to educate in social and citizenship competencies; the Internet and video games are also social spaces where it is possible to foster these abilities. Indeed Chaux (2009) pointed out that education directed to cognitive and socioemotional process can help to break cycles of violence. In consideration of the conflict-ridden context entrenched in Colombian society, *Civia* project intends to foster social and citizenship competencies based on play and digital technology nurturing knowledge structures in Colombian children and youths.

The research team decided to work with children and youths because they experience a critical age stage to develop social competence. According to Semrud-Clikeman (2007) childhood and adolescence are equally essential for the building of social skills. Specifically, preadolescents and adolescents where targeted since they are most common users of the type of games with social tools and interaction. Based on the importance of game content and mechanics for players stated by the previously mentioned research findings, this project emphasizes on conceptual game design to take advantage of the game language possibilities and audiovisual metaphors. The project had a special challenge because it requires the use of non-violent game incentives. The core content is based on the Colombian National Program of Citizenship Competencies (MEN, 2004a), which is a guide of standards for the education system in citizenship and social competencies. Chaux (2009), who led the team that developed this program, pointed out that these competencies can be learned in real-life simulations instead of traditional coursework with civic content.

The project had three phases. The first was analysis of children's values and the conceptual game design according to the actual context. In this phase the basic design method of divergence-convergence was applied, generating many ideas in the divergent stage and synthesizing a concept in the convergence stage. The second phase included the interactive design based on the concept and software development. The interactive design required detail creation from the concept to define the core mechanics of the game and the visualization of interfaces. The model proposed by Rollings and Adams (2003) to document the game design structure was used, consisting of three basic elements: narrative, core mechanics, and interactivity. The software development was blended from the beginning, and engineers supported the conceptual design. The team used the iterative and incremental method suggested by Jacobson, Booch and Rumbaughl (1999) called Unified Process (UP), which is guided by use cases and centric architecture. The final phase of evaluation of the learning efficiency of the video game was conducted using mixed methods embedding complementary qualitative data into quantitative data.

VIRTUAL GAMES FOR SOCIAL CHANGE

As stated above virtual games have shown their ability to teach children and youths issues like social, strategy, or spatial competencies. Video games, especially online games, include social interaction that can be targeted to teach social competence and foster social change. Recent reports (Sánchez Buron, Rodriguez, & Fernández Martín, 2009; Manders, 2008) reveal that video game use among children and youths aids processes of social interaction and learning. Manders (2008) contended that video games are predominantly harmless and can be used in children and youths for learning and essential competencies of the twenty-first century. He reported an expansion of online video games, in which the participation of individuals in forums and social communities is allowed. In addition, Kahne, Lenhart, and colleagues (Kahne, Middaugh, & Evans, 2008; Lenhart et al., 2008) have conducted research about civic experiences included in video games that can aid the educational process in the classroom. Their work shows that adolescent users of video games participate with more frequency and commitment that those who do not play video games, even though there is no correlation between video game use and civic and political commitments. According to this research, the best practices to foster civic responsibility are experiences that enable users to:

- Simulate civic and political activities (present in games like *Civilization* or *SimCity*)
- Learn how governmental, political, economic, and legal systems work (present in games like *The Oregon Trail, Carmen San Diego, Zoo Tycoon*, or *Lemonade Stand*)

- Voluntarily help others (in games that allow participation and help)
- Help guide or direct a given organization or group (trough Massive Multiuser Online Games [MMOGs] like *World of Warcraft*, *Everquest*)
- Participate in clubs or organizations where young people have the opportunity to practice productive group norms and to form social networks (in games with social tools like chat and forums)
- Take part in open discussions of ethical, social, and political issues (visible in games that relate to communication and social issues)

Therefore, Kahne, Lenhart, and colleagues suggested that parents and schools should promote video game use in adolescents with the aim of increasing social participation. Further, they recommend the use of civic-content video games to increase social and citizenship competencies in youths. This is possible not only because some commercial video games have civic content but also because there are serious games options that focus on social and civic change. This named category of serious games is a relatively new genre "with purpose beyond just providing entertainment" (Michigan State University, 2009, para. 1). These games use simulation and entertainment to aid learning of social issues such as social conflicts, military situations, and governability. Even though many organizations and initiatives are promoting and developing serious video games for social change, there is still lack of evidence that supports their effectiveness in encouraging learning and behavioral changes. More research is needed to establish principles of efficacy in design for this genre of games. Some examples of serious games for social change are *A Force More Powerful* from the International Center on Nonviolent Conflict, which teaches activism against political dictators and corruption; *Darfur is Dying* from the MTV channel, which helps players to understand Darfur's humanitarian crisis; *Under Siege / Under Ash* from Radwan Kasmiya (Afkarmedia), which explains the Arab-Israeli conflict from the Palestinian viewpoint; *Matari 69200* from Rolando Sánchez, which reproduces episodes of armed war in Peru; *Food Force* from United Nations World Food Program, which describes how the program executes their food projects against hunger; *Estrecho Adventure* from Valeriano López, which animates and narrates adventures of a Maghrebi immigrant traveling to Spain; and *Peace Maker* from Impact Games, which challenges players to solve the Palestinian-Israeli conflict through role playing as an Israeli prime minister or Palestinian president. These games have a diverse use of video game language. Some of them are short interactions with a clear message, while others include complex role-playing games with strategies. Surprisingly, none of the games described above for social change includes multiuser interaction. (for a comprehensive list of social serious games visit http://www.socialimpactgames.com)

CIVIA MULTIUSER VIRTUAL GAME

Game Concept

The conceptual design problem required the use of a straightforward and understandable model for children and youths and also the use of an engaging narrative. It was decided to create a game for non-formal education to facilitate these aims. The narrative proposes a metaphor of community with social challenges, which looks for the creation of emotional links with the Colombian reality. The video game was named *Civia* to create an analogy with the concepts of civism and civility. *Civia* was designed as a multiuser online role-playing game that intends to create a virtual world to let children engage in conflict with others. The videogame proposes a fantastic virtual world in a future age of the earth after the planet has undergone a severe environmental crisis. The few survivors gather in a hive called *Civia*. Figure 1 shows a general view of the hive plan used in the design process. The concept of a hive is used to reinforce the idea of collaborative and communal housing. This game scenario was modeled as an agglomerated community prone to social conflict. Each avatar has complete dependence from the hive and the others to survive. The player challenge consists of achieving the survival that is possible through individual and collective actions. This metaphor presents a desired model of social reality, where it is necessary to behave in a collective way, using peaceful conflict solutions to achieve social stability. As well, the players can perform activities in groups to reach collective survival, which lets players increase their understanding of living in community with social participation. It was expected that the combination of social interaction and collective challenge would lead to an increase of social competence and prosocial behaviors.

Core Mechanics

Before the first interaction, a video is presented that explains history of *Civia* and how the hive is organized in four subgroups, which have specific strengths for the survival of the hive: fishing, farming, energy production, and construction/recycling. Players join a subgroup, pick an avatar, and enter the virtual world (See Figure 2). The community has a collective challenge that consists of the continuous production and preservation of the hive's resources: food, energy and housing. There are also individual challenges. Players can find and exchange power cards to perform better in certain activities. Good players may gain specific powers to be more effective and the best players are part of a governance committee that makes decisions about the management of resources and access to private spaces. Avatars of committee members are also identified with visual elements. To achieve these benefits players must get points by taking care of the avatar and supporting duties in subgroups. Avatars have need of food, physical exercise, learning, sleeping and diversion. Players are not limited in these individual actions and subgroup duties. For example, a player can make his or her avatar eat more or less or do farming instead of fishing. The player challenge consists of figuring out what is the best balance and right actions to support the avatar and its community. Players also have social tools like chat and forum, where they can discuss game issues. It was expected that the player would realize what are the best way to live together with others, which is essential for social competence. Figure 3 shows one scenario with some avatars of players.

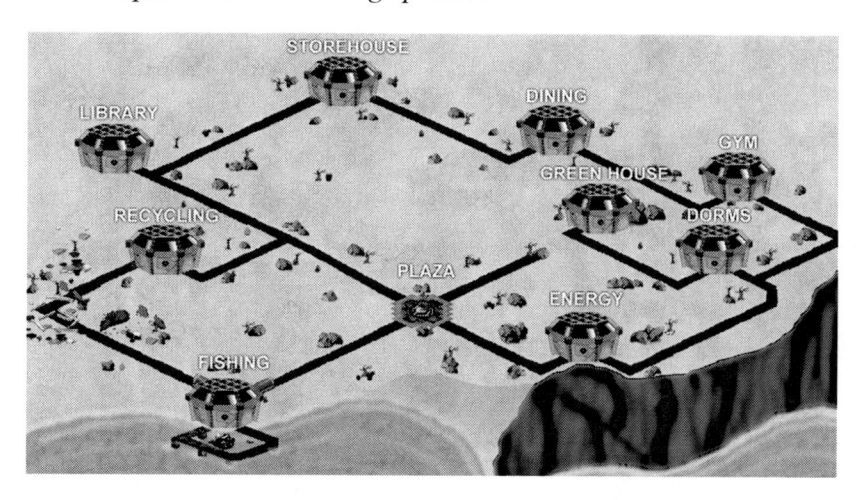

Figure 1. General Civia plan used in the design process

Figure 2. Interface of avatar selection

Figure 3. Plaza scenario

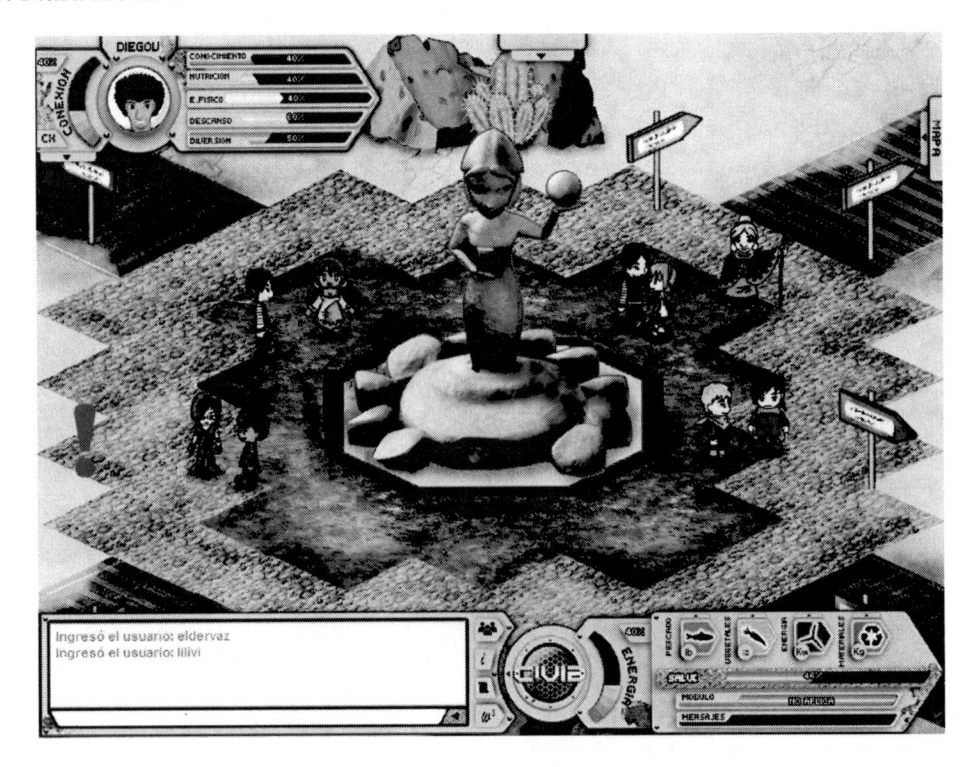

METHOD, OUTCOMES AND FINDINGS

Research Design of Evaluation

The main goal of evaluation was to measure and analyze the citizenship competencies before and after the game experience by comparing attitudes, behaviors, and abilities in these two moments. The first instrument used everyday conflict examples and the second used game conflict examples in the questions. The subjects consisted of 70 children and youths in central western Colombia between twelve and sixteen years old, which is the age range targeted in the design process. The instruments to measure citizenship competencies were developed according to the guidelines and standards of MEN (2004a). It was evaluated the level of competence in the three type of competencies: (1) peaceful social interaction, (2) democratic participation, and (3) respect and diversity. The pretest used standards of MEN (2004a) through closed-ended questions that asked subjects to predict their behavior in various everyday situations by selecting one of five reaction possibilities. (Example of question: Daisy and Johan in a math exam cheated and the instructor applied them a disciplinary action according to the school ethics' manual ¿How would you react to this situation? (1) You get irritated with Daisy and Johan, (2) You consider unjust the instructor's decision, (3) You consider that there would not be ethics's manual, (4) You agree with the application of the ethics' manual, (5) You agree with the instructor. In this question answer one has the lowest level of competence and answer five the highest.)

This allowed researchers to see a 1 to 5 scale in levels of competence. Also, participants answered questions about their video game preferences, frequency of play, and use of technology. The questionnaire included 16 items: nine competence items (three for each type), one overall competence item, and six video game and technology items. Before giving out the questionnaire, researchers explained the clarity needed in the items. Afterwards, researchers explained the game basics, subjects played the game a session that lasted between 1.5 and 2 hours, and then a second questionnaire was administered. It used the same standards of MEN (2004a) with levels and types of competence but using *Civia* situations instead of everyday situations. *Civia* items about interaction and usability replaced general video game and technology use items. Subjects also had the opportunity to include opinions about the game. In addition, a chat log recorded behavior patterns within the game interaction. A complementary qualitative analysis was done based on the chat log, open-ended opinions in the questionnaire, non-structured observations, non-structured interviews, and video records.

Findings

The data were coded according to each type of competence group for both pre and post-video game interaction. Results of the range from 1 to 2 were classified as a low level of competence, 2 to 3 low medium, 3 to 4 medium, and 4 to 5 high. Next, a frequency descriptor was used to determine percentages of increase and decrease in the 3 types of competence. Table 1 shows the results of the citizenship competence items of the pre and post questionnaires.

According to these results, the first type of competence called *peaceful social interaction* was higher after the video game interaction. As well, in the second type of competence called *democratic participation* there was a considerable increase in the level of competence after the game. By contrast, in the competence *respect and diversity* there was a decrease. The analysis of variance (see Table 2) shows an overall increase in citizenship competencies after the game interaction.

The qualitative data provided further insight into the participants' attitudes, behaviors and abilities. First, many players expressed the need for weapons or possibilities of game aggression

Table 1. Results in levels of competence

Type of Competence	Level of Competence			
	Low	Low medium	Medium	High
Pretest				
Peaceful social interaction	2.9%	12.9%	44.2%	40.%
Democratic participation		7.1%	51.4%	41.%
Respect and diversity		2.9%	34.3%	62.%
Posttest				
Peaceful social interaction		1.4%	14.5%	81.%
Democratic participation		7.2%	33.4%	59.%
Respect and diversity		5.7%	42.9%	51.4%

Table 2. Analysis of variance, t-test

Type of competence	mean
	Pretest $\bar{X}$
Peaceful social interaction	3.23
Democratic participation	3.35
Respect and diversity	3.60

within the game, mainly players that showed a preference for shooter games. Second, female subjects showed more fascination for the game, its social tools, and its avatars. Third, although the majority of subjects were familiar with video games, multiuser games were not popular among them. Fourth, there was an uneven learning curve of the game interaction, particularly in those who were not familiarized with video games. Finally, the overall analysis of the interaction pattern followed the MUD (multi-user dungeon) interaction categories proposed by Bartle (2000): players with a preference for socializing and creating strategies to chase similar paths and be together in the game were the most common, followed by those who prefer exploration of the possibilities of the virtual space and enjoy finding hidden elements. Next came players interested in scoring, looking for strategies to improve, and achieving a better ranking; and last, a small group with aggressive behavior that tried to win through harm.

Because *Civia* does not have aggression options, the latter group used the chat tool to express aggressions.

Discussion

Civia achieved the goal of increasing overall citizenship competencies after the game interaction. Taking the General Learning Model (Buckley & Anderson, 2006) as the theoretical basis to observe learning and behavioral outcomes from the virtual game experience of *Civia*, it can be said that the game provided situational variables that affected cognitive reasoning, emotions and arousal in the players. As a result, the game increased competencies for players to appraise situations and make decisions about social and citizenship behavior. *Civia* also adds to the evidence that violence is not the only option to motivate and engage children and youths in game play. However, some issues raise questions about whether games like this may either help or worsen learning and behavior. The researchers are wary about the uneven increase in competencies, the limited game experience and the lack of long-term data, and details on how the outcomes seen are actually transferred to everyday life. Future design and applied research projects are needed to address the design questions raised by these conclusions.

The first issue concerns the uneven change in the measured citizenship competencies. It was stronger in *peaceful social interaction*, which suggests that *Civia* fostered this competence through collective participation of the proposed survival challenge. This case leads the researchers to conclude that the central game concept has a strong effect on behavior. *Democratic participation* also moderately increased, which also supports the positive upshot of the game. In contrast *respect and diversity* decreased. This might be explained because the game let some players feel freer to insult others when their usernames were anonymous and then the game caused a counter effect. These conclusions suggest that anonymity is a critical issue in serious game design for social change. The philosophy of serious games propounds the use of known game principles, which in the case of multiuser games is founded strongly in anonymity of usernames and avatars. Game design is then challenged to find a balance between respecting the engagement given by anonymity and the aggressive behaviors that it allows.

The second issue is the research design based on short-term effects, which raises other questions. Most of the research in effects of media and video games has looked at short-term effects. The game experience did support prosocial behavior and show positive citizenship competencies outcomes. But a longer game interaction and a long-term study may show different results. For example, the players did not use some democratic participation game mechanics in depth. A particular case for this is that those players who gained specific powers and made decisions about the hive's resources were few. These game mechanics, when used in several game interactions, were designed to metaphorically explain the relevance of participation and governance in the Civia's hive. As a result the measure of short-term effects did not include mechanics designed for long-term game experience. As well, some of the conditions for long-term learning contended by Gee (2008), that require constant interpretation, discussion, and opportunities to apply previous game experiences, are not included in short-term outcomes because they need long-term game experience. The researchers hypothesize that long-term interaction with the game would improve even more prosocial behaviors and citizenship competencies.

Besides uneven results and a lack of long-term data, it is still pending how learning outcomes and antisocial or prosocial behaviors are actually transferred to everyday life and even how future adults traits can be modified. As cited above, Prensky (2006) argued that because children interact with complex factors in life, research has not identified a general negative influence of video games on children's behavior. Thus, the same argument applies for general positive influences. Further in-depth research should include some of those complex factors that Prensky mentioned. Otherwise, the only option is build new findings on the General Learning Model (Buckley & Anderson, 2006), which provides a stable foundation to argue that games like Civia contribute to long-term situational appraisals and decisions. Prosocial behaviors found in games might be successfully transferred to real life situations in the future. Therefore, the model suggests the possibility that the game will contribute to knowledge structures that support prosocial behavior and citizenship competence, although within the scope of this research it was not assessed if these competencies and behaviors were successfully transferred to everyday life.

For all these reasons the researchers conclude that the virtual game of Civia has the potential to contribute to social change in Colombia. Even though further evidence is needed to clearly identify outcomes, the theoretical foundations and initial findings suggest that the game affected behavior by creating prosocial cognitions. Within the context of Colombia this has foundations in research done by Chaux and colleagues (Chaux, 2009), which contended that educational initiatives targeting cognitive and emotional competencies can promote peaceful interaction, even in a violent environment. In addition, the Ministry of Education (MEN, 2004b) has also extensively promoted multiples educational experiences of citizenship competencies learning and encouraged educators and researcher to innovate and create diverse programs and projects for this purpose. Civia is a good example of how learning with technology can take advantage of what children and youths like and then speak in their symbolic language.

ACKNOWLEDGMENT

This project was supported by a grant from the Departamento Administrativo de Ciencia, Tecnología e Innovación de Colombia (Colombian Administrative Department of Science, Technology, and Innovation –COLCIENCIAS-) code 112740520334, and Universidad de Caldas, Colombia code 0043508.

REFERENCES

Anderson, C. A. (2004). An update on the effects of playing violent video games. *Journal of Adolescence*, *27*(1), 113–122. doi:10.1016/j.adolescence.2003.10.009

Anderson, C. A., & Bushman, B. (2002). Human Aggression. *Annual Review of Psychology*, *53*, 27–51. doi:10.1146/annurev.psych.53.100901.135231

Bartholow, B. D., Sestir, M. A., & Davis, E. B. (2005). Correlates and Consequences of Exposure to Video Game Violence: Hostile Personality, Empathy, and Aggressive Behavior. *Personality and Social Psychology Bulletin*, *31*(11), 1573–1586. doi:10.1177/0146167205277205

Bartle, R. (2000). *Hearts, Clubs, Diamonds, Spades: Players who Suit MUDs*. Colchester, UK: Muse.

Buckley, K. E., & Anderson, C. A. (2006). A theoretical model of the effects and consequences of playing video games. In Vorderer, P., & Bryant, J. (Eds.), *Playing video games—Motives, responses, and consequences* (pp. 363–378). Mahwah, NJ: Lawrence Erlbaum.

Bushman, B. J., & Anderson, A. (2002). Violent Video Games and Hostile Expectations: A Test of the General Aggression Model. *Personality and Social Psychology Bulletin*, *28*(5), 1679–1686. doi:10.1177/014616702237649

Chaux, E. (2009). Citizenship Competencies in the Midst of a Violent Political Conflict: The Colombian Educational Response. *Harvard Educational Review*, *79*(1), 84–93.

Departamento Administrativo Nacional de Estadística [DANE] (2005). *Informe Especial Censo General 2005 Colombia – Educación* [Specia Report General Census 2005 Colombia – Education]. Retrieved from http://www.dane.gov.co/censo.

Departamento Administrativo Nacional de Estadística [DANE] (2008). *Encuesta de Consumo Cultural 2008* [Cultural Consumption Survey 2008] [Data file]. Retrieved from http://www.dane.gov.co/.

Departamento Administrativo Nacional de Estadística [DANE] (2009a). *Contador de Población* [Population Counter]. Retrieved from http://www.dane.gov.co/reloj/reloj_animado.php

Departamento Administrativo Nacional de Estadística [DANE] (2009b). *Encuesta de Ingresos y Gastos 2006/2007* [Survey of Incomes and Expenses 2006/2007]. Retrieved from: http://www.dane.gov.co/

Departamento Administrativo Nacional de Estadística [DANE] (2009c). *Indicadores Básicos de Tecnologías de la Información y la Comunicación TIC* [Basic Indicators of Information and Communication Technology ICT]. Retrieved from http://www.dane.gov.co/

Flusser, V. (2000). *Towards a Philosophy of Photography*. London: Reaktion.

Gee, J. P. (2003). *What Video Games Have to Teach us About Learning and Literacy*. New York: Palgrave Macmillan.

Gee, J. P. (2007). *Good Video Games and Good Learning: Collected Essays on Video Games, Learning and Literacy*. New York: Peter Lang Publishing.

Gee, J. P. (2008). Video games, learning, and "content". In Miller, C. T. (Ed.), *Games: Purpose and Potential in Education* (pp. 43–53). New York, NY: Springer.

Gentile, D. A., Anderson, C. A., Yukawa, S., Ihori, N., Saleem, M., & Shibuya, A. (2009). The Effects of Prosocial Video Games on Prosocial Behaviors: International Evidence From Correlational, Longitudinal, and Experimental Studies. *Personality and Social Psychology Bulletin, 35*(6), 752–763. doi:10.1177/0146167209333045

Goldstein, J., Buckingham, D., & Brougére, G. (2004). Introduction: Toys, games, and media. In Goldstein, J., Buckingham, D., & Brougére, G. (Eds.), *Toys, Games, and Media* (pp. 1–8). Mahwah, NJ: Lawrence Erlbaum.

Greitemeyer, T., & Osswald, S. (2009). Prosocial video games reduce aggressive cognitions. *Journal of Experimental Social Psychology, 45*(4), 896–900. doi:10.1016/j.jesp.2009.04.005

Hartman, W., & Brougère, G. (2004). Toy culture in preschool education and children's toy preferences. In Goldstein, J., Buckingham, D., & Brougère, G. (Eds.), *Toys, Games, and Media* (pp. 37–53). Mahwah, NJ: Lawrence Erlbaum.

Huizinga, J. (1955). *Homo ludens. A Study of the Play-element in Culture*. Boston: Beacon Press.

Jacobson, I., Booch, G., & Rumbaugh, J. (1999). *The Unified Software Development Process*. Reading, Massachusetts: Addison-Wesley.

Kahne, J., Middaugh, E., & Evans, C. (2008). *The Civic Potential of Video Games*. MacArthur Foundation, Digital Media and Learning Program.

Kerr, A. (2006). *The Business and Culture of Digital Games: Gamework/Gameplay*. Thousand Oaks, CA: Sage Publications.

Lenhart, A., Kahne, J., Middaugh, E., Macgill, A. R., Evans, C., & Vitak, J. (2008) *Teens, Video Games, and Civics. Teens' gaming experiences are diverse and include significant social interaction and civic engagement*. Retrieved from: http://www.pewinternet.org/

Linderoth, J., Lindström, B., & Alexandersson, M. (2004). Learning with computer games. In Goldstein, J., Buckingham, D., & Brougère, G. (Eds.), *Toys, Games, and Media* (pp. 157–176). Mahwah, NJ: Lawrence Erlbaum.

Manders, T. (2008). *Proyecto de Informe sobre la Protección de los Consumidores, en Particular de los Menores, por lo que se refiere al Uso de Juegos de Vídeo* [Project of Report about Consumers' Protection, Particularly Minors, About Video Game Use]. Bruselas: Parlamento Europeo, Comisión de Mercado Interior y Protección del Consumidor.

Michigan State University. (2009), *Serious Game Design Program*. Retrieved from: http://woz.commtechlab.msu.edu/sgd/index.php.

Ministerio de Educación Nacional. (2004a). *Estándares Básicos de Competencias Ciudadanas. Formar para la Ciudadanía... ¡sí es posible! Lo que necesitamos saber y saber hacer* [Basic Standards of Citizenship Competencies... it is possible! What we Need to Know and Know How to Do]. Bogotá, Colombia: Ministerio de Educación Nacional.

Ministerio de Educación Nacional. (2004b). *Proceedings from Foro Nacional de Competencias Ciudadanas* [National Forum of Citizenship Competencies]. Bogotá, Colombia: Ministerio de Educación Nacional.

Mockus, A. (2004, February-March). Por qué competencias ciudadanas en Colombia? [Why Citizenship Competencies in Colombia?] *Altablero, 27*. Retrieved from: http://www.mineducacion.gov.co/1621/article-87299.html

Murray, J. H. (1997). *Hamlet on the Holodeck: the Future of Narrative in Cyberspace*. New York: Free Press.

Prensky, M. (2001). *Digital Game-based Learning*. New York, NY: McGraw-Hill.

Prensky, M. (2006). *"Don't bother me Mom, I'm learning!" How Computer and Video Games are preparing your Kids for Twenty-first Century Success and how you can Help!* St. Paul, Minnesota: Paragon House.

Przybylski, A. K., Ryan, R. M., & Rigby, C. S. (2009). The Motivating Role of Violence in Video Games. *Personality and Social Psychology Bulletin, 35*(2), 243–259. doi:10.1177/0146167208327216

Rollings, A., & Adams, E. (2003). *Andrew Rollings and Ernest Adams on Game Design*. Indianapolis: New Riders.

Sanchez Buron, A., Rodríguez, L., & Fernández Marín, M. P. (2009). *Los Adolescentes en la Red. Estudio sobre sus Hábitos de Comportamiento en Internet* [Adolescents in the Net. Study about their Behavior Habits in Internet]. Retrieved from http://www.ucjc.edu

Semrud-Clikeman, M. (2007). *Social Competence in Children*. New York, NY: Springer.

Singer, D., & Singer, J. (2005). *Imagination and Play in the Electronic Age*. Cambridge, MA: Harvard University Press.

KEY TERMS AND DEFINITIONS

Citizenship and Social Competencies: Human abilities to behave peacefully, participate, and respect others within communities.

Game Design: Activities of conceptualization, planning, creation and production of games.

Multiuser Virtual Game: Online game that can be played by several users at the same time.

Serious Games: Games that have main aims other than entertainment such as learning, marketing, or social change.

Social Change: Transformation of social behaviors and culture.

Compilation of References

Abraham, W. C., & Bear, M. F. (1996). Metaplasticity: the plasticity of synaptic plasticity. *Trends in Neurosciences*, *19*(4), 126–130. doi:10.1016/S0166-2236(96)80018-X

Ackoff, R. L. (1989). From Data to Wisdom. *Journal of Applies Systems Analysis*, 3-9.

Ait Kaci, Y., & Mestaoui, N. (2006). *Electronic Shadow*. Retrieved September 15th, 2009 from http://www.electronicshadow.com/

Alexander, C., & Ishikawa, S. (1977). *A pattern language: towns, buildings, construction*. New York: Oxford University Press.

Allen, R. (2005). Sensitive Spaces, in *Living Objects, Sensitive Spaces*, catalog of the *ArtFutura 2005* exhibiton, Barcelona: ArtFutura. Retrieved December 22, 2008 from http://www.artfutura.org/v2/artthought.php?idcontent=10&idcreation=37&mb=6&lang=En.

Museums in Education. Towards the End of the Century. In Ambrose, T. (Ed.), *Education in Museums, Museums in Education*. Edinburgh, Scotland: HMSO.

Anderson, C. A. (2004). An update on the effects of playing violent video games. *Journal of Adolescence*, *27*(1), 113–122. doi:10.1016/j.adolescence.2003.10.009

Anderson, C. A., & Bushman, B. (2002). Human Aggression. *Annual Review of Psychology*, *53*, 27–51. doi:10.1146/annurev.psych.53.100901.135231

Arrabales, R., Ledezma, A., & Sanchis, A. (2009), Establishing a Roadmap and Metrics for Conscious Machines Development, *The 8th IEEE International Conference on Cognitive Informatics*: 94-101

Artsmachine Media Lab. (2010). Retrieved from http://www.artsmachine.com

Ascott, R. (1994). "The Architecture of Cyberception" in Shanken, E. (ed.) *Telematic Embrace*, Los Angeles, CA: Unversity of California Press, 2003.

Ascott, R., & Shanken, E. (Ed). (2003). *Telematic Embrace: Visionary Theories of Art, Technology, and Consciousness*, University of California Press. Berkeley, CA.

Ashby, R. (1956). *Introduction to Cybernetics*. London, England: Chapman & Hall.

Atkinson, R. L., & Hilgard, E. R. (1996). *Hilgard's Introduction to Psychology*. Fort Worth: Harcourt Brace College.

Auslander, P. (1999). *Liveness: Performance in a mediatized culture* (2nd ed.). New York: Routledge.

Auslander, P. (2002). Live from cyberspace: or, I was sitting at my computer this guy appeared he thought I was a bot. *PAJ a Journal of Performance and Art*, *24*(1), 16–21. doi:10.1162/152028101753401767

Ayiter, E. (2008), Syncretia: A Sojourn in the Uncanny Valley, *New Realities: Being Syncretic*, Springer, Vienna, AT

Babeli, G. (2007). *Save your skin*. Retrieved on October 4, 2009 from http://gazirababeli.com/saveyourskin.php

Babeli, G. *Gazira Babeli: Code performer*. [Artist website]. Retrieved on October 4, 2009 from http://gazirababeli.com

Bachelard, G. (1966). *The poetics of space* (p. 203). Boston, MA: Beacon Press.

Ball, G., & Breese, J. (2000). Emotion and Personality in a Conversational Agent. pp. 189–219.

Balla, G., & Depero, F. (1958). Ricostruzione Futurista dell'Universo. In *Archivi del Futurismo*. Roma: Manifesto.

Barbaud, P. (1966). *Introduction à la composition musicale automatique*. Paris, France: Dunod.

Barfield, W., Zelter, D., Sheridan, T. B., & Slater, M. (1995). Presence and Performance within Virtual Environments. In W. Barfield & T.A. Furness (Eds,), *Virtual Environments and advanced interface design*. Oxford: Oxford University Press

Barthes, R. (1953/1977). *Writing Degree Zero (Jonathan Cape Ltd., Trans.)*. New York: Hill & Wang.

Bartholow, B. D., Sestir, M. A., & Davis, E. B. (2005). Correlates and Consequences of Exposure to Video Game Violence: Hostile Personality, Empathy, and Aggressive Behavior. *Personality and Social Psychology Bulletin*, *31*(11), 1573–1586. doi:10.1177/0146167205277205

Bartle, R. (2000). *Hearts, Clubs, Diamonds, Spades: Players who Suit MUDs*. Colchester, UK: Muse.

Bates, J., Loyall, B. A., & Reilly, S. W. (1992). An architecture for Action, Emotion, and Social Behavior. *Modelling Autonomous Agents in a Multi-Agent World*, pp. 55-68.

Baudrillard, J. (1981/1994). *Simulacra and simulation (Glaser, S., Trans.)*. Ann Arbor, MI: University of Michigan.

Bellasi, P., Corà, B., Fiz, A., Hayek, M., & Magnagnagno, G. (2004). *Tinguely e Munari. Opere in Azione*. Milano, Italy: Mazzotta Editore.

Benedict, M. (Ed.). (1991). *Cyberspace First Step*. Cambridge, MA: MIT Press.

Benjamin, W. (1936/1969). The work of art in the age of mechanical reproduction. In Arendt, H. (Ed.), *Illuminations: Essays and reflections*. New York: Schocken.

Bertalanffy, V. L. (1983). *Teoria generale dei sistemi*. Milano, Italy: Mondadori.

Besant, A. (1899). *The Ancient Wisdom: An outline of Theosophical Teachings, 2nd edition, 146*. London: Theosophical Publishing Society.

Beyeler Museum, A. G. (2008). *Action Painting, Jackson Pollock*. Catalogue. Beyeler Museum AG.

Bilal, W., & Lydersen, K. (2008), *Shoot an Iraqi, Art, Life and Resistance Under the Gun*. City Lights Publisher, San Franscico, California. from: http://www.wafaabilal.com

Biocca, F., & Levy, M. R. (1995). *Communication in the age of virtual reality*. Hillsdale, NJ: Lawrence Erlbaum.

Bir, B. Lee S. (1994) Genetic Learning for Adaptive Image Segmentation *The Springer International Series in Engineering and Computer Science, 287*

Black, M. (1937). Vagueness: An exercise in logical analysis in *Science, n. 4, pp 427-455*.

Blais, J., & Ippolito, J. (2006). *At the Edge of Art*. London: Thames & Hudson.

Bless, H., & Schwarz, N. (1999). *Mood, its Impact on Cognition and Behavior. Blackwell Encyclopedia of Social Psychology*. Oxford, England: Blackwell.

Bloomer, K. C., & Moore, C. W. (1977). *Body, Memory and Architecture*. New Haven, CT: Yale University Press.

Blyth, F. M., March, L. M., Brnabic, A. J., Jorm, L. R., Williamson, M., & Cousins, M. J. (2001). Chronic pain in Australia: A prevalence study. *Pain*, *89*(2-3), 127–134. doi:10.1016/S0304-3959(00)00355-9

Boccioni, U. (1971). *Manifesto tecnico della scultura futurista*. Milano, Italy: Feltrinelli.

Boden, M. (1997). *The Creative Mind: Myths and Mechanism*. London: Weidenfeld & Nicholson.

Bogost, I. (June 16, 2007). *Bloomsday on Twitter: A performance of "Wandering rocks" on Twitter, and a commentary on both*. Retrieved on October 4, 2009 from http://www.bogost.com/blog/bloomsday_on_twitter.shtml

Bolter, J., & Gromala, D. (2005). *Windows and mirrors: Interaction design, digital art and the myth of transparency*. Cambridge, MA: MIT Press.

Compilation of References

Bolter & Grusin. (1999). *Remediation: Understanding New Media*. Cambridge, MA: MIT Press.

Borges, J. L. (1946/1975). A new refutation of time. In R. Simms (Trans.), *Other inquisitions: 1937–1952*. Austin, TX: University of Texas.

Bosse, T., Pointer, M., & Treur, J. (2007). A Dynamical System Modelling Approach to Gross' Model of Emotion Regulation. *Proceedings of the 8th International Conference on Cognitive Modeling (ICCM'07)*, Oxford, UK: Taylor & Francis/Psychology Press, pp. 187-192.

Botella, C. M., Juan, M. C., Baños, R. M., Alcañiz, M., Guillén, V., & Rey, B. (2005). Mixing realities? An application of augmented reality for the treatment of cockroach phobia. *Cyberpsychology & Behavior*, *8*(2), 162–171. doi:10.1089/cpb.2005.8.162

Bret, M. (2000) Virtual Living Beings, in Lecture Notes in Artificial Intelligence. *Virtual Worlds*, Ed. Jean-Claude Heudin, Springer. 1834: 119-134

Bridges, A., & Charitos, D. (1998). On architectural design in virtual environments. *Design Studies*, *2*(18), 143–154.

Brinkman, W. van der Mast, C., & de Vliegher, D. (2008). Virtual reality exposure therapy for social phobia: A pilot study in evoking fear in a virtual world. *Proceedings of the 22nd HCI 2008 Workshop - HCI for Technology Enhanced Learning* (pp. 29-35), Liverpool, UK.

Brown, R. (1989). *Handbook of Creativity* (Glover, B., Eds.). London, New York: Plenum Press.

Bru, C. (1955). *L'estetique de l'abstraction*. Paris: Presses Universitaires de France.

Buckley, K. E., & Anderson, C. A. (2006). A theoretical model of the effects and consequences of playing video games. In Vorderer, P., & Bryant, J. (Eds.), *Playing video games—Motives, responses, and consequences* (pp. 363–378). Mahwah, NJ: Lawrence Erlbaum.

Burden, E. (1999). *Visionary Architecture: Unbuilt Works of the Imagination*. USA: McGraw-Hill Professional.

Burkhart, C. (1979) *Anthology for Musical Analysis*, Holt, Rinehart and Winston, Toronto, Ontario, Canada: H B Holt.

Bushman, B. J., & Anderson, A. (2002). Violent Video Games and Hostile Expectations: A Test of the General Aggression Model. *Personality and Social Psychology Bulletin*, *28*(5), 1679–1686. doi:10.1177/014616702237649

Callaghan, M.-J., Harkin, J., Scibilia, G., Sanfilippo, F., McCusker, K., & Wilson, S. (2008) Experiential based learning in 3D Virtual Worlds: Visualization and data integration in Second Life, *Proceedings of Remote, Engineering and Virtual Instrumentation* (REV 2008) Conference, Dusseldorf, Germany.

Calvino, I. (1997). *De osynliga städerna (invisible cities)* (Swedish edition). Stockholm: Bonniers Förlag.

Camfield, W. (1979). *Francis Picabia his art, life and times*. Princeton, NJ, USA: Princeton Univ. Press.

Canadian Pain Coalition (2007). Retrieved November 7, 2005, from http://www.rsdcanada.org/parc/english/resources/coalition.htm.

Carver, C. S., & Scheier, M. (1998). *On the self-regulation of behavior*. London: Cambridge University Press.

Castranova, E. (2007), Exodus to the Virtual World, Palgrave MacMillan, NY, NY.

Catanese, P. (2008). *The Small Magnetism: Interrogating the Nature of Collection*. Appeared in *Imaging in the Digital* Fall 2008 edition of *Media-N Journal of the New Media Caucus*, Retrieved December 1, 2009, from: http://www.newmediacaucus.org/journal/issues.php?f=papers&time=2008 winter&page=catanese

Causey, M. (1999). The screen test of the double: The uncanny performer in the space of technology. *Theatre journal 51(4)*, 383–394.

Centre for Metahuman Exploration. (1998). *Project paradise*. Retrieved on October 4, 2009 from http://cultronix.eserver.org/pparadise/happinessflows.html

Chaux, E. (2009). Citizenship Competencies in the Midst of a Violent Political Conflict: The Colombian Educational Response. *Harvard Educational Review*, *79*(1), 84–93.

Chomsky, N.(1959). Three models for the description of language in Information and Control. *On certain formal properties of grammars, n. 2*, IRE Transactions on Information Theory.

Cingolani, P. (2009). jfuzzylogic: An Open Source Fuzzy Logic Library and FCL Language Implementation. http://jfuzzylogic.sourceforge.net/html/index.html.

Coates, G. (1992). *A virtual show. A multimedia performance work*. San Francisco.

Colace, F., Santo, M., Vento, M., & Foggia, P. (2004): A Semi-Automatic Bayesian Algorithm for Ontology Learning. In *International Conference on Entreprise Information* pp. 191-196.

Coleridge, S. T. (1985). *Biographia literaria: Biographical sketches of my literary life & opinions*. Princeton, NJ: Princeton University. (Original work published 1817)

Conradi, I., & Vasudevan, J. (2010). *Internal External, Ina Conradi*. (J. Vasudevan, Ed.) Sinagpore: Ina Conradi.

Costa, P. C., & Laskey, K. B. (2005). PR-OWL: A Bayesian Ontology Language for the Semantic Web. *Proceedings of the ISWC Workshop on Uncertainty Reasoning for the Semantic Web*.

Couchot E.Bret, M & Tramus M-H., (1998) Art virtuel, créations interactives et multisensorielles, *Beaux-arts Magazine*, hors série:11-18.

Cramer, F. (2005). *Words Made Flesh: Code, Culture, Imagination*. Piet Zwart Institute: Willem De Kooning Academy Hogeschool Rotteradam, Retrieved December 1, 2009, from: http://pzwart.wdka.hro.nl/mdr/research/fcramer/wordsmadeflesh/

Creed, C. and Beale, R. (2008) Simulated Emotion in Affective Embodied Agents Affect and *Emotion in Human-Computer Interaction, From Theory to Applications. Lecture Notes in Computer Science Systems And Apllications*. Springer Berlin / Heidelberg. 4868: 163-174

Cruz-Diez, C. (n.d). *My Reflexions on Color*. Retrieved January 9, 2007 from Carlos Cruz-Diez Official Site, www.cruz-diez.com

d'Escriván, J. (2006). To sing the body electric: Instruments and effort in the performance of electronic music. *Contemporary Music Review*, 25(1–2), 183–191. doi:10.1080/07494460600647667

Daden (2009). Automated Avatars. http://www.youtube.com/watch?v=9hte2MJ54CA

Damer, B. (2008). A New Virtual World Winter? http://terranova.blogs.com/terra_nova/2008/06/possibility-of.html

Damer, B., & Judson, J. Dove, j., (1997) Avatars! Exploring and Building Virtual Worlds on the Internet, Peachpit Press Berkeley, CA, USA

Damisch, H. (1984). *Fenêtre jaune cadmium: ou les dessous de la peinture*. Paris, France: Gallimard.

Davies, C. (2001). *Multimedia: from Wagner to Virtual Reality* (pp. 293–300). New York: W.W. Norton & Company.

De Boeck, J., Raymaekers, C., & Coninx, K. (2006). Exploiting proprioception to improve haptic interaction in a virtual environment. *Presence (Cambridge, Mass.)*, 15(6), 627–636. doi:10.1162/pres.15.6.627

De Certeau, M. (1990). *L'Invention du Quotidien. 1. Arts de Faire*. Paris: Gallimard.

De Mauro, T. (Ed.). (1967). *Corso di linguistica generale*. Roma-Bari, Italy: Laterza.

De Nicola, A. Missikoff, M. Navigli R. (2009). "A Software Engineering Approach to Ontology Building". *Information Systems*, 34(2), Elsevier, 2009, pp. 258-275.

Debono, M.-W. (1996). *L'Ere des plasticiens: De nouveaux hommes de science face à la poésie du monde*. Coll. Sciences et Spiritualité, Editions Aubin.

Deger, J. (2006). *Shimmering Screens, Making Media in an Aboriginal Community*. Minneapolis: University of Minnesota Press.

Deleuze, G. (2006). *The Fold. Translation by Tom Conley*. Minneapolis, MN: University of Minnesota Press.

Departamento Administrativo Nacional de Estadística [DANE] (2005). *Informe Especial Censo General 2005 Colombia – Educación* [Specia Report General Census 2005 Colombia–Education]. Retrieved from http://www.dane.gov.co/censo.

Departamento Administrativo Nacional de Estadística [DANE] (2008). *Encuesta de Consumo Cultural 2008* [Cultural Consumption Survey 2008] [Data file]. Retrieved from http://www.dane.gov.co/.

Departamento Administrativo Nacional de Estadística [DANE] (2009a). *Contador de Población* [Population Counter]. Retrieved from http://www.dane.gov.co/reloj/reloj_animado.php

Departamento Administrativo Nacional de Estadística [DANE] (2009b). *Encuesta de Ingresos y Gastos 2006/2007* [Survey of Incomes and Expenses 2006/2007]. Retrieved from: http://www.dane.gov.co/

Departamento Administrativo Nacional de Estadística [DANE] (2009c). *Indicadores Básicos de Tecnologías de la Información y la Comunicación TIC* [Basic Indicators of Information and Communication Technology ICT]. Retrieved from http://www.dane.gov.co/

Dinkla, S. (2002). The Art of Narrative. In Rieser, M., & Zapp, A. (Eds.), *New Screen Media*. London: Brittish Film Institute.

DiPaola, S. (1989) Implementation and use of a 3d parameterized facial modeling and animation system, *ACM Siggraph'89 Course Notes, State of the Art in Facial Animation. SIGGRAPH Conference 16th 1989 Boston, Mass 22:20-33*.

Dixon, S. (2007). *Digital performance: A history of new media in theater, dance, performance art, and installation*. Cambridge, MA: MIT.

Donaldson, N., Haque, U., Hasegawa, A., & Tremmel, G. (2008). *Remote*. Retrieved on October 4, 2009 from http://turbulence.org/Works/remote

Dorfles, G. (2002). *Il divenire delle Arti. Ricognizione nei linguaggi artistic*. Bologna, Italy: Bompiani.

Dormor, C. (2008). *Skin, Textile, Film, Textile. The Journal of Cloth and Culture,* November, 2008.

Doyle, D. (2008). *Kritical Works in SL. Exhibition Catalogue*. Morrisville, North Carolina: Lulu Publishing.

Doyle, D & Kim, T. (2007). Embodied Narrative: the virtual nomad and the metadreamer. *International Journal of Performance Arts and Digital Media* 3: 2 & 3, pp.209-222, doi: 10.1386/padm.3.2&3.209/1.

Doyle, D. (2009). *Kritical Works in SL ii*. Exhibition Catalogue. Wolverhampton: CADRE publications.

Dunne, A. (2006). *Hertzian Tales. Electronic Products, Aesthetic Experience and Critical Design*. Cambridge, MA: MIT Press.

Eco, U. (1976). *A theory of Semiotics*. Bloomington, IN: Indiana University Press.

Eco, U., & Munari, B. (1962). *Arte Programmata (catalogo Olivetti)*. Milano, Italy: Olivetti.

Eco, U. (1989). *The Open Work*. Cambridge, MA: Harvard University Press. Translation by Anna Cancogni.

Eco, U. (2004). On Beauty, A History of a Western Idea. In *The Contemporary Re-assessment of Material* (p. 438). Secker & Warburg.

Edschmid, K. (1920), Paul Klee Schopferische Konfession. In *Tribune der Kunst und Zeit* (Vol. XIII). Berlin, Germany.

Efrat, A., Indyk, P., & Venkatasubramanian, S. (2004). Pattern Matching for Sets of Segments *Algorithmica. Springer New York, 40*(3), 147–160.

Egges, A., Kshirsagar, S., & Magnenat-Thalmann, N. (2004). Generic Personality and Emotion Simulation for Conversational Agents. *Journal of Visualization and Computer Animation, 15*(1), 1–13.

Einstein, A. (1905). Zur Elektrodynamik bewegter Körper [On the electrodynamics of moving bodies]. *Annalen der Physik und Chemie, 17*, 891–921.

Ekman, P. (1994). *Moods, Emotions, and Traits, the Nature of Emotion: Fundamental Questions*. New York, NY, USA: Oxford University Press.

Elkins, J. (1999). *Pictures of the body: Pain and metamorphosis*. Stanford: Stanford University Press.

El-Nasr, M. S., Yen, J., & Ioerger, T. R. (2000). FLAME: Fuzzy Logic Adaptive Model of Emotions. *Autonomous Agents and Multi-Agent Systems, 3*(3), 219–257. doi:10.1023/A:1010030809960

Emmelkamp, P. (2006). Post-traumatic stress disorder: Assessment and follow-up. In Roy, M. J. (Ed.), *Novel approaches to the diagnosis and treatment of posttraumatic stress disorder* (pp. 309–320). Washington: IOS Press.

Emmerson, S. (2007). *Living electronic music*. Burlingon, VT: Ashgate.

Feulner, J., & Hörnel, D. (1994). *Neural Networks that Learn Harmony-Based Melodic Variations." Proceedings. ICMC, Århus* (pp. 121–124). MELONET.

Fink, C. A. (1979). *Searching for the most powerful behavioral theory: The whole of the Behavior Systems Research Institute and the Behavioral Model Analysis Center*. Falls Church, Va: Fink.

Flusser, V. (2000). *Towards a Philosophy of Photography*. London: Reaktion.

Foucault, M. (1973/2008). *This is not a pipe: With illustrations and letters by René Magritte* (2nd ed.). Los Angeles: University of California.

Foucault, M. (1984). Dits et écrits 1984, Des espaces autres (conférence au Cercle d'études architecturales, 14 mars 1967), in *Architecture, Mouvement, Continuité*, n°5, october 1984, 46-49. Retrieved September 20th, 2009 from http://foucault.info/documents/heteroTopia/foucault. Hetero Topia.en.html. Translation by Jay Miskowiek.

Freeman, J. C. (2008). *Imaging Beijing*. Retrieved on October 4, 2009 from http://turbulence.org/Works/ImagingBeijing

Freud, S. (1955). *The interpretation of Dreams*, Standard edition, Vol. V, 1900 reprint. London : Hogarth press.

Fridja, N. (1987). *The Emotions: Studies in Emotion and Social Interaction*. New York: Cambridge Universty Press.

Front, S. (2007). *The gate*. Retrieved on October 4, 2009 from http://secondfront.org/Performances/The_Gate.html

Front, S. *Second Front: The pioneering performance art group in Second Life*. [Artist website]. Retrieved on October 4, 2009 from http://secondfront.org

Fusi, S., & Abbott, L. (2007). Limits on the memory storage capacity of bounded synapses. *Nature Neuroscience, 10*, 485–493.

García-Rojas, A., Gutiérrez, M., & Thalmann, D. (2008). Simulation of Individual Spontaneous Reactive Behavior. In *AAMAS '08: Proceedings of the 7th International Joint Conference on Autonomous Agents and Multiagent Systems*, Richland, SC. International Foundation for Autonomous Agents and Multiagent Systems, pp. 143–150.

Gartner (2009). Hype Cycle Special Report. http://www.gartner.com/it/page.jsp?id=1124212

Gatchel, R., Peng, Y., & Peters, M. (2007). The biopsychosocial approach to chronic pain: Scientific advances and future directions. [Washington: American Psychological Association.]. *Psychological Bulletin, 133*(4), 581–624. doi:10.1037/0033-2909.133.4.581

Gatchel, R. (2009). Biofeedback as an adjunctive treatment modality in pain management. *American Pain Society Bulletin, 14* (4). Retrieved March 10, 2009, from http://www.ampainsoc.org/pub/bulletin/jul04/clin1.htm

Geary, A. (2009, February). *Digital Physicality - Exploring Hybrid Practices in Drawing and Printmaking*, Paper presented at the College Art Association conference, Los Angeles, CA.

Gee, J. P. (2003). *What Video Games Have to Teach us About Learning and Literacy*. New York: Palgrave Macmillan.

Gee, J. P. (2007). *Good Video Games and Good Learning: Collected Essays on Video Games, Learning and Literacy*. New York: Peter Lang Publishing.

Gee, J. P. (2008). Video games, learning, and "content". In Miller, C. T. (Ed.), *Games: Purpose and Potential in Education* (pp. 43–53). New York, NY: Springer.

Gentile, D. A., Anderson, C. A., Yukawa, S., Ihori, N., Saleem, M., & Shibuya, A. (2009). The Effects of Prosocial Video Games on Prosocial Behaviors: International Evidence From Correlational, Longitudinal, and Experimental Studies. *Personality and Social Psychology Bulletin*, *35*(6), 752–763. doi:10.1177/0146167209333045

Georgeff, M. P., Pell, B., Pollack, M. E., Tambe, M., & Wooldridge, M. (1999). The Belief-Desire-Intention Model of Agency. *In Proceedings of the 5th international Workshop on intelligent Agents, Agent theories, Architectures, and Languages*. In J. P. Müller, M. P. Singh, and A. S. Rao, Eds. Lecture Notes In Computer Science, Springer-Verlag, London, 1555, pp. 1-10.

Georges Poulet, b. (1961). *The Matamorphoses of the Circle*. Baltimore: Johns Hopkins Press.

Gershenfeld, N. (2005). *FAB: The coming Revolution on Your Desktop, From Personal Computers to Personal Fabrication*. New York: Basic Books.

Gerstner, W., & Kistler, W. (2002). *Spiking neuron models: single neurons, populations, plasticity*. Cambridge: Cambridge University Press.

Ghasem-Aghaee, N., & Oren, T. (2007). Cognitive Complexity and Dynamic Personality in Agent Simulation. *Computers in Human Behavior*, *23*(6), 2983–2997. doi:10.1016/j.chb.2006.08.012

Goldberg, K. (1995). *The Telegarden website*. Retrieved on October 4, 2009 from http://goldberg.berkeley.edu/garden/Ars/

Goldstein, J., Buckingham, D., & Brougére, G. (2004). Introduction: Toys, games, and media. In Goldstein, J., Buckingham, D., & Brougére, G. (Eds.), *Toys, Games, and Media* (pp. 1–8). Mahwah, NJ: Lawrence Erlbaum.

Gooding, M. (2001). Abstraction: An Introduction. In Gooding, M. (Ed.), *Movements in Modern: Abstract Art* (pp. 6–7). Cambridge: Cambridge University Press.

Goodman, N. (2003). *I Linguaggi dell'Arte. L'esperirenza estetica: rappresentazione e simboli, trad. I* (Saggiatore, I., Ed.). Milano, Italy.

Gratch, J., & Marsella, S. (2001). Tears and Fears: Modeling Emotions and Emotional Behaviors in Synthetic Agents. *In Proceedings of the Fifth international Conference on Autonomous Agents* (Montreal, Quebec, Canada). AGENTS '01. ACM, New York, NY, pp. 278-285.

Gray, W. D. (Ed.). (2007). *Integrated Models of Cognitive Systems*. USA: Oxford University Press.

Gredler, M. E. (2004). Games and Simulations and their Relationships To Learning. In Jonassen, D. H. (Ed.), *Handbook of Research on Educational Communications and Technology*. Mahwah, NJ: Lawrence Erlbaum Associates.

Greenaway, P. and Akademie der Bildenden Künste in Wien. (1992) *Hundert Objekte zeigen die welt = Hundred objects to represent the world: 100*. Verlag erd Hatje, Stuttgart.

Greimas, A. J., & Courtés, J. (1979). *Sémiotique. Dictionnaire raisonné de la théorie du langage*. Paris, France: Hachette.

Greimas, A. J., & Courtés, J. (1982). *Semiotics and Language: An Analytical Dictionary*. Bloomington, Indiana, USA: Indiana University Press.

Greitemeyer, T., & Osswald, S. (2009). Prosocial video games reduce aggressive cognitions. *Journal of Experimental Social Psychology*, *45*(4), 896–900. doi:10.1016/j.jesp.2009.04.005

Grinder, J., & Bandler, R. (1975). *The Structure of Magic II: A Book About Communication and Change*. Palo Alto, CA: Science & Behavior Books.

Gromala, D. (2000). Pain and subjectivity in virtual reality. In Bell, D. (Ed.), *The cybercultures reader* (pp. 598–608). London: Routledge.

Gromala, D., & Sharir, Y. (1996). Dancing with the virtual dervish: Virtual bodies. In Moser, M., & MacLeod, D. (Eds.), *Immersed in technology: Art and virtual environments* (pp. 281–285). Cambridge, MA: MITPress.

Groupe μ, (1992). *Traité du signe visuel. Pour une rhétorique de l'image.* Paris: Le Seuil.

Hamilton, A., & Simon, J. (2006). *An Inventory of Objects.* New York, New York: Gregory R. Miller & Company.

Hang, H. P., & Alessi, N. E. (1999). Presence as an emotional experience. In *Medicine meets virtual reality: The Convergence of Physical and Information Technologies options for a new era in Healthcare* (pp. 148–153). Amsterdam: IOS Press.

Hansen, M. (2006). *Bodies in Code: Interfaces with Digital Media.* London: Routledge.

Haque, U. (2006). Architecture, interaction, systems. First developed for an article in *AU: Arquitetura & Urbanismo,* N° 149 August 2006, Brazil. Retrieved Februray 25th, 2009 from Usman Haque Official Site, www.haque.co.uk.

Harrison, D. (1997). Hypermedia as Art System. In Drucker, J. (Ed.), *Digital Reflections: The Dialogue of Art and Technology. Art Journal, Fall, 56(3)* (pp. 55–59).

Harrison, D. (2009). The Writing on the Wall. In *ISEA2009 15th International Symposium on Electronic Art.* Belfast: University of Ulster.

Hartman, W., & Brougère, G. (2004). Toy culture in preschool education and children's toy preferences. In Goldstein, J., Buckingham, D., & Brougère, G. (Eds.), *Toys, Games, and Media* (pp. 37–53). Mahwah, NJ: Lawrence Erlbaum.

Hauswirth, M., & Decker, S. (2007). Semantic Reality – Connecting the Real and the Virtual World. *Microsoft SemGrail workshop.* Redmond, WA.

Hautop Lund, H., & Ottesen, M. (2008). RoboMusic: a behavior-based approach. *Artificial Life and Robotics Computer Science. Springer Japan., 12*(1-2), 18–23.

Heim, M. (1993). *The Metaphysics of Virtual Reality.* Oxford: Oxford University Press.

Hiiro, M., Kojima, K., Iwasaki, K. (1996, August). *The Concept of the Space Block and its Achievable Media: Design Method Using the Space Block,* Japan Association of Architecture Academic Lecture Series, Vol. 1996(19960730) (pp. 463-464). E-1 Architectural Plan I.1996, Japan Assciation of Architecture.

Hill, J. R., Wiley, D., Miller-Nelson, L., & Han, S. (2003). Exploring Research on Internet-based Learning: From Infrastructure to Interactions. (pp: 433 – 461) In Jonassen, D. H. (Ed), Harris, P. (Ed). *Handbook of Research on Educational Communications and Technology.* Lawrence Erlbaum Associates. Mahwah, NJ.

Hoberg, C. (2006) *Komplexitets Max,* Dialoger Förlag, Stockholm, Sweden Jackson, Steve, (2002) *GURPS basic set third edition,* Steve Jackson games, USA.

Hofer, S. (2006). I am they: Technological mediation, shifting conceptions of identity and techno music. In *Convergence: The international journal of research into new media technologies 12(3),* 307–324.

Hoffman, H. (2009). Virtual reality as an adjunctive pain control during burn wound care in adolescent patients. *Pain, 85*(1), 305–309. doi:10.1016/S0304-3959(99)00275-4

Hoffman, H. G., Garcia-Palacios, A., Carlin, C., Furness, T. A. III, & Botella-Arbona, C. (2003). Interfaces that heal: Coupling real and virtual objects to cure spider phobia. *International Journal of Human-Computer Interaction, 16*(2), 283–300. doi:10.1207/S15327590IJHC1602_08

Hoffman, H. G., Patterson, D. R., & Carrougher, G. J. (2000). Use of virtual reality for adjunctive treatment of adult burn pain during physical therapy: A controlled study. *The Clinical Journal of Pain, 16*(3), 244–250. doi:10.1097/00002508-200009000-00010

Hoffman, H. G., Richards, T., Coda, B., Bills, A., Blough, D., Richards, A., & Sharar, S. (2004a). Modulation of thermal pain-related brain activity with virtual reality: Evidence from fMRI. *Neuroreport, 15*(8), 1245–1248.

Compilation of References

Hoffman, H. G., Sharar, S., & Coda, B. (2004b). Manipulating presence influences the magnitude of virtual reality analgesia. *Pain*, *111*(1–2), 162–168. doi:10.1016/j.pain.2004.06.013

Hoffman, H. G., & Patterson, D. (2005). Virtual reality pain distraction. *American Pain Society Bulletin 15* (2). Retrieved March 31, 2009, from http://www.ampainsoc.org/pub/bulletin/spr05/inno1.htm.

Hofstadter, D. (1979/1999). *Gödel, Escher, Bach: An eternal golden braid* (20th anniversary ed.). New York: Basic Books.

Hong, J. (2008). *From Rational to Emotional Agents: A Way to Design Emotional Agents*. Saarbrucken, Germany: VDM Verlag.

Huizinga, J. (1955). *Homo Ludens: A study of the play-element in culture*. Boston, USA: Beacon Press.

Huyghe, P. D. (2006). *Éloge de l'aspect*. Paris: Éditions Mix. Translation by Carola Moujan.

Hwang, K. (2009). Artist Statement from Artist's personal website. Retrieved December 1, 2009, from http://www.dancekk.com/dance2/welcome.html

IEC 1131 (1997). Programmable Controllers, Part 7: Fuzzy Control Programming, Fuzzy Control Language. http://www.fuzzytech.com/binaries/ieccd1.pdf.

Iglesias, A., & Luengo, F. (2007). *AI Framework for Decision Modeling in Behavioral Animation of Virtual Avatars* (pp. 89–96). Berlin, Heidelberg: Springer-Verlag.

IJsselsteijn, W., deRidder, H., Freeman, J., & Avons, S. E. (2000). Presence: Concept, determinants and measurement. *Proceedings of the SPIE, Human Vision and Electronic Imaging*, San Jose, USA.

Imbert, R., & de Antonio, A. (2005). An Emotional Architecture for Virtual Characters. *ICVS'05, Third International Conference on Virtual Storytelling*. Lecture Notes in Computer Science, Springer. Strasbourg, France, 3805, 63-72.

Inoue, C., Kojima, K., Hiiro, M., Watanave, H. (1999, August). *Construction of the Space Block Database, Space Block as the Interactive Spatial Design Tool, Part 1*, Japan Association of Architecture Academic Lecture Series, Vol. 1999(19990730) (pp. 531-532). E-1 Architectural Plan I, Japan Assciation of Architecture.

Itmedia (2009, May). *Second Life Will Not End: Increasing number of users, Growing Economy*, from http://www.itmedia.co.jp/news/articles/0905/18/news037.html

Jacobson, I., Booch, G., & Rumbaugh, J. (1999). *The Unified Software Development Process*. Reading, Massachusetts: Addison-Wesley.

Jakobson, R. (1973). *Questions de poetique, Paris, France: Seuil, Kuhn, A. (1974) The Logic of Social systems*. San Francisco, USA: Jossey-Bass.

Jakobson, R. (1960). Linguistics and poetics. In Sebeok, T. A. (Ed.), *Style in Language* (pp. 350–377). Cambridge, MA: MIT Press.

Jaques, P. A., & Viccari, R. M. (2006). Considering students' emotions in computer-mediated learning environments. In Ma, Z. (Ed.), *Web-based intelligent e-learning systems: Technologies and applications* (pp. 122–138). Hershey, PA: Information Science Publishing.

Jiang, H., & Vidal, J. M. (2006). From Rational to Emotional Agents. *Proceedings of the American Association for Artificial Intelligence (AAAI). Workshop on Cognitive Modeling and Agent-based Social Simulation*.

Jiang, H., Vidal, J. M., & Huhns, M. N. (2007). EBDI: An Architecture for Emotional Agents. *In Proceedings of the 6th international Joint Conference on Autonomous Agents and Multiagent Systems*. AAMAS '07. ACM, New York, NY, pp. 1-3.

Joerissen, B., (2008) The Body is the Message. Avatare als visuelle Artikulationen, soziale Aktanten und hybride Akteure. *Paragrana, Volume: 17, Issue: 1*.

Joyce, J. (2008). *Ulysses*. New York: Oxford University. (Original work published 1922)

Kabat-Zinn, J. (2006). *Coming to our senses: Healing ourselves and the world through mindfulness*. New York: Hyperion.

Kafai, Y. (Ed.). Resnick, M. (Ed). (1996) Constructionism in Practice: Designing, Thinking, and Learning in a Digital World, (pg: 11) Lawrence Erlbaum Associates. Mahwah, NJ.

Kahne, J., Middaugh, E., & Evans, C. (2008). *The Civic Potential of Video Games*. MacArthur Foundation, Digital Media and Learning Program.

Kandinsky, W. (1968). *Punto Linea Superficie*. Milano, Italy: Adelphi Edizioni.

Kerr, A. (2006). *The Business and Culture of Digital Games: Gamework/Gameplay*. Thousand Oaks, CA: Sage Publications.

Khan, K. S., & Al-Khatib, W. G. (2006). Machine-learning based classification of speech and music. *Multimedia Systems. Computer Science, 12*(1), 55–67.

Khoo, Y. H., & Conradi, I. (2009). *Painting Using Experimentall Animation*. Singapore: NTU URECA Proceedings.

Khulood, A. M., & Raed, A. Z. (2007). Emotional Agents: A Modeling and an Application. *Information and Software Technology, 49*(7), 695–716. doi:10.1016/j.infsof.2006.08.002

Kildall, S., & Scott, V. (2008). *No Matter*. Sponsored by Turbulence, a project of New Radio and Performing Arts, Inc. for the exhibition, Mixed Realities—An International Networked Art Exhibition and Symposium. Retrieved on October 4, 2009 from http://turbulence.org/Works/nomatter/

Kirriemuir, J. Measuring" the Impact of Second Life for Educational Purposes. Retrieved on 29/03/2008. http://www.eduserv.org.uk/upload/foundation/sl/impactreport032008/impactreport.pdf/

Kiss J. & all. (2006) Voice interaction system for video games used within "virtual singer" computer interface. *Proceedings of CGAMES Conference. 8th International Conference on Computer Games: AI, Animation, Mobile, Educational & Serious Games*. Dublin, Ireland. Louisville. USA

Koda, T., Ishida, T., Rehm, M., & André, E. (2009). Avatar culture: cross-cultural evaluations of avatar facial expressions. *AI & Society. Computer Science. Springer London., 24*(3), 237–250.

Kowalski, R., & Sergot, M. (1986). A Logic-Based Calculus of Events. *New Generation Computing, 4*(1), 67–95. doi:10.1007/BF03037383

Krijn, M., Emmelkamp, P., Ólafsson, R., Bouwman, M., van Gerwen, M. L., & Spinhoven, P. (2007). Fear of flying treatment methods: Virtual reality exposure vs. cognitive behavioral therapy. *Aviation, Space, and Environmental Medicine, 78*(2), 121–128.

Kshirsagar, S., & Magnenat-Thalmann, N. (2002). A Multilayer Personality Model. *In SMARTGRAPH '02: Proceedings of the 2nd International Symposium on Smart Graphics*, New York, NY, USA. ACM, pp. 107–115.

Kuhn, A. (1974). *The logic of social systems*. San Francisco, USA: Jossey-Bass

Kunii, T. L. (2005). *Cyberworld modeling – integrating Cyberworld, the Real World and Conceptual World. CyberWorlds 05, CW2005* (pp. 3–11). IEEE.

Kupiec, Jean-Jacques [coordinator] (2009) *Le hasard au cœur de la cellule, Probabilités, déterminisme, génétique*, Syllepse Edition

Kushner, D. (Producer), & Lisberger, S. (Director). (1982). *Tron*. [Motion picture]. United States: Lisberger Studios.

Lagorio, C. (2007) The Ultimate Distance Learning, The New York Times. Retrieved on 13/06/2007 http://www.nytimes.com/2007/01/07/education/edlife/

Lakoff, G. (1987). *Women, fire and dangerous things: what categories reveal about the mind*. Chicago, IL, USA: University of Chicago Press.

Laliberté, M. (1993). Informatique musicale: utopies et réalités (1957-90) *Les Cahiers de l'Ircam no 4. Utopies*, *4*, 155–172.

Lambert, D., & Rezsöhazy, R. (2004). *Comment les pattes viennent au serpent Essai sur l'étonnante plasticité du vivant, Nouvelle bibliothèque scientifique*. Flammarion.

Lanier, J. (1989). Interview with Jaron Lanier. *Whole Earth Review*, *64*, 108–119.

Larman, C., & Basili Victor, R. (2003). Iterative and Incremental Development: A Brief History. *IEEE Computer*, *36*(6), 47–56.

Latour, B. (2005). *Reassembling the Social: An Introduction to Actor-Network Theory*. Oxford: Oxford University Press.

Laurel, B. (1990). *The Art of Human-Computer Interface Design*. Reading, MA: Addison-Wesley.

LeDoux, J. E. (1996). *The Emotional Brain: The Mysterious Underpinnings of Emotional Life*. New York: Simon and Schuster.

Lenhart, A., Kahne, J., Middaugh, E., Macgill, A. R., Evans, C., & Vitak, J. (2008) *Teens, Video Games, and Civics. Teens' gaming experiences are diverse and include significant social interaction and civic engagement*. Retrieved from: http://www.pewinternet.org/

Levi-Strauss, C. (2007). *Anthropology and Aesthetics (Ideas in Context)*. Cambridge, UK: Cambridge University Press.

Lévy, P. (1997). *Qu'est-ce que le virtuel? Il Virtuale, it.translation*. Milano, Italy: Raffaello. Cortina Editore.

Linden, T. (2009), The Second Life Economy - First Quarter 2009 in Detail, Retrieved on 09/26/2009 https://blogs.secondlife.com/community/features/blog/2009/04/16/the-second-life-economy--first-quarter-2009-in-detail

Linderoth, J., Lindström, B., & Alexandersson, M. (2004). Learning with computer games. In Goldstein, J., Buckingham, D., & Brougère, G. (Eds.), *Toys, Games, and Media* (pp. 157–176). Mahwah, NJ: Lawrence Erlbaum.

Liu, Z., & Pan, Z.-G. (2005). An Emotion Model of 3D Virtual Characters. In Picard, R. W. (Ed.), *Intelligent Virtual Environments. In Affective Computing and Intelligent Interaction, Tao, JH* (pp. 629–636). Tan, TN: Springer-Verlag. doi:10.1007/11573548_81

Liu, Z., & Lu, Y.-S. (2008). A Motivation Model for Virtual Characters. *In Proceedings of the Seventh International Conference on Machine Learning and Cybernetics*, *5*, 2712–2717.

Lobard, M., & Ditton, T. (1997). At the heart of it all: The concept of presence. *Journal of Computer-Mediated Communication*, Indiana, USA.

Loeser, J. D., Butler, S. H., Chapman, C. R., & Turk, D. C. (Eds.). (2001). *Bonica's Management of Pain* (3rd ed.). Philadelphia: Lippincott Williams & Wilkins.

Lopez, X., Annamalai, M., Banerjee, J., Ihm, J., Sharma, J., & Steiner, J. (2009). *Oracle Database 11g Semantic Technologies Semantic Data Integration for the Enterprise. Oracle Database 11g Semantic Technologies Semantic Data Integration for the Enterprise*. Redwood Shores, CA: Oracle Corporation.

Lovejoy, M. (1990). Art, Technology and Postmodernism: Paradigms, Parallels, and Paradoxes. *Art Journal*. Fall, 257 ff.

Lutz, A., Slagter, H., Dunne, J. D., & Davidson, R. J. (2008). Attention regulation and monitoring in meditation. *Trends in Cognitive Sciences*, *12*(4), 163–169. doi:10.1016/j.tics.2008.01.005

MacLean, P. (1973). *A Triune Concept of the Brain and Behaviour*. Toronto: University of Toronto Press.

Magritte, R. (1928–1929). *La trahison des images* [The treachery of images]. [Painting].

Maher, M. L., Simoff, S., Gu, N., & Lau, K. H. (2000). *Designing virtual architecture*, Proceedings of the Fifth Conference on Computer Aided Architectural Design Research in Asia (pp. 481- 490), Singapore.

Mahrer, N., & Gold, J. (2009). The use of virtual reality for pain control: A review. *Current Pain and Headache Reports*, *13*(2), 100–109. doi:10.1007/s11916-009-0019-8

Manders, T. (2008). *Proyecto de Informe sobre la Protección de los Consumidores, en Particular de los Menores, por lo que se refiere al Uso de Juegos de Vídeo* [Project of Report about Consumers' Protection, Particularly Minors, About Video Game Use]. Bruselas: Parlamento Europeo, Comisión de Mercado Interior y Protección del Consumidor.

Manovich, L. (2008). "Understanding MetaMedia" Opening Keynote at *Computer Art Congress 2008 [CAC.2]*. Toluca, Mexico. March 26, 2008.

Manovich, L. *The Death of Computer Art.* http://www-apparitions.ucsd.edu/~manovich/text/death.html

Mansoor, J. (2003) Kurt Schwitters' Merzbau: The Desiring House, in Invisible Culture, Issue 4, Retrieved October 12, 2009 http://www.rochester.edu/in_visible_culture/Issue4-IVC/Mansoor.html

Mantovani, G., & Riva, G. (1999) "Real" Presence: How Different Ontologies Generate Different Criteria for Presence, Telepresence, and Virtual Presence, *Presence: Teleoperators & Virtual Environments*; Vol. 8 Issue 5, p540, 11p.

Marsciani, F., & Zinna, A. (1991) *Elementi di Semiotica Generativa. Processi e sistemi della significazione*, Progetto Leonardo, Bologna, Italy: Ed. Esculapio

Martinez-Miranda, J., & Aldea, A. (2005). Emotions in Human and Artificial Intelligence. *Computers in Human Behavior*, *21*(2), 323–341. doi:10.1016/j.chb.2004.02.010

Mase, K., Yasuyuki Sumi, Y., & Fels, S. (2007). Welcome to the special issue on memory and sharing of experience for the Journal of Personal and Ubiquitous Computing *Memory and Sharing of Experiences Personal and Ubiquitous Computing. Springer London.*, *11*, 213–328.

Mattes, E., & Mattes, F. (2007). *Synthetic performances*. Retrieved on October 4, 2009 from http://www.0100101110101101.org/home/performances

Maturana, H., & Varela, F. (1980). *Autopoiesis and Cognition: the Realization of the Living*. Robert S. Cohen and Marx W. Wartofsky (Eds.), Boston Studies in the Philosophy of Science 42. Dordecht: D. Reidel Publishing

Mayer, J. D., & Geher, G. (1996). Emotional Intelligence and the Identification of Emotion. *Intelligence*, *22*, 89–113. doi:10.1016/S0160-2896(96)90011-2

Mayer, J. D., & Salovey, P. (1995). Emotional Intelligence and the Construction and Regulation of Feelings. *Applied & Preventive Psychology*, *4*, 197–208. doi:10.1016/S0962-1849(05)80058-7

Mayer, J. D., Salovey, P., & Caruso, D. R. (2000). *Models of Emotional Intelligence. Handbook of Intelligence*. Cambridge, UK: Cambridge University Press.

McCaul, K. D., & Malott, J. M. (1984). Distraction and coping with pain. *Psychological Bulletin*, *95*(3), 516–533. doi:10.1037/0033-2909.95.3.516

McCauley, L., Franklin, S., & Bogner, M. (2000). An Emotion-Based "Conscious" Software Agent Architecture. *Affective Interactions*, LNAI, Springer-Verlag. Berlin Heidelberg, 1814, 107-120.

McCrae, R., & John, O. (1992). An Introduction to the Five-Factor Model and its Application. *Journal of Personality*, *60*(2), 175–215. doi:10.1111/j.1467-6494.1992.tb00970.x

McCullough, M. (1996). *Abstracting Craft: The Practiced Digital Hand*. Boston, MA: MIT Press.

McKenzie, J. (1994). Retrieved March 2010 from http://www.bham.wednet.edu/muse.htm

McPhail, T. L. (2006). *Global communication: Theories, stakeholders, and trends* (2nd ed.). Malden, MA: Blackwell.

Mcpherson, M., & Nunes, M. B. (2004). *Developing Innovation in Online Learning: An Action Research Framework* (pp. 46, 47, 54–60). London, UK: RoutledgeFalmer.

Melzack, R. (1990). The tragedy of needless pain. *Scientific American*, *262*(2), 27–33. doi:10.1038/scientificamerican0290-27

Melzack, R., & Wall, P. D. (1996). *The challenge of pain* (2nd ed.). London: Penguin Books.

Mezirow, J. (1991). *Transformative Dimensions of Adult Learning*. Jossey Bass Higher and Adult Education Series.

Mezur, K. (2009, February). *Invisible Intimacies and Cold Burn: Haptic Migrations in 3D Tele-Immersion Choreography*. Paper presented at the College Art Association conference, Los Angeles, CA.

Michigan State University. (2009), *Serious Game Design Program*. Retrieved from: http://woz.commtechlab.msu.edu/sgd/index.php.

Milekic, S. (2002, April). *Towards Tangible Virtualities: Tangialities*. Paper presented at the meeting of Archives & Museum Informatics: Museums and the Web conference, Boston, MA.

Mills, B., & Walker, Wm. (2008). *Memory Work, Archaeologies of Material Practices*. Santa Fe, New Mexico: School for Advanced Research Press.

Mingers, J. (1994). *Self-Producing Systems*. Kluwer Academic/Plenum Publishers.

Ministerio de Educación Nacional. (2004a). *Estándares Básicos de Competencias Ciudadanas. Formar para la Ciudadanía... ¡sí es posible! Lo que necesitamos saber y saber hacer* [Basic Standards of Citizenship Competencies... it is possible! What we Need to Know and Know How to Do]. Bogotá, Colombia: Ministerio de Educación Nacional.

Ministerio de Educación Nacional. (2004b). *Proceedings from Foro Nacional de Competencias Ciudadanas* [National Forum of Citizenship Competencies]. Bogotá, Colombia: Ministerio de Educación Nacional.

Minsky, M. (2000). Commonsense Based Interfaces: To Build a Machine that Truly Learns by Itself will Require a Commonsense Knowledge Representing the Kinds of Things Even A Small Child Already Knows. *Communications of the ACM, 43*(8), 66–73. doi:10.1145/345124.345145

Minsky, M. (1979). *Web of Thoughts and Feelings*. Sound Recording. Cornell University, Ithaca, NY. April 18th.

Miranda, E. R., & Matthias, J. (2009). Music Neurotechnology for Sound Synthesis: Sound Synthesis with Spiking Neuronal Networks. *Leonardo, 42*(5), 439–442. doi:10.1162/leon.2009.42.5.439

MMOX. (2009), Massively Multi-participant Online Games and Applications http://wiki.secondlife.com/wiki/MMOX

Mockus, A. (2004, February-March). Por qué competencias ciudadanas en Colombia? [Why Citizenship Competencies in Colombia?] *Altablero, 27*. Retrieved from: http://www.mineducacion.gov.co/1621/article-87299.html

Moholy-Nagy, L., & Kemény, A. (1922). *The Constructive Dynamic System of Forces: Manifesto of Kinetic Sculpture*. Berlin, Germany: Der Sturm.

Monroe, D. (2009, August). Just for You. *Communications of the ACM, 52*(Issue 8). doi:10.1145/1536616.1536622

Montie Roland. (2008) Product Design Speak 101: Linear versus Iterative Design. Retrieved October 12, 2009, from http://carolinanewswire.com/news/News.cgi?database=pipeline.db&command=viewone&id=8

Mori, M. (1970). The uncanny valley(K.F. MacDorman & T. Minato, Trans.). *Energy, 7*(4), 33–35.

Morris, J. M. (2009). Ontological substance and meaning in live electroacoustic music. In *Computer Music Modeling and Retrieval, Genesis of meaning in sound and music*. New York: Springer. doi:10.1007/978-3-642-02518-1_15

Morris, J. M. (2008). Embracing a mediat[is]ed modernity: An approach to exploring humanity in posthuman music. In *Performance paradigm 4*.

Morris, J. M. (2008). Structure in the dimension of liveness and mediation. In *Leonardo music journal 18: Why live? Performance in the age of digital reproduction*, 59–61.

Munster, A. (2006). *Materializing New Media, Embodiment in Information Aesthetics*. Dartmouth College Press, University Press of New England.

Mura, G. (2006). *The red and black semantics: a fuzzy language in Visual Computer, n.23* (pp. 359–368). Berlin, Germany: Springer.

Mura, G. (2009). *Cyberworld Cybernetic Art Model for Shared Communications in CyberWorlds 2009, Bradford University*. UK: IEEE.

Mura, G. (2007). The Metaplastic Virtual Space. In Wyeld, T. G., Kenderdine, S., & Docherty, M. (Eds.), *Virtual Systems and Multimedia* (pp. 166–178). Berlin, Germany: Springer.

Mura, G. (2005) *Virtual Space Model for Industrial Design,* Unpublished doctoral dissertation, Politecnico di Milano University, Italy

Mura, G. (2006) Conceptual artwork model for virtual environments in *JCIS journal* (ISSN 1553-9105), Vol.3, n.2, pp.461-465, BinaryInfoPress

Mura, G. (2007) The Metaplastic Virtual Space in Wyeld, T.G., Kenderdine,S.,Docherty,M.(Eds) *Virtual Systems and Multimedia* (pp.166-178), Berlin,Germany: Springer

Mura, G. (2008). The meta-plastic cyberspace: A network of semantic virtual worlds. In *ICIWI 2008* WWW/Internet International Conference, Freiburg, IADIS.

Murakami, Y., Sugimoto, Y., & Ishida, T. (2005). Modeling Human Behavior for Virtual Training Systems. *In Proceedings of the 20th National Conference on Artificial intelligence* – Vol. 1. A. Cohn, Ed. Aaai Conference On Artificial Intelligence. AAAI Press, pp. 127-132.

Murray, J. H. (1997). *Hamlet on the Holodeck: the Future of Narrative in Cyberspace*. New York: Free Press.

Murtaugh, M. (2008). *Interaction* in M. Fuller (Ed.), *Software Studies: A Lexicon* (pp. 143-148). Cambridge, MA: MIT Press.

Nelson, T. H. (1965). *A File Structure for the Complex, The Changing and the Indeterminate*. Cambridge, MA: MIT Press.

Nelson, G. (2003). *I Linguaggi dell'Arte. L'esperienza estetica: rappresentazione e simboli, trad. it* (Saggiatore, I., Ed.). Milano, Italy.

Newman, W., & Sproul, R. F. (1987). *Computer Graphics principles*. New York, USA: McGraw Hill.

Nierhaus, G. (2009). *Algorithmic Composition, Paradigms of Automated Music Generation*. Vienna: Springer.

Nomura Sougou Research Lab, Inc. (2007, July). *IT Road Map,* from http://www.nri.co.jp/news/2007/070525.html

Ok, A. (2009). Virtual Worlds Timeline, http://www.dipity.com/xantherus/Virtual_Worlds

Oliver, J. (2008). *Optical Illusion Art as Radical Interface*. Retrieved February 18th, 2009, from http://julianoliver.com.

Olsen, H. F., & Belar, H. (1961). Aid to Music Composition Employing a Random Probability System. *The Journal of the Acoustical Society of America*, *33*, 1163–1170. doi:10.1121/1.1908937

Ören, T., & Yilmaz, L. (2004). Behavioral Anticipation in Agent Simulation. *Winter Simulation Conference,* pp. 801-806.

Ortony, A., Clore, G. L., & Collins, A. (1988). *The Cognitive Structure of Emotions*. New York: Cambridge Universty Press.

Osberg, T. M., Haseley, E. N., & Kamas, M. M. (2008). The MMPI-2 Clinical Scales and Restructured Clinical (RC) Scales: Comparative Psychometric Properties and Relative Diagnostic Efficiency in Young Adults. *Journal of Personality Assessment*, *90*, 81–92.

Pamminger, W. *(2008).* 3delyxe, Transdisciplinary Approaches to Design. In *Serial Machine, Generative Systems* (p. 180). Frame Publishers.

Papert, S. (1980). *Mindstorms: Children, computers, and powerful ideas*. Brighton, England: Harvester Publications.

Papert, S. (1990). Introduction. In Harel, I. (Ed.), *Constructionist learning* (pp. 1–8). Cambridge, MA: Media Laboratory Publication.

Parker, S. P. (Ed.). (2002). *McGraw-Hill Dictionary of Scientific & Technical Terms*. New York, NJ, USA: McGraw-Hill.

Pereira, D., Oliveira, E., Moreira, N., & Sarmento, L. (2005). Towards an Architecture for Emotional BDI Agents. *In EPIA'05: Proceedings of 12th Portuguese Conference on Artificial Intelligence*, pp. 40-47.

Peretti, J. (2001). Contagious Media Project. Retrieved on October 4, 2009 from http://www.contagiousmedia.org/

Perrot, X. (1993). Applications in Museums. In D. Lees (Ed.), *Museums and Interactive Multimedia. Proceedings of the Sixth International Conference of the MDA and the Second International Conference on Hypermedia and Interactivity in Museums (ICHIM '93)*. Cambridge, UK, September 20-24, Archives & Museum Informatics, Pittsburgh, PA, USA.

Perrot, X. (1999). L'avenir du musée à l'heure des médias interactifs. In B. Darras, D. Chateaux (p.149) *Arts et multimédia*. Paris: Publication de la Sorbonne.

Perrot, X., & Mura, G. (2005). VR Workshop 2005 "Virtuality in Arts and Design" Virtual Exhibition Projects. Archives&Museum Informatics. Retrieved from http://www.archimuse.com/publishing/ichim_05.html

Phelan, P. (1993). *Unmarked: The politics of performance*. New York: Routledge. doi:10.4324/9780203359433

Pierce, C., & Weiss, P. (1935). *Collected papers of Charles Sanders Peirce, Cambridge*. Ma, USA: Harvard University Press.

Pinto, H, S, and Martins, J, P. (2004). Ontologies: How can They be Built? *Knowledge and Information Systems*, 6, 441–464. doi:10.1007/s10115-003-0138-1

Popper, K. R., & Eccles, J. C. (1977). *The self and its brain*. New York: Springer International. QBIC (Query by Image Content). Retrieved March 2010 from http://www.hermitagemuseum.org

Poznanski, M., & Thagard, P. (2005). Changing Personalities: Towards Realistic Virtual Characters. *Journal of Experimental & Theoretical Artificial Intelligence*, 17(3), 221–241. doi:10.1080/09528130500112478

PRAMnet. (2009) Journey to Abadyl. Retrieved October 12, 2009, from http://www.pramnet.org/PRAMnet_JourneyToAbadyl_web/JOURNEY_TO_ABADYL.pdf

Prensky, M. (2006). *"Don't bother me Mom, I'm learning!" How Computer and Video Games are preparing your Kids for Twenty-first Century Success and how you can Help!* St. Paul, Minnesota: Paragon House.

Prestinenza Puglisi, L. (2005). *Hyperarchitecture: Spaces in the Electronic Age*. Basel: Birkhauser.

Pridmore, S. (2002). *Managing chronic pain A biopsychosocial approach*. London: Martin Dunitz Ltd.

Przybylski, A. K., Ryan, R. M., & Rigby, C. S. (2009). The Motivating Role of Violence in Video Games. *Personality and Social Psychology Bulletin*, 35(2), 243–259. doi:10.1177/0146167208327216

Ralston, A., & Reilly, E. D. (1983). *Encyclopedia of Computer Science and Engineering*. New York: Van Nostrand Reinhold.

Ramsey, A. (2009) Three Reasons to Start Designing Iteratively Retrieved October 12, 2009, from http://www.andersramsay.com/2009/03/01/three-reasons-to-start-designing-iteratively

Reilly, E. D. (2004). *Concise Encyclopedia of Computer Science*. Hoboken, NJ, USA: Wiley&Sons.

Reyes, E. (2009). "Hypermedia as Media" in 20th. *ACM Conference on Hypertext and Hypermedia*. New York: ACM Press.

Reyes, E., & Zreik, K. (2008). "A Social Hyperdimension of Media Art" in Zreik & Reyes (eds.) *Proceedings of the Second International Congress Computer Art Congress 2008 [CAC.2]*. Paris: Europia.

Richards, C. (2009). *Artist Statement* Retrieved December 30, 2009 from: http://www.catherinerichards.ca

Riemschneider, B., & Grosenick, U. (Eds.). (1999). *Art at the turn of the millennium*. Los Angeles: Taschen.

Ringbom, S. (1970). *The Sounding Cosmos: A study of the Spiritualism of Kandinsky and the Genesis of Abstract Painting*. Acta Academiae Aboensis.

Ringbom, S. (1986). *Transcending the Visible: The Generation of the Abstract Pioneers*. In M. Tuchman, *The Spiritual in Art: Abstract Painting 1890-1985* (p. 137). Los Angeles: Abbeville Press Publishers.

Rizzo, P., Veloso, M., Miceli, M., & Cesta, A. (1997). Personality-Driven Social Behaviors in Believable Agents. *AAAI, Fall Symposium on Socially Intelligent Agents*.

Robillard, G., Bouchard, S., Fournier, T., & Renaud, P. (2003). Anxiety and presence during VR immersion: A comparative study of the reactions of phobic and non-phobic participants in therapeutic virtual environments derived from computer games. *Cyberpsychology & Behavior*, *6*(5), 467–476. doi:10.1089/109493103769710497

Rollings, A., & Adams, E. (2003). *Andrew Rollings and Ernest Adams on Game Design*. Indianapolis: New Riders.

Ropars-Wuilleumier, M. C. (2002). *Écrire l'espace. St Denis: Presses Universitaires de Vincennes*. Translation by Carola Moujan.

Rosch, E. (1978) Principles of categorization, in Rosch E. and Lloyd B.B.(eds) *Cognition and categorization*, (pp.27-48):New York,NJ,USA: Erlbaum, Hillsdale

Roseman, I. J., Jose, P. E., & Spindel, M. S. (1990). Appraisals of Emotion-Eliciting Events: Testing a Theory of Discrete Emotions. *Journal of Personality and Social Psychology*, *59*(2), 899–915. doi:10.1037/0022-3514.59.5.899

Rothbaum, B., Hodges, L., Smith, S., Lee, J. H., & Price, L. (2000). A controlled study of virtual reality exposure therapy for the fear of flying. *Journal of Consulting and Clinical Psychology*, *68*(6), 1020–1026. doi:10.1037/0022-006X.68.6.1020

Rousseau, D., & Hayes-Roth, B. (1997). A Social-Psychological Model for Synthetic Actors. *Technical Report KSL-9707*, Knowledge Systems Laboratory, Stanford University.

Russo, C., & Brose, W. (1998). Chronic pain. *Annual Review of Medicine*, *49*, 123–133. doi:10.1146/annurev.med.49.1.123

Salovey, P., & Mayer, J. D. (1990). Emotional Intelligence. *Imagination, Cognition and Personality*, *9*, 185–211.

Salovey, P., & Sluyter, D. J. (1997). *Emotional Development and Emotional Intelligence*. New York: Basic Books.

Sanchez Buron, A., Rodríguez, L., & Fernández Marín, M. P. (2009). *Los Adolescentes en la Red. Estudio sobre sus Hábitos de Comportamiento en Internet* [Adolescents in the Net. Study about their Behavior Habits in Internet]. Retrieved from http://www.ucjc.edu

Schacter, D. L. (1989). On the Relation Between Memory and Consciousness: Dissociable Interactions and Conscious Experience. In Roediger, H., & Craik, F. (Eds.), *Varieties of Memory and Consciousness: Essays in Honour of Endel Tulving* (pp. 22–35). Hillsdale, NJ: Erlbaum.

Schatman, M., & Campbell, A. (2007). *Chronic pain management: Guidelines for multidisciplinary program development*. New York: Informa Healthcare.

Schiphorst, T. (2009). *Artist Statement* Retrieved January 15, 2010 from: www.siat.sfu.ca/faculty/thecla-shiphorst/

Schleiner, A.-M., Leandre, J., & Condon, B. (2003). *Velvet-strike*. Retrieved on October 4, 2009 from http://www.opensorcery.net/velvet-strike

Schmidt, B. (2002). How to Give Agents a Personality. *In Proceeding of third Workshop on Agent-Based Simulation*, Passau, Germany.

Schroeder, R. (2008) Defining Virtual Worlds and Virtual Environments. In *Virtual Worlds Research: Past, Present & Future*. 1(1). 2-3. ISSN: 1941-8477. http://journals.tdl.org/jvwr/article/viewFile/294/248

Seaman, W. (1999). *Recombinant poetics: Emerging meaning as examined and explored within a specific generative virtual environment*. Unpublished doctoral dissertation, Centre for Advanced Inquiry in the Interactive Arts, University of Wales.

Semrud-Clikeman, M. (2007). *Social Competence in Children*. New York, NY: Springer.

Senior, J. (1959). The Way Down and Out: The Occult in Symbolist Literature. In Tuchman, M. (Ed.), *The Spiritual in Art* (pp. 39–41). New York: Greenwood Press.

Shanahan, M. (1999). The Event Calculus Explained. *Springer Verlag, LNAI, 1600*, 409–430.

Shaw, C., Gromala, D., & Fleming Seay, A. (2007). The Meditation Chamber: Enacting autonomic senses. *Proc. of ENACTIVE/07, 4th International Conference on Enactive Interfaces*, Grenoble, France, 19-22 November 2007, 405-408.

Sheared, V., Sissel, P. A., & Cunningham, P. M. (2001). *Making Space: Merging Theory and Practice in Adult Education* (p. 250). Westport, CT: Bergin and Garvey.

Sherman, W. R., & Craig, A. (2002). *Understanding Virtual Reality: Interface Application, and Design*. San Francisco, CA: Morgan Kaufmann Publishers.

Shreve, J. (2009). *All the Labs a Stage*: Publicity release on the website for the Center for Information Technology Research in the Interest of Society (CITRIS), Retrieved November 14, 2009, from: http://ucberkeley.citris-uc.org/publications/article/all_lab%2526%2523039%3Bs_stage

Simms, L. J., Casillas, A., Clark, L. A., Watson, D., & Doebbeling, B. I. (2005). Psychometric Evaluation of the Restructured Clinical Scales of the MMPI-2. *Psychological Assessment, 17*, 345–358. doi:10.1037/1040-3590.17.3.345

Singer, D., & Singer, J. (2005). *Imagination and Play in the Electronic Age*. Cambridge, MA: Harvard University Press.

Slater, M., Pertaub, D., & Steed, A. (1999). Public speaking in virtual reality: Facing an audience of avatars. [USA.]. *IEEE Computer Graphics and Applications, 19*.

Smirnov, A., Shilov, N., Levashova, T., Sheremetov, L., & Contreras, M. (2007). Ontology-driven intelligent service for configuration support in networked organizations. *Knowledge and Information Systems Springer London., 12*, 229–253. doi:10.1007/s10115-007-0067-5

SommererC.MignonneauL. (2009) from http://www.interface.ufg.ac.at/christa-laurent/

Song, M. (2009). *Virtual reality for cultural heritage applications*. Saarbrücken: VDM-Verlag Dr. Muller.

Soto, M., Allongue, S., Pierre, U., Curie, M., & Jussieu, P. (1997). A Semantic Approach of Virtual Worlds Interoperability. *Proc. IEEE IEEE International Workshops on Enabling Technologies: Infrastructure for Collaborative Enterprises WET-ICE '97* (pp. 173-178). IEEE Press.

Steele, E., Grimmer, K., Thomas, B., Mulley, B., Fulton, I., & Hoffman, H. (2003). Virtual reality as a pediatric pain modulation technique: A case study. *Cyberpsychology & Behavior, 6*(6), 633–638. doi:10.1089/109493103322725405

Steffe, L. P. (Ed.). Gale, J. (Ed), (1995) Constructivism in Education. (pg: 351). Lawrence Erlbaum Associates. Hillsdale, NJ.

Steuer, J. (1992). Defining virtual reality: Dimensions determining Telepresence. *The Journal of Communication, 42*.

Suler, J. (2007). "The Psychology of Avatars and Graphical Space in Multimedia Chat Communities", Retrived on 05/05/2008 The Psychology of Cyberspace, www.rider.edu/suler/psycyber/basicfeat.html (article orig. pub. 1996)

Sutherland, I. (1965). *The ultimate Display*. Paper presented at the IFIP Congress, *65, 506-508,* Information Processing.

Tellegen, A., Ben-Porath, Y., Arbisi, P., Graham, J., & Kaemmer, B. (2003). *The MMPI-2 Restructured Clinical Scales: Development, Validation, and Interpretation*. Minneapolis, MN, USA: University of Minnesota Press.

The Ensemble. (2009). http://www.the-ensemble.com/content/netflix-prize-movie-similarity-visualization

The Museum of Modern Art. New York. (1999). *Jackson Pollock, New Approcahes*. New York: The Museum of Modern Art, New York.

ThinkBalm. (2009). ThinkBalm Data Garden. http://www.youtube.com/watch?v=-GBxafGzZsY

Thürlemann, F. (1982) *Paul Klee. Analyse sémiotique de trois peintures.* Lausanne,Suisse: L'Age d'Homme

Tinguely, J. (1982). *Catalogue Raisonné, Sculptures and Reliefs 1954-1968.* Zurich, Switzerland: Edition Galerie Bruno Bischofberger.

Tisseron, S. (2003). *Le bonheur dans l'image.* Paris: Les Empêcheurs de penser en rond/Le Seuil. Translation by Carola Moujan.

Transforming Pain Research Group (2010). Retrieved April 01 2010 from http://www.transformingpain.org.

Truckenbrod, J. (1992). Integrated Creativity, Transcending the Boundaries of Visual Art, Music & Literature. *Leonardo Music Journal, 2*(1), 89–95. doi:10.2307/1513214

Truckenbrod, J. (2008). *Digital Raw Materials*, in Imaging in the Digital Fall 2008 edition of *Media-N Journal of the New Media Caucus*,http://www.newmediacaucus.org/journal/issues.php?f=papers&time=2008_winter&page=truckenbrod

Tsichritzis, D., & Gibbs, S. (1991, October). Virtual Museums and Virtual Realities. In *Proc. Intl. Conf. on Interactivity and Hypermedia in Museums,* Pittsburgh, USA.

Tuchman, M. (1986). *Hidden Meaning in Abstract Art.* In M. Tuchman, *The Spiritual in Art: Abstract Painting 1890-1985* (p. 19). Los Angeles: Abbeville Press Publishers.

Turk, D. C., & Nash, J. M. (1993). Chronic pain: New ways to cope. In Goleman, D., & Gurin, J. (Eds.), *Mind body medicine.* New York, NY: Consumer Reports Books.

Turkle, S. (2008). *Falling for Science, Objects in Mind.* Boston: MIT Press.

Ushiba,J. (2009, February).*Operation through the Characters of the Virtual World Second Life*, The Brain Science to Open the Future (No. 5) Technology Challenging the Interaction between the Brain and the Environment (2), Bio-industry 26 (2) (299), (pp. 89~93), CMC Publishing.

Valéry, P. (1928). La Conquete de l'ubiquite. In *De La Musique avant toute chose.* Paris: Editions du Tambourinaire.

Varela, F., Thompson, E., & Rosch, E. (1992). *The embodied mind: Cognitive science and human experience.* Cambridge, MA: MIT Press.

Vélasquez, J. D. (1997). Modeling Emotions and Other Motivations in Synthetic Agent. *In Proceedings of the AAAI Conference,* AAAI Press and the MIT Press.

Villanueva, L. (2009). Diffuse Noxious Inhibitory Control (DNIC) as a tool for exploring dysfunction of endogenous pain modulatory systems. *Pain, 143*(3), 161–162. doi:10.1016/j.pain.2009.03.003

Watanave, H. (2008, December). *"SIGGRAPH Asia Archive in Second Life".* Artwork exhibited in SIGGRAPH ASIA 2008 Art Gallery (curated works). Suntec Singapore International Convention & Exhibition Centre. Singapore.

Watanave, H. (2008, November). *"Contents Oriented Space"* ACM International Conference Proceeding Series; Vol. 352. Proceedings of the 2008 International Conference on Advances in Computer Entertainment Technology. Posters. (pp. 418-418), ACM.

Watanave,H. (2008, August). *"3D image database of Oscar Niemeyer"- NIPPAKU 100 (Memorial of Japan and Brazil exchange year) Official Project,* Artwork exhibited in FILE - Electronic Language International Festival in Sao Paulo 2008. SESI' Cultural Centre. Sao Paulo. Brazil.

Watanave,H. (2009, April). *"Laval VRchive",* Laval Virtual France Virtual Reality International Converence 2009 Proceedings (pp. 397-398), Laval Virtual.

Watanave,H. (2009, April), *"Archidemo - Architecture in Metaverse",* Artwork exhibited in 9th edition of FILE - Electronic Language International Festival in Rio de Janeiro, in Media Art category, 2009

Weiner, I. B., Freedheim, D. K., Schinka, J. A., Gallagher, M., Nelson, R. J., & Velicer, W. F. (2003). *Handbook of Psychology.* Hoboken, NJ: John Wiley and Sons. doi:10.1002/0471264385

Weinhart, M. (2007). In The Eye of The Beholder. In *Op Art (*pp. 18-38). Catalogue of the exhibition held at the Schrim Kunsthalle, Francfort, february 17th, 2007 - may 20th, 2007. Köln: Walter König.

Weiser, M. (2001). The Computer for the Twenty-First Century, *Scientific American*, September 1991, 94-100. Retrieved from http://www.ubiq.com/hypertext/weiser/SciAmDraft3.html.

Weiser, M., & Brown, J. S. (1996). The Coming Age of Calm Technology. Revised version of *Designing Calm Technology*. PowerGrid Journal, v 1.01, july 1996. Retrieved March 10, 2009. Website: http://www.ubiq.com/hypertext/weiser/acmfuture2endnote.htm.

Whorf, B. L., & Carroll, J. B. (1964). *Language, Thought and Reality: selected writings*. Cambidge, MA, USA: MIT Press.

Whyte, J. (2002). *Virtual Reality and the Built Environment*. USA: Architectural Press.

Wiener, N. (1948). *Cybernetics or Control and Communication in the Animal and the Machine*. New York, Cambridge, MA, USA: MIT Press.

Wiener, N. (1988). *The Human Use of Human Beings: Cybernetics and Society*. Cambridge, MA, USA: Da Capo Press.

Williams, N., Wilkinson, C., Stott, N., & Menkes, D. B. (2008). Functional illness in primary care: Dysfunction versus disease. *BMC Family Practice*, *9*(30).

Winkler, T. (2004). *GPS:Tron*. Retrieved on October 4, 2009 from http://gps-tron.datenmafia.org/?page_id=2

Wolfram, S. (2002). *A new kind of science*. Champaign, IL: Wolfram Media.

Xu J, Yang Zhao Y., Chen, Z. and Liu, Z. (2009) Music snippet extraction via melody-based repeated pattern discovery. *Science in China Series F: Information Sciences*. Science in China Press, co-published with Springer-Verlag GmbH. 52(5): 804-812.

Yee, N., & Bailenson, J. N. (2007). The Proteus Effect: The Effect of Transformed Self-Representation on Behavior. (271-290). *Human Communication Research*, 33.

Young, P.T.(1967) Affective Arousal in *American psychologist*, n.22, pp.32-40

Zadeh, L. (1969). *Biological application of the theory of fuzzy sets and systems in the proceedings of an international symposium on BioCybernetics of the central nervous system* (pp. 199–206). Boston: Little Brown.

Zegher, C. d. (2005). *3 x abstraction: New Methods of Drawing, Hilma Klint, Emma Kunz, Agnes Martin 1862-1944*. The Drawing Center and Yale University Press.

Zimmerman, E. (2003) Play as Research: The Iterative Design Process. Retrieved October 12, 2009, from http://www.ericzimmerman.com/texts/Iterative_Design.htm

About the Contributors

Gianluca Mura is a transdisciplinary media researcher, architect, digital artist and designer. Researcher at the Politecnico di Milano University and Universidade Catolica Portuguesa, Porto. He holds a PhD in Industrial Design and Multimedia Communication from the Politecnico di Milano University. His research area is within interrelations among Design, Architecture, Art, Science, Technology. Founder and Editor of the International Scientific Journal of Art, Culture and Design Technologies (IJACDT), IGI Global. He is a member of several International Scientific Committees: CAE Computational Aesthetics, ARTECH, CYBERWORLDS, WEBSTUDIES, IEEE Italian Committee, MIMOS (Italian Movement on Modeling and Simulation). He organized the International Workshop "Virtuality in Arts and Design" between Politecnico di Milano and Ecole du Louvre, Paris with the High Patronage of Italian Ministry of Foreign Affairs, Rome, Italy. His recent digital artwork "The Metaplastic Constructor" has been exhibited on the Museum of Modern Art of Touluca, Mexico and into the Rhizome Artbase digital art collection.

* * *

Elif Ayiter, aka Alpha Auer is an artist, designer and researcher specializing in the development and implementation of hybrid educational methodologies between art & design and computer science, teaching full time at Sabanci University, Istanbul, Turkey. She has presented creative as well as research output at conferences including Siggraph, Consciousness Reframed, Creativity and Cognition, ISEA, ICALT, Computational Aesthetics (Eurographics) and Cyberworlds. She is also the chief editor of the forthcoming journal Metaverse Creativity with Intellect Journals, UK and is currently studying for a doctoral degree at the Planetary Collegium, CAiiA hub, at the University of Plymouth with Roy Ascott.

Paul Catanese is a hybrid media artist, an Associate Professor in the Department of Interdisciplinary Arts at Columbia College Chicago and the President of the New Media Caucus, a College Art Association Affiliate Society. His artwork has been exhibited widely including at the Whitney Museum of American Art, the New Museum of Contemporary Art, SFMOMA Artist's Gallery, La Villette-Numerique and Stuttgarter Filmwinter among others. Paul is the recipient of numerous grants and awards, including commissions for the creation of new artwork from Turbulence.org as well as Rhizome.org.

Ina Conradi, Assist.Prof, received M.F.A. in art in 1989 from University of California at Los Angeles. She is a member of Singapore-China Association for Advancement of Science and Technology (SCAAST), Union of Slovene Fine Art Arts Association (ZDSLU), and is Japan Foundation fellow. She has been teaching in the School of Art Design and Media at Nanyang Technological University Singapore

About the Contributors

since 2007. Her research explores experimental and immersive abstract computer animation, responsive and reactive painted surfaces with imagery integrating experimental 3d animation, oversized image creation using algorithmic paint strokes, high-resolution computer rendering technique, advanced print prototyping and finishing technique.

Denise Doyle is a Researcher, Artist Curator, and Senior Lecturer in Digital Media at the University of Wolverhampton. With a background in Fine Art Painting (BA Hons) from Winchester School of Art, and Design and Digital Media (MA), she is completing her PhD research at SMARTlab Digital Media Institute, University of East London, under the directorship of Professor Lizbeth Goodman. Denise's research investigates the Artist's experience of the Imaginary in Virtual Worlds, and she is developing a framework for a new theory of the Imagination that incorporates experiences of mediated spaces created through interdisciplinary practice in Art and Technology.

Adel S. Elmaghraby is Professor and Chair of the Computer Engineering and Computer Science Department at the University of Louisville. He has also held appointments at the Software Engineering Institute - Carnegie-Mellon University, and the University of Wisconsin-Madison. He advised approximately 60 master's graduates and 20 doctoral graduates. His research contributions and consulting spans the areas of Intelligent Multimedia Systems, Networks, HPC, Visualization, and Simulation. He is a well-published author (over 200 publications), a public speaker, member of editorial boards, and technical reviewer. He has been recognized for his achievements by several professional organizations including a Golden Core Membership Award by the IEEE Computer Society. He is a senior member of the IEEE, a member of ACM and ISCA. He served a term as an elected ISCA Board member and currently is a Senior Member and an Associate editor for ISCA Journal.

Paula Andrea Escandón is industrial designer from Universidad Autónoma de Manizales, in Colombia. She is finishing her Master of Interactive Design and Creation at Universidad de Caldas, in Manizales, Colombia. She is adjunct professor in assisted design at Universidad Nacional de Colombia at Manizales. She is also research assistant in the research group DICOVI - Design and cognition in visual and virtual environments and her research interests are interaction design issues.

Victor Fernández Cervantes is Computer Engineer from the University Center of the Cienega (CUCI) at the University of Guadalajara in Mexico. He received his Master in Sciences degree in 2009 from the Center for Research and Advanced Studies of National Polytechnic Institute (CINVESTAV of I.P.N.) in Guadalajara, Mexico. He is a PhD Student at the CINVESTAV of I.P.N. in Guadalajara, Mexico. His research interests include Virtual Reality, Human Behavior Simulation, Human-Computer Interaction Systems, Multi-Agent Systems, Distributed Systems and Computer Vision.

J. Octavio Gutierrez-Garcia received his PhD in Electrical Engineering and Computer Science from both CINVESTAV, Mexico and Grenoble Institute of Technology, France, in 2009. Since then, he has been a computer science lecturer and technical editor of several computer science related journals, as well as a program committee member of various international conferences. His research interests focus on multi-agent systems.

About the Contributors

Diane Gromala is the Canada Research Chair at Simon Fraser University's School of Interactive Art & Technology. Her work has been at the forefront of emerging forms of technology, from the earliest multimedia application to one of the very first instances of Virtual Reality. Gromala heads the Transforming Pain Research Group, whose research focuses on innovative uses of technologies for managing chronic pain. Gromala is co-author of MITPress' Windows and Mirrors: Electronic Art, Design, and the Myth of Transparency, with Jay David Bolter.

Dew Harrison is a researcher and practitioner in virtual and computer-mediated art currently working as Associate Dean in the School of Art & Design at the University of Wolverhampton. With a BA in Fine Art, an MSc in Computer Science and a PhD from the Planetary Collegium, CAiiA, in Interactive Art, her practice undertakes a critical exploration of Conceptual Art, semantic media and intuitive interfaces where she often work's collaboratively and considers virtual curation an art practice.She continues to show internationally and has over 40 publications to date spanning digital art, consciousness studies, interactive games, art history and museology.

Michael Johansson, Artist, Lecturer, Researcher. Born 1962, Gothenburg, Sweden. Educated at the Royal College of Fine Arts in Copenhagen 1984 -1990. Worked with digital media as part of my work practice for over 25 years. Done about 50 exhibitions both in Sweden and abroad since 1999. Been involved in research, first at the interactive institute, space and virtuality studio, and between 1998-2007 at Malmo University arts and communication. planned and taught, design, animation, architectural visualization, interaction design, pervasive gaming and digital prototyping on the master level. In 2005 one of the founders of the research network PRAMNET.

Jocelyne Kiss iscurrently Associate Professor at University Paris East (France) in Numerical Art and Multimedia, her research focuses on exploitation of connectionism methods within artistic immersive-interactive interfaces. Her research interests lie in developing concepts and computer games in the area of synaesthetic sensations. Author of five interactive installations, especially "sing with me, avatar singer". She published "composition and cognitive sciences", about thirty papers about art & technology.

Martin Laliberté composer, Musicologist, Professor at U. of Paris-Est. Was born in Quebec in 1963. He studied musical composition and computer music in Quebec, California and France, 1982-1990. After working freelance as a composer in Hollywood and in Quebec, he moved to Paris in 1988 where he finished a PhD at IRCAM in 1994. Appointed Professor in 1995, he teaches composition, music theory and history. His research centers on the aesthetics of contemporary music and the emergence of electronic and computer musical instruments.

Felipe César Londoño is an architect and PhD in Multimedia Engineering from the Universitat Politècnica de Catalunya, in Spain. He is professor in visual design and experimental research in art, design and new media, and director of the PhD in Design at Universidad de Caldas in Manizales, Colombia, where he co-founded the Department of Visual Design. He is also participant professor of the PhD in Multimedia Engineering at Universitat Politècnica de Catalunya. He is head of the research group DICOVI - Design and cognition in visual and virtual environments and director of the International Image Festival.

About the Contributors

Diane J. Love is a systems engineer staff in Lockheed Martin's Information Systems and Global Services, specializing in usability. She graduated from the University of Edinburgh with a Bachelor of Sciences in Physics, and Imperial College, University of London, with a PhD in Physics. Her professional interests include human-computer interaction, virtual worlds and collaboration, and she is a Certified Human Factors Engineering Professional (CHFEP). During her career in Lockheed Martin, Diane has worked on air traffic control systems in the UK and USA. She previously worked for Logica UK on a variety of systems for UK government.

Germán Mauricio Mejía is industrial designer from Universidad Autónoma de Manizales, in Colombia. With a Fulbright grant, he is finishing his Master of Design at University of Cincinnati, in United States. He is assistant professor in digital design at Universidad de Caldas, in Manizales, Colombia. He is also research fellow of the group DICOVI - Design and cognition in visual and virtual environments and his research interests are visual design for social empowerment and the role of human differences in the interaction with visual information.

Jeffrey M. Morris is an Instructional Assistant Professor in computer music and coordinator of technology facilities for the Department of Performance Studies at Texas A&M University. Dr. Morris gives improvised performances with interactive electronics in addition to composing for traditional instruments and electronic media. His creative works have been performed internationally and include intermedia works and collaborations with dance artists. His scholarly writings explore mediation by technology in live performance and its implications for the human experience.

Carola Moujan Designer & Artist. Lives and works in Paris at the Université Paris 1 – Pantheon-Sorbonne. She was born in 1969 in Montevideo (Uruguay). As an architecture and fine arts student, she was awarded in 1991 with a study and travel grant from the University of Minnesota, where she took part in early virtual reality and 3D-modeling research. In 1995 she moved to Paris and begun a career as a graphic designer & art director for magazines and design studios. From 1997 on, her work shifted to the emerging field of interaction design. She started her own design studio in 2000, focusing on global brand design with a strong emphasis on interactive media, while carrying on personal work around the notion of virtual and hybrid environments. Her interactive installation project "Jour d'angoisse" was awarded with the SCAM New Media Grant (Bourse d'aide à l'art numérique) in 2008. From 2006 on, she pursues theoretical and practice-based research on ubiquitous computing.

Catherine Nyeki, Pluridisciplinary artist,developed her research for several years around the concept of virtual biology in the field of digital arts, music and song. She has taken part in many individual and collective exhibits in France and abroad and has received several awards. 2010 "Plastika" Cité des Sciences de La Villette, 2009 Espace d'Art Contemporain Camille Lambert, Videoformes, 2008 eArts Shanghai, 2007 Ars Electronica Linz, Kiev, Talents video, Slick, 2006 Galerie Fraich'attitude, 2005 Galerie Hors Sol, 2004 ICHIM Berlin, Art-Metz, Festival Nemo, 2003 WRO International Media Art Biennale, 2000 @rt Outsiders, ISEA 10ème symposium international arts électroniques, Paris, 1998, Galerie Donguy.

Héctor Rafael Orozco Aguirre is Computer Engineer from the University Center for Science and Engineering (CUCEI) at the University of Guadalajara in Mexico. He is Professor of Computer Sciences at the CUCEI at the University of Guadalajara in Mexico. He received his Master in Sciences degree in 2006 from the Center for Research and Advanced Studies of National Polytechnic Institute (CINVESTAV of I.P.N.) in Guadalajara, Mexico. He is a PhD Student at the CINVESTAV of I.P.N. in Guadalajara, Mexico. His research interests include Virtual Reality, Human Behavior Simulation, Human-Computer Interaction Systems, Multi-Agent Systems, Distributed Systems, Robotics and Computer Vision.

Félix Francisco Ramos Corchado, received his PhD in Computer Sciences from Technology University Of Compiègne France in 1997, a MSc. In Computer Science from the Centre of Advanced Researches of IPN in Mexico, a MSc. in Distributed Systems from the CNAM in Paris in 1994. Since 1997 he works as full time Professor in the Computer and Electric Engineering Department of the Centre of Advanced Researches in Guadalajara CINVESTAV. He has been working on projects on Distributed Virtual Reality, Multi-agent Systems, Distributed Systems and Self-organization. His main research topics are Multi-agent systems, co-operative work and Networks.

Everardo Reyes-García received his doctorate in Information and Communication Sciences at the University of Paris VIII, France. Since 2003 is researcher member of the CiTu group at Paragraphe Lab. From 2007 is program director of the BA in Animation and Digital Art at Monterrey Tech, Toluca, Mexico. He teaches, lectures, and publishes extensively on design, production, usages, sociality and epistemology of hypermedia, media art, and digital media. He was distinguished as member of the National System of Researchers, Conacyt, Mexico, in 2008.

Chris Shaw is an Associate Professor in the School of Interactive Arts & Technology at Simon Fraser University in Surrey, British Columbia, Canada. His research interests are in Virtual Environments and Visual Analytics. Prior to joining SFU in 2005, Shaw held a faculty position at the College of Computing at the Georgia Institute of Technology. Shaw Received his PhD in Computing Science from the University of Alberta in 1997.

Vadim Slavin is a senior software engineer, a principal investigator, and the head of Human-Systems Interaction group within Modeling Simulation and Information Sciences Department at Lockheed Martin Space Systems' Advanced Technology Center. He graduated from Brown University with a BS in Math-Physics, a BA in Computer Science, and a MS in Computer Science. Having joined Lockheed Martin, Vadim has worked on a number of modeling and simulation projects for space industry customers as well as premier government research organizations. His research interests include human-computer interaction, immersive collaborative visualization environments, and semantic technologies for reasoning under uncertainty. Vadim serves on the board of several internal Lockheed Martin-wide working groups and has participated in NASA proposal reviews as a lead reviewer.

Meehae Song is currently a PhD student at Simon Fraser University's School of Interactive Art & Technology. She has a BSc in Computer Science from Ewha Womans University, South Korea and a M.Eng. in Virtual Reality applications from Nanyang Technological University, Singapore. She worked for the Centre for Advanced Media Technology (CAMTech), a joint research and development centre between NTU and Fraunhofer IGD for 6 years specializing in various immersive VR applications. Her current interests lie in applying Art Theory and looking at how abstractions and connotations/denotations affect the experience in immersive and mixed Virtual Reality environments.

About the Contributors

Sidi Soueina was born in North Africa, educated in the middle east and Japan. His doctorate AI/Intelligent agents. Sidi worked as Research Engineer in Sydney Australia for few years then moved to the US where he currently teaches and does research. His research interest are Personality Based Recognition Tools and Creativity.

Olga Sourina received her MSc in Computer Engineering from Moscow Engineering Physics Institute (MEPhI) in 1983, and her PhD in Computer Science from NTU in 1998. Dr. Sourina was awarded of the honorary diploma of the Academy of Sciences of USSR, the Silver Medal of the National Exhibition Centre of USSR, and the Medal of the Ministry of Education of USSR. Since 1 December 2001 she is Assistant Professor in NTU. Her research interests are in interactive digital media IDM (particularly in visual data mining, virtual reality, and visual and haptic interfaces), and Biomedical Engineering (visual analysis and quantification of brain responses, virtual surgery).She collaborates with SBS, MAE, SCE, and medical doctors from SGH (Singapore General Hospital),NUH (National University Hospital),and NNI (National Neuroscience Institute). She is a member of program committee of international journal and conferences of: Cyberworlds, Journal "Computer Graphics & Geometry", IEEE Computer Society and Biomedical Engineering Society.

Daniel Thalmann is Director of EPFL VRlab, Switzerland. He is a pioneer in Virtual Humans. He is coeditor-in-chief of the Journal of Computer Animation and Virtual Worlds. He has published numerous papers in Graphics, Animation and VR. He is coauthor of several books including" Crowd Simulation" (2007). He received his PhD in CS from University of Geneva and an Honorary Doctorate from University Paul-Sabatier in Toulouse, France.

Joan Truckenbrod exhibits her artwork internationally in Chicago, New York, London, Paris and Berlin. An article about her video sculpture was featured in the September 2007 issue of SCULPTURE magazine. ARTnews has featured a Review of one of Truckenbrod's exhibitions. A book about her artwork titled ³Portfolio: Joan Truckenbrod² has been published by Telos Publishing. Ms.Truckenbrod published one of the early books about computer art titled Creative Computer Imaging in 1988 (Prentice Hall). She is a Professor in the Art and Technology Department at The School of the Art Institute of Chicago.

Hidenori Watanave is researching the arts in the Virtual Worlds. He is interested in collaborative work in the realms of architecture and environmental design in tele-existence. Spatial design in the Virtual Worlds was established through the Archidemo project (2007-2008), which was selected to be part of FILE2008, SIGGRAPH 2008, SIGGRAPH ASIA 2009 and the 13th Japan Media Arts Festival. His current experiment focuses GPS and GIS, using techniques like those developed by Hidenori in the NetAIBO project (2004-2005, Honorary Mention, Prix Ars Electronica) and the ObaMcCain project (2008) of 3Di-chatterbots-space, which was exhibited in Mission Accomplished at the Location One gallery, New York.

Index

Symbols

2D canvas 223
3D animation 268
3D cloud 208
3D environment 138, 199
3D imensional sets 223
3D interactive facial animated avatar 158
3D models 90, 253
3D object 90, 93
3D programming 5
3D sculptures 232
3D tools 230
3D virtual world 185, 240, 241, 242, 243, 244, 245, 246, 248, 249

A

Abadyl 251, 252, 253, 254, 256, 257, 258, 259, 260, 261, 262, 263, 265
abstract art 218, 219, 221, 227
abstract figure 38
abstract imagery 218, 221
abstract painting 218, 221
Academic Research Fund (AcRF) 218
active learning 268, 270
actorialization 46
acute pain 126, 132
aggressive behavior 267, 268, 269, 276
algorithm 217, 219, 220, 227
animated articulation 160
animated avatar 158
antisocial 268, 269, 277
archidemo 241, 246, 249, 250
architectonic manifestations 253
architectural space 240, 241, 242, 243, 244, 245, 246, 247
architectural spatial design 241
armed conflict 266, 267
arrhythmic machines 29
Artifacts of Touch 63
artificial agent 99
artificial context 40
artificial environment 40, 41, 43
artificial graft 239
artificial intelligence 98, 99, 137, 138, 154, 155, 203
artificial neural network 158, 160
artistic activity 193
artistic choice 160
artistic happenings 171
artistic languages 168
artistic material 258
artistic value 211
augmented reality 178, 181
Aura 120
autonomous cells 234
autonomous evolution 158
avatar 70, 71, 79, 91, 92, 93, 94, 96, 98, 99, 158, 159, 160, 161, 163, 164, 165, 182, 183, 185, 188, 190, 191, 193, 194, 196, 197, 198, 199, 200, 201, 205, 206, 207, 208, 209, 210, 212, 213, 214, 216, 242, 250, 273, 274, 276, 277
avatar engine 159

B

Baroque 167, 168, 169, 171, 172, 177, 179, 181
Baroque aesthetics 169
Baroque architecture 167, 171, 172, 173
Baroque art 179

Baroque cities 178
Baroque spaces 171
behavioral effects 266
biochemistry 229
biofeedback 121, 122, 123, 125, 127, 128, 129
biologic model 159
biology 188, 203, 228, 229, 230, 235, 237
biopsychosocial approach 122, 129, 130, 131
biotechnology 228, 231, 235
black boxes 114
blogs 104, 107
body-as-data 67
body graft 228
body-in-code 192
brain-body connection 110
Bridging Virtual to Real Spaces 215
bud cell 228
built environment 261, 264

C

Carmen San Diego 271
Cartesian space 253, 254
Cave Automatic Virtual Environment (CAVE) 7
celibate machines 30
cell biology 228
cell division 236, 237
Cellula 231
cellular buds 229, 236
Center for Information Technology Research in the Interest of Society (CITRIS) 67, 76
centric architecture 271
Cerberus 158, 159, 160, 161, 162, 164
Cerberus engine 159, 162
CERN (European Council of Nuclear Research) 4
chalk and talk 93
chaotic brain plans 229
chatbot 96, 106
chat log 275
chronic pain 121, 122, 125, 126, 127, 128, 129, 130, 131, 132
Citizenship and Social Competencies 280
citizenship competence 266, 267, 268, 270, 272, 275, 276, 277, 278
City of Abadyl 251, 252, 253, 254, 256, 258, 262, 263

city planning 246, 261
Civia 266, 270, 272, 273, 275, 276, 277, 278
Civia project 270
civilization 271
coarse materials 217
co-creation 265
cognitive development 187, 197
collaboration 251, 257
collaborative creativity 257
collective cinema 84
color attraction 86, 87
communication 105, 110, 118, 119
communicative learning 187
computational environment 189
computer art 2, 204
computer-based media 82
computer data 82
computer-enhanced 83
computer games 185, 197
computer graphics 2, 10
computer graphics imagery (CGI) 78, 82
computer-mediated 83, 88, 197, 204, 263
conceptual art 204
conceptual design 271, 272
concretization 80
constructionism 182, 187, 196
constructive competition 187
constructivist educational theory 186
constructivist theories 187
contemporary art 204, 206
contents oriented space 240, 241, 245, 246, 247, 248, 249, 250
convergence stage 271
conversation theory 171, 181
Costa Del Numero 246, 247
creative enablement 182, 186
creative identity 194
creative learning 190
critical learning 270
cultural data 82
cultural diversity 266, 267
cultural domains 257
cultural exchange 269
cultural identity 267
cultural perspective 167
cultural production 251, 257

culture artists 203
cyberception 83
cybernetic art 204
cybernetic era 203
cybernetic platonic 191
cybernetics 2, 3, 14, 26, 30, 31, 32, 34, 40, 42, 186, 187, 281, 291, 299
cyberpsychology 182, 186, 195
cyberspace 7, 9, 24, 196, 280, 282, 294, 297
cyborg 6
Cysp1 project(Cybernetic and Space Dynamics) 30

D

dancing avatar 161
data silos 95
defuzzified values 43
design material 253, 258
design theory 251, 253
dialogue system 39, 40, 42, 57
digital animation 78
digital art 2, 3, 6, 203, 215, 217
digital art 256
digital art history 1
digital artists 59, 74, 203, 230
digital artwork 58, 64, 73, 83, 231
digital content 78
digital creations 230
digital creativity 183
digital culture 203
digital design 251, 258, 259
digital domain 88
digital embodiment 192
digital environment 189
digital experience 192
digital format 205
digital games 270
digital imagery 217, 223
digital images 221, 226
digital interfaces 59
digital material 258, 260
digital media 14, 17, 23, 26, 58, 59, 64, 70, 78, 79, 80, 82, 83, 88, 199, 204, 252, 267
digital medium 183, 218, 223
digital museum 20, 26
digital narrative 202

digital painting 220, 221, 226
digital planning 223
digital plunder 107
digital prototyping 221, 222, 223
digital raw material 69, 70
digital society 207
digital space 251, 257
digital studio practice 58, 67, 70, 74
digital technology 61, 72, 263
digital versions 107
divergent cultures 209
divergent stage 271
diverse hybrid methodologies 219
diversity 228, 230
DNIC (diffuse noxious inhibitory controls) 129, 132
Do-Fi-Do 258
dynamic system 34, 38, 40, 57, 102, 103

E

earthbound hybrid carrier 191
economic activity 192
economic systems 40
educational research 270
educational theory 186, 187
electroencephalograms 31
electromagnetic spectrum 61
electronic communications 102, 110
electronic media 105, 107, 116, 204, 257
electronic processing 114
electronic reality 9
Electronic Visualization Laboratory (EVL) 7
Elektrodynamik 115, 118, 119
embryology 229
emotion 147, 153, 154, 155, 156, 157, 164
emotional agents 137
Emotional-Belief-Desire-Intention (EBDI) 134, 135, 137, 147, 148, 150, 153, 154
emotional cognitive structural system 40
emotional competence 267, 268, 277
emotional competence framework 134, 136
emotional expression 187
emotional influence 135, 141, 143, 145, 152
emotional intelligence 134, 135, 136, 137, 140, 146, 153
emotional intelligence model 134, 136

emotional regulation 135, 146
emotional state 161
emotion-related feelings 136, 157
emotions sharing 159
emotion synthesis 136
empathy 111
ENIAC (Electronic Numerical Integrator and Computer) 4
equilibrium 28, 29, 30, 31, 32, 34, 35, 36, 39, 44, 45, 57, 86, 172
Everquest 272
evolutionary biology 229
experiential 225
experiential learning 187

F

female virtual human 136
fieldasy 251, 253, 258
field studies 258, 265
figuration theory 41
finite state machine (FSM) 34, 37, 38, 57
flat web 91, 94
Form_Rhythm 36, 38, 48, 57
Form-Rhythm Complexity 39
Frijda's theory of emotion 137
frivolous play 190
fuzzy agent 138
fuzzy control language (FCL) 141, 153
fuzzy dynamic system 34, 38, 57
fuzzy functions 42
fuzzy inference 43
fuzzy inference system (FIS) 141
fuzzy limits 141, 152
fuzzy logic 27, 32, 33, 34, 54, 57, 137, 141, 152
fuzzy logic adaptive model of emotions (FLAME) 137, 153
fuzzy rules 134, 135, 137, 141, 143, 144, 145, 146, 147, 152
fuzzy sets 135, 141, 143, 144, 145, 146, 147

G

galvanic skin response (GSR) 123, 124, 125, 129
game core mechanics 270
game design 271, 277
game design 279, 280
game designer 270
game interaction 270, 275, 276, 277
game language 271, 272
game mechanics 277
gaming industry 183
general aggression model (GAM) 268, 269, 278
general learning model (GLM) 269, 276, 277
general systems theory 203
generative algorithm 217
generic universal role playing system (GURPS) 261, 263
genetic manipulation 228, 235
genetics 228, 229, 237
Gleich Unendlich 44
global capacity 157
global space 66
GPS-enabled mobile phones 110
Grand Vacance 246
Green Hand 234, 235, 236
Groundcourse 182, 185, 186, 187, 194
GUI (Graphic User Interface) 4

H

haptic interaction 60
haptic visuality 73
head-mounted display (HMD) 122, 123, 125, 133
heterotopia 174, 175
holistic approach 191
homeostatic system model 40
homo ludens 269
human behavior determination model 40
human connection 110, 112
human craftsmanship 183
human element 106, 109, 115, 116
Humanities, Arts, Science and Technology Advanced Collaboratory (HASTAC) 68
human-machine interaction 40, 42
human monopoly 116
humanness 102, 103, 106, 107, 110, 117
human presence 116
human systems 40
hybrid creations 251, 257
hybrid genre 205, 206

hybrid harvest 239
hybrid reality 238
hybrid world 234
hyperreal 104
Hyperreal 120

I

ICOM (International Council of Museums) 19, 20, 21
image creation 218, 221
imaginary incubator 228, 232
imaginary museum 20, 21, 26
imaginary objects 71, 72
immersive learning 182
immersive technologies 91, 92
immersive virtual reality 121, 132
in-between dimensions 169
Industrial Revolution 184
infinite 219, 222, 225
information age 267
information-avatar 23
information technology (IT) 87
information theory 203
intelligent agents 136, 137, 203
interactive design 271
interactive movie 172
interactive painting 172
interactive public video 207
interactivity 3, 7, 9, 14, 25, 26
Inter-Society of Electronic Art (ISEA) 199, 204, 213, 214, 215
IRCAM 238
ISEA2008 198, 200
ISEA2009 198, 210, 215

J

Journey to Abadyl 259, 261, 262, 263

K

kaleidoscopes 1
knowledge base 101
Kritical Works 198, 199, 203, 204, 210, 211, 213, 214, 215

L

learning activity 184, 185, 190
learning machines 158
learning objectives 270
learning outcome 266, 270, 277
learning platform 185
learning strategy 182
Lemna Luciola 234, 235
Lemonade Stand 271
Le Rouge et le noir 46
Life Squared project 205
limbic system 135
Linden Lab 199
linear pattern abstraction 219
live cinema 69
longitudinal aspects 129
long-term efficacy 128
long-term learning 270, 277
ludology 182, 195

M

Main Verte 234
male virtual human 136
mammalian brain 135, 139
mammalian traits 95
man as media 89
massive multiuser online games (MMOG) 272
materiality 58, 59, 60, 61, 72, 73, 74, 75
material world 169, 172, 174, 176, 178
media artwork 64
mediated environment 190
mediation 103, 106, 107, 110, 114, 118, 119, 120
mediatization 103, 104, 106, 110, 113, 115, 116, 117
mediatize 103
meditation 121, 122, 123, 124, 125, 126, 127, 128, 129, 131, 132, 133
Meditation Chamber 121, 122, 123, 124, 125, 126, 128, 129, 131
meiosis 237
melody analysis 159
mental impulse 217
meta-archive 205
metadata 90, 93, 94, 95, 96, 97

metalanguage 34, 35, 38
meta-matic 30
metanomics 182, 192
metaplastic 27, 31, 34, 36, 38, 39, 40, 48, 49, 50, 51, 52, 54, 57
metaplastic discipline 1, 14, 27
metaplastic entities 15
metaplastic interactivity 26
metaplasticity 1, 14, 18, 26, 158, 159, 161, 162, 164
metaplastic language 26, 34, 48, 49, 57
metaplastic media entities 15
metaplastic metalanguage 34, 35
metaplastic metaspace 15, 26
metaplastic virtual media 1, 16
metaplastic virtual museum 22, 23, 26
metaplastic virtual worlds 13, 15, 26, 27, 38, 39, 48, 52, 53, 54
metaverse 182, 184, 185, 186, 188, 189, 190, 192, 195, 197
microorganisms 1, 232, 235
micros univers 232
migratory population 189
Ministerio de Educación Nacional [MEN] 268
Minnesota Multiphasic Personality Inventory (MMPI) 134, 135, 140, 141, 155, 156, 157
mitosis 237
mixed reality 107, 109, 167, 168, 169, 170, 171, 172, 173, 174, 175, 176, 177, 178, 179
mixed reality installation 167, 168, 169, 170, 172, 173, 174, 175, 176, 177, 178, 179
mixed reality installation/space 181
mixed reality spaces 167, 172, 175, 177, 178, 179
molecular biology 203, 229, 235, 237
mood influence 135, 143, 144, 145
mood regulation 135, 147
MOO (Muds Object-Oriented) 5
mortal physicality 191
Motoo Viridis 236, 237
multi-dimensionality 172
multilingual semantic zoo 228
multiple-loop interaction 171
multi-user dungeon (MUD) 276

multi-user virtual environment 8
multiuser virtual game 266, 280
multi-user virtual world environment 94
museums without walls 19, 26
musical intentions 159
music creation 159
Mμ herbarium 232, 233, 234, 235

N

nano evolutions 228, 233
nano-looping-sound 233
neocortex 135
neural network 158, 159, 160, 161, 162, 163
new media 79, 82, 83, 87, 88, 265
nodal concept 229
noms-de-plume 194
non-digital artists 203, 215
Numbers Beach 246

O

object oriented paradigm 95, 101
OCEAN model 138, 139
online communities 194
online economy 192
online educational systems 184
online environment 70, 194
online game 66, 205, 271
online game communities 205
online gameplay 214
online role playing games 190
online social play 208
online synthetic worlds 182, 191
online video games 271
online world space 210
ontological treasures 107
ontology 38, 39, 57, 100, 101
open source 240
open systems 2
open world 251
optional environment 251
organic abstract geometry 225
organic tissues 235
Oslen method 159
OULIPO (Ouvroir de Littérature Potentielle) 2, 8

P

pain distraction 122, 126, 127, 128, 130, 133
pain modulation 131, 133
pain self-modulation 127, 128
panel debate 215
paradigm shifts 212
parallel botany 228
pattern language 255, 263, 264
pattern matching 160
peaceful social interaction 267, 268, 275, 277
perennial mutation 45
periphery 181
personality 134, 135, 136, 137, 138, 139, 140, 141, 142, 143, 144, 145, 146, 147, 150, 152, 153, 156, 278, 279, 280
personality change 138
personality theories 136, 137
personality traits 135, 137, 138
pervasive computing 172
pervasiveness 78, 83
pervasiveness of digital media 78
phenomenology 122
physical world 63, 72, 73, 251, 257
pictorial artwork 82
ping space 201
plastic anticipation 231
plastic art 29, 31, 45
plasticity 228, 229, 230, 231, 232, 234, 235, 236, 237, 238
plasticity 229, 239
plasticity process 231, 234
plastic language 34, 43
plastic plasticity 235
polymorphic properties 90, 94, 100
post-Cartesian shift 191
post-digital materiality 58, 75
post-humanistic world 191
post-traumatic stress disorder (PTSD) 126
Project paradise 110, 111, 112, 116, 117
prosocial behaviors 269, 270, 273, 277
prosocial cognitions 277
prosocial content 269
Proteus Effect 194, 196
prototypes 257, 258, 265
psychological traits 135
psychometric practice 135
pure play 190

Q

QBIC (Query by Image Content) 21, 26
qualitative data 271, 275
quantitative data 271
quasi-permanent feeling 157

R

rational agents 137
R-complex 135
realistic virtual humans 134, 152
reality jam 200
real reality 9
real world 105, 107, 109, 110, 116, 241, 242, 243, 244, 246
real world campus 184
real-world galleries 199, 215
real-world viral communications 110
reconstructed reality 6
red and black semantics 57
reptilian brain 135, 139, 140
RFID reader 258
RFID tag 258

S

sample cell 236
scientific awareness 177
scientific community 229
sculpture networks 233
Second Life® 183, 184, 185, 188, 189, 190, 191, 192, 193, 194
Second Life (SL) 71, 72, 74, 96, 107, 108, 109, 113, 118, 185, 192, 196, 198, 199, 200, 201, 202, 203, 204, 205, 206, 207, 208, 209, 210, 211, 212, 213, 214, 215, 240, 242, 248, 249, 250
selective memory 159
Self-actualization 13
self-conscious 159
self-consciousness 136
Self-consciousness 13
self media networks 22
self-modulation 122, 126, 127, 128, 129

self-motivation 136
self-regulation 136
semantic knowledge 90, 93, 94, 95, 96, 99, 101
semantic model 47, 54, 57
semantic space 54
semantic technologies 93, 94, 95, 96, 97, 99
semantic web 98, 100, 101
semantic zoo 228
semi-machine 39
semiotics 45, 55
sensitive microscope 228, 232
sensorial experience 169, 172
sensory microscope 233
serious games 272, 277
Serious Games 280
serious virtual game 266
short-term effects 269, 277
SIGGRAPH 204, 248, 250
silent epidemic 128
SimCity 271
simstim 9
single-loop interaction 171
SL community 198, 199
social awareness 136
social balance 267
social bonding 187
social change 267, 269, 270, 271, 272, 277, 280
social change processes 267
social competence 267, 270, 271, 273
social education 267
social ethics 54
social factor 191
social interaction 137, 184, 190, 266, 267, 268, 271, 273, 275, 276, 277, 279
socialization experiences 159
social landscape 267
social network 205
social presence 191
social processes 267
social-psychological model 138
social skills 136
social spaces 270
social standpoint 78
social systems 40
social tools 271, 272, 273, 276
social world 191

socio-cultural milieu 183
socio-cultural web 191
socio-economic impact 182
socio-political structures 183
software agents 136, 137
sound plasticity 235
space block 243, 246, 249
space dynamics 31, 43
space model 243
space-time dimension 171
space-time events 171
spatial competencies 271
spatial experience 240, 241, 243, 244, 245, 246, 247, 248
spatial gaps 241
spatialization 46
spatial model 240, 241, 245, 246, 247, 249
spectral bodies 62
spiritual 217, 219, 225, 227
spores 229, 234
static equilibrium 32
STDP concept 159, 163
stem cell division 236
stem cells 228, 229, 231, 234, 236, 237, 239
stem cell videos 238
stem drawing 231
Strip Hit Music 262
structuralism 203
sub-neural network 162
suspension of disbelief 103, 105, 107
symbolic approach 137
symbolic emotional rule-based systems 137
symbolic vocabulary 159
synaesthesia 57
synaptic modification 159
synaptic pruning 163
synaptic spikes 159
synthetic reality 9
synthetic virtual worlds 195
synthetic world 182, 183, 184, 189, 190, 191, 192, 194, 195, 197
Synthetic World 197

T

tactile laboratory 228, 232
tangiality 58, 59, 63, 64, 66, 69, 70, 71, 72, 73, 74, 75

tangible 59, 60, 61, 62, 63
Tappatappatappa 115, 118
technological developments 102, 103
telegarden 107, 118
tele-immersion 67
tele-immersive 68
telepistemology 107
telepresence 6, 8, 59, 107, 120, 196
telepresent 107
temporalization 46
The Actual 80, 81, 83, 89
The Memory Stairs 210
The Oregon Trail 271
theoretical biology 230
The Real 80, 81
The Self 106, 110, 112, 117
The Uncanny 116
The Virtual 80, 81, 82, 83, 88, 102, 103, 107, 108, 109, 110, 112, 116
three dimensional (3D) 90, 91, 93, 94, 96, 98, 123
three dimensional (3D) synthetic worlds 182, 192
three dimensional avatar 194
totipotency 229, 234, 238
totipotent stem cells 229, 231
traceroutes 209, 210
traditional aesthetic 171
transformative learning 182, 187
Traora Jukebox 262
triune brain model 134, 135, 139
two dimensional (2D) 91, 94

U

ubiquitous computing 167, 172, 175, 176, 180, 181
ubiquitous technologies 183
uncanny 103, 116, 117, 118, 120, 194, 195
Unified Process (UP) 271
University of Urbana-Champaign Computer Science Department (UIUC) 67
unorthodox media 203

V

vantage point 173, 179, 189
video art 207
video game 191, 266, 267, 268, 269, 270, 271, 272, 275, 276, 277, 278, 279
video game interaction 275
video stem cell 234, 235, 236, 237, 239
virtools software 158
virtual agents 90, 93
virtual artifacts 183
virtual artists 210, 211
virtual assemblage 182
virtual assets 110
virtual avatars 138
virtual body 194
virtual building 160
virtual camera 161
virtual campus 184
virtual cell plasticity 237
virtual character 159
virtual content 184
virtual counterpart 212
virtual creations 238
virtual creativity 183
virtual creatures 138
virtual curation 216
virtual dancer 161
virtual data 59, 72
virtual economies 183, 184
virtual elements 234
virtual embodiment 192
virtual environment (VE) 3, 5, 6, 7, 8, 9, 16, 17, 18, 19, 23, 24, 39, 40, 41, 44, 49, 50, 51, 62, 71, 90, 93, 101, 121, 122, 123, 127, 129, 130, 148, 149, 182, 186, 190, 191, 197, 199, 200, 209, 210, 211, 214
virtual experience 109, 116, 190, 192
virtual extensions 184
virtual fabric 199
virtual gallery 198
virtual game 266, 271, 276, 277
virtual geography 203
virtual haptic interface for printmaking (VHIP) 60
virtual human 134, 135, 136, 138, 139, 140, 141, 143, 144, 145, 146, 147, 148, 150, 151, 152, 153
virtual interactions 105, 110
virtuality 1, 4, 10, 27, 54, 58, 59, 66, 67, 73, 89, 102, 103, 105, 106, 117, 120

Index

virtualization 78, 79, 80, 81, 82, 83, 85, 87, 88
virtualized self 89
virtual materiality 58, 73
virtual media 1, 8, 14, 15, 16, 24, 27
virtual meditative walk 121, 128
virtual medium 9, 183
virtual museum 20, 21, 22, 23, 26
virtual object 90, 192, 251, 257
virtual organic plasticine 232, 238
virtual participation 66
virtual pastures 183
virtual performance 109
virtual personas 200
virtual Petri dish 233
virtual plasticine 239
virtual presence 66
virtual property 199
virtual proposals 222
virtual reality archetypes 57
virtual reality (VR) 1, 4, 5, 6, 7, 8, 9, 10, 15, 17, 18, 19, 24, 25, 26, 62, 63, 64, 67, 76, 78, 121, 122, 125, 126, 127, 128, 129, 131, 161, 199, 203, 215, 241
virtual realms 191
virtual renderings 222
virtual self representations 191
virtual singer 158, 159, 165
virtual situation 109
virtual skin 234
virtual social interaction 190
virtual sound 201
virtual space 7, 11, 12, 14, 18, 21, 23, 25, 28, 34, 35, 38, 39, 41, 43, 54, 56, 168, 174, 203, 204, 209, 211, 212, 213, 214
virtual spores 234
virtual state 80
virtual tissue 228, 239
virtual tools 253

virtual world 27, 38, 39, 48, 49, 50, 51, 52, 53, 54, 56, 57, 63, 64, 70, 72, 73, 78, 79, 80, 83, 88, 90, 91, 93, 94, 95, 96, 98, 99, 102, 210, 211, 212, 214, 215, 216, 229, 235, 238, 239, 240, 241, 242, 243, 244, 245, 246, 248, 249, 250, 272, 273
virtual world metadata 95
virtual world platform 95, 96, 198, 214
visual alphabet 38
visual communication 190
visual elements 193, 194, 273
visual field 225
visual identity 182, 194
visual language 34, 35, 36, 38, 39
visual language model 15
visual poetics 68
visual syntax 39, 45, 51
vivarium 228
VRML (Virtual Reality Modeling Language) 5
VR system 9, 10

W

web 2.0 domains 184
web environment 78
web interface 201
weblog 104
web museum 26
Wikipedia 82
World 1 5
World 2 5
World 3 5, 6
World of Warcraft 272
World Wide Web 101

Z

zoomable 79, 83
zoomable media 89
Zoo Tycoon 271
zygotes 229